VOLUME SEVENTY ONE

ADVANCES IN CATALYSIS

Advances in the Synthesis and Catalytic Applications of Boron Clusters: A Tribute to the Works of Professor Francesc Teixidor and Professor Clara Viñas

VOLUME SEVENTY ONE

ADVANCES IN CATALYSIS

Advances in the Synthesis and Catalytic Applications of Boron Clusters: A Tribute to the Works of Professor Francesc Teixidor and Professor Clara Viñas

Edited by

MONTSERRAT DIÉGUEZ
Departament de Química Física i Inorgànica, Universitat Rovira i Virgili, Tarragona, Spain

ROSARIO NÚÑEZ
Institut de Ciència de Materials de Barcelona (ICMAB-CSIC), Barcelona, Spain

ELSEVIER

ACADEMIC PRESS
An imprint of Elsevier

Academic Press is an imprint of Elsevier
50 Hampshire Street, 5th Floor, Cambridge, MA 02139, United States
525 B Street, Suite 1650, San Diego, CA 92101, United States
The Boulevard, Langford Lane, Kidlington, Oxford OX5 1GB, United Kingdom
125 London Wall, London, EC2Y 5AS, United Kingdom

First edition 2022

Notices
Knowledge and best practice in this field are constantly changing. As new research and experience broaden our understanding, changes in research methods, professional practices, or medical treatment may become necessary.

Practitioners and researchers must always rely on their own experience and knowledge in evaluating and using any information, methods, compounds, or experiments described herein. In using such information or methods they should be mindful of their own safety and the safety of others, including parties for whom they have a professional responsibility.

To the fullest extent of the law, neither the Publisher nor the authors, contributors, or editors, assume any liability for any injury and/or damage to persons or property as a matter of products liability, negligence or otherwise, or from any use or operation of any methods, products, instructions, or ideas contained in the material herein.

ISBN: 978-0-323-98831-5
ISSN: 0360-0564

For information on all Academic Press publications
visit our website at https://www.elsevier.com/books-and-journals

Publisher: Zoe Kruze
Acquisitions Editor: Sam Mahfoudh
Developmental Editor: Cindy Angelita Gardose
Production Project Manager:
Sudharshini Renganathan
Cover Designer: Victoria Pearson

Typeset by STRAIVE, India

Contents

Contributors

Fayaz Ali
Department of Chemistry, Abbottabad University of Science and Technology, Abbottabad, Pakistan

Sergey A. Anufriev
A. N. Nesmeyanov Institute of Organoelement Compounds, Russian Academy of Sciences, Moscow, Russia

Simon Duttwyler
Department of Chemistry, Zhejiang University, Hangzhou, China

Tanzeela Fazal
Department of Chemistry, Abbottabad University of Science and Technology, Abbottabad, Pakistan

Narayan S. Hosmane
Department of Chemistry and Biochemistry, Northern Illinois University, DeKalb, IL, United States

Chun-Xiao Li
School of Chemistry and Chemical Engineering, Nanjing University, Nanjing, Jiangsu, People's Republic of China

Rosario Núñez
Institut de Ciència de Materials de Barcelona (ICMAB-CSIC), Barcelona, Spain

José Giner Planas
Institut de Ciència de Materials de Barcelona (ICMAB-CSIC), Barcelona, Spain

Isabel Romero
Departament de Química, Universitat de Girona, Girona, Spain

Akim V. Shmal'ko
A. N. Nesmeyanov Institute of Organoelement Compounds, Russian Academy of Sciences, Moscow, Russia

Igor B. Sivaev
A. N. Nesmeyanov Institute of Organoelement Compounds, Russian Academy of Sciences, Moscow, Russia

Francesc Teixidor
Institut de Ciència de Materials de Barcelona (ICMAB-CSIC), Barcelona, Spain

Clara Viñas
Institut de Ciència de Materials de Barcelona (ICMAB-CSIC), Barcelona, Spain

Zuowei Xie
Department of Chemistry and State Key Laboratory of Synthetic Chemistry, The Chinese University of Hong Kong, Shatin, New Territories, Hong Kong, China

Hong Yan
School of Chemistry and Chemical Engineering, Nanjing University, Nanjing, Jiangsu, People's Republic of China

Jie Zhang
Department of Chemistry and State Key Laboratory of Synthetic Chemistry, The Chinese University of Hong Kong, Shatin, New Territories, Hong Kong, China

Yinghuai Zhu
Sunshine Lake Pharma Co., Ltd., Songshan Lake Industrial Park, Dongguan, China

Preface

Icosahedral boron clusters are molecular clusters mainly made of B—H units. Among them, carboranes with B—H and C—H in their structures show unique properties such as 3D aromaticity, rigid geometry, high thermal and chemical stability, low toxicity, and hydrophobicity, among others. All these properties make them and their derivatives attractive for a wide variety of applications in various fields of research from medical to materials science. In this regard, significant progress in the functionalization of boron clusters has been made by research scientists over the past few decades. In the particular case of carboranes, since the properties of the CH and BH groups differ significantly, very different methods have been used for their functionalization. Recent advances in the transition metal-catalyzed B—H activation of neutral and anionic boron clusters have been achieved, in particular the regioselective functionalization of B—H that has led to a large number of structures. The direct conversion of B—H bonds to substituted vertices constitutes an attractive strategy to boron cluster compounds with desired properties. Moreover, the design of new transition metal catalysts that incorporate ligands containing boron clusters with the aim to obtain highly efficient catalytic species has been a major topic in the last few years. Carborane and metallacarborane complexes have been widely used as catalysts for reactions as different as hydrogenation, cyclopropanation, olefin polymerization, and olefin metathesis and, most recently, as photoredox catalysts in water.

Following these considerations, this volume of *Advances in Catalysis*, titled "Advances in the Synthesis and Catalytic Applications of Boron Clusters," contains six top-quality chapters written by experts in the field with the common aim to show the developed strategies and latest advances in the field of boron chemistry. When putting together this special issue, we also aimed to dedicate this volume to honor the work of Professors Francesc Teixidor and Clara Viñas as major contributors and internationally recognized experts in the field. We decided to do so by identifying fellow colleagues in the field we truly admire and simply asking them to share their chemistry.

Prof. Dr. Francesc Teixidor

Prof. Dr. Clara Viñas

In Chapter 1, Núñez and coworkers describe the advances in the use of rhoda and ruthenacarboranes containing *o*-carborane-based ligands as efficient catalysts in diverse reactions such as hydrogenation, cyclopropanation, or Kharasch addition. Besides, Pd complexes containing *m*-carborane ligands have been shown to be good catalyst precursors in Suzuki coupling reactions in water. Furthermore, recent advances in the photocatalytic activity of metallacarboranes toward the oxidation of alcohols and alkenes under UV light in water have also been reported. In Chapter 2, Sivaev and coworkers review the current state of functionalization of polyhedral boron hydrides to form B—N, B—P, B—O, and B—S bonds using transition metal-catalyzed BX and BH activation reactions. In Chapter 3, Xie and Zhang describe recent advances in the catalytic regio- and enantioselective cage BH functionalization of *o*-carborane to provide new derivatives for applications in medicine, materials, and coordination/organometallic chemistry. In Chapter 4, Zhu and coworkers provide an overview of the contributions of boron-based compounds as catalysts, i.e., the application of boron-based compounds such as frustrated Lewis pair catalysts with the focus on heterogeneous catalysis to produce products ranging from good chemicals to selective pharmaceuticals. In Chapter 5, Yan and Li collect the progress on carboranylation approaches in order to link organic fragments to specific vertices of *closo*- and *nido*-carboranes by functionalization via B—H bond activation, transformation of pre-functionalized carborane, and functionalization on the carbon vertex. Finally, in Chapter 6, Duttwyler describes advances in the derivatization of other types of anionic boron clusters such as the monocarborane and the dodecaborate clusters by controlled activation of the B—H vertex.

We thank Mr. Sam Mahfoudh and Tara A. Nadera, who helped us initiate this project, and Ms. Cindy Angelita P. Gardose, whose help during the editing and production process was invaluable. We also thank the Spanish Ministry of Science and Innovation (PID2019-106832RB-I00) and the Catalan Government (2017/SGR/1720). Finally, we specially thank all the authors who contributed to this project.

Rosario Núñez
Montserrat Diéguez

CHAPTER ONE

Advances in the catalytic and photocatalytic behavior of carborane derived metal complexes

Francesc Teixidor[a], Clara Viñas[a], José Giner Planas[a], Isabel Romero[b], and Rosario Núñez[a,*]
[a]Institut de Ciència de Materials de Barcelona (ICMAB-CSIC), Barcelona, Spain
[b]Departament de Química, Universitat de Girona, Girona, Spain
*Corresponding author: e-mail address: rosario@icmab.es

Contents

Abstract

Boron clusters whose geometrical shape is that of an icosahedron in combination with a transition metal have been used as homogeneous phase catalysts for significant reactions of practical interest. Key features include the formation of B—M—H or B—H—M bonds, their arrangement at the periphery of the cluster or with the metal forming part of the cluster, their interaction with light without photoluminescence losses, their great capacity for SET reactions, a fantastic catalyst-substrate ratio and water as a solvent.

Advances in Catalysis, Volume 71
ISSN 0360-0564
https://doi.org/10.1016/bs.acat.2022.04.001

1. Introduction

Among the carbaborane cluster families, the 12-vertex icosahedral *closo*-dicarbadodecaborane or *closo*-carborane (1-2-$C_2B_{10}H_{12}$) and its derivatives are the most extensively studied (Chart 1) *(1,2)*. These clusters possess delocalized electron density *(3)*, are considered analogues to arenes and possess an exceptional stability, three-dimensional (3D) geometry and σ-aromaticity *(4,5)*. The *closo-o*-carborane can be thermally isomerized to give *closo-m*-carborane (Chart 1) and *closo-p*-carborane *(1)*. These neutral carboranes are remarkably robust with unique material-favorable properties such as hydrophobicity, electronic and thermal stability, chemical versatility and low toxicity in biological systems *(6)*. Moreover, despite their chemical stability in front of acids and oxidizing agents, the *closo-o*-carborane (*o*-carborane) can be partially deboronated in basic media, by removing formally a BH^+ vertex to give the corresponding anionic species *nido-o*-carborane (Chart 1). All these properties make these boron clusters promising building blocks for a wide variety of applications *(7)*, such as optoelectronic materials *(8–11)*, coordination chemistry *(12,13)*, metal-organic frameworks *(14,15)*, polymers *(16)*, and biomedicine *(17–21)*, among others.

Apart from the above applications, carborane-based metal complexes (metallacarboranes), in particular half- and mixed sandwich metal complexes, have been applied in different catalytic areas, such as olefin hydrogenation *(22,23)*, and asymmetric hydrogenation *(24,25)*, among others *(26–29)*.

Further to the properties described for carboranes, in anionic sandwich metallacarboranes the presence of a metal center (Co or Fe, Chart 1) confers to this cluster unique redox properties, that can be tuned upon dehydrohalogenation *(30,31)*, as well as extraordinary high chemical, thermal and

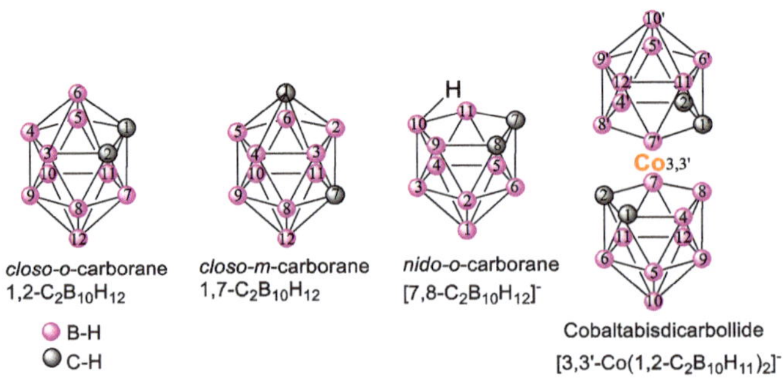

Chart 1 Main neutral and anionic icosahedral carboranes with their vertex numbering.

photochemical stability *(1,32,33)*. These anions have a high molecular volume, low nucleophilic character and low charge densities because the negative charge is distributed throughout 45 atoms *(34)*. The protonated form and sodium salt of the cobaltabis(dicarbollide) anion (namely COSAN), $[3,3'\text{-Co}(1,2\text{-C}_2\text{B}_{10}\text{H}_{11})_2]^-$, have shown exceptional physicochemical properties such as amphiphilicity, leading to supramolecular interactions and self-assembling in water *(35–41)*. All these properties and their biocompatibility have made metallacarboranes promising scaffolds in a range of applications from switchable luminescent materials *(42)*, to biomedical applications *(43,44)*. Nevertheless, to the best of our knowledge, the use of sandwich metallacarboranes in catalysis has not been previously explored, being Teixidor and Viñas's group the first to develop the use of metallacarboranes, mainly COSAN, in sustainable photoredox catalysis.

In this chapter, we have focused mainly on the catalytic activity of Rhoda and ruthenacarboranes developed in our laboratories. Furthermore, the more recent advances on the photocatalytic activity of some metallacarborane derivatives toward the oxidation of alcohols and alkenes under UV light in water will be also reported.

2. Ligands and complexes for catalysis

The first work on the use of rhodacarboranes as precursor catalysts for the hydrogenation and isomerization of alkenes was published by Hawthorne *(45,46)*. After several years of research, it was reported a mechanistic study that attributed the catalytic activity to the formation of B-Rh^{III}-H species by the oxidative addition of terminal B—H bonds to Rh centers in *exo-nido*-rhodacarboranc derivatives *(47–51)*. In these complexes, a tautomerism equilibrium between the catalytic active structure *exo-nido* to the catalytic inactive *closo* one, which diminishes the catalytic activity of the metallacarbone complexes was observed. Later, a great advance in the targeted synthesis of metallacarboranes that fixed the catalytic *exo-nido-* structures and consequently increase the catalytic potential of carborane-based transition metal complexes was achieved by Teixidor and Viñas's group *(52–54)*. A large number of Rh and Ru complexes derived from *nido*-dithioether-, *nido*-monothioether-, *nido*-diphosphino- and *nido*-monophosphino-*o*-carborane ligands were developed to be applied as catalytic precursors. Taking Hawthorne's results as a landmark, our main aim was to force the formation of complexes with stable B–H–Rh and B–H–Ru agostic bonds, to avoid the *exo-nido* to *closo* tautomerism. This implies the deactivation of the hydrogenation process. For *nido*-dithioether and *nido*-diphosphino-*o*-carborane ligands, they usually act as chelating ligands through the S or P atoms when they coordinate to Rh(I) or Pd(II) metals yielding complexes with square-planar stereochemistry. Moreover, when these ligands coordinate to Ru(II), octahedral complexes with B–H–M agostic bonds in addition to the S–M or P–M bonds are formed *(55–59)*.

For this purpose, the synthesis of *nido*-monothioether- and *nido*-monophosphino-*o*-carborane ligands derived from the *ortho*-carborane was performed. The hypothesis in mind was that these kind of ligands would lead to the formation of the expected B–H–Rh and B–H–Ru bonds when coordinated to Rh and Ru metals, which contribute to a better activity in the catalytic reaction because the S–M or P–M bond would fix the catalytic active B–H–Rh and B–H–Ru bonds. The first step for the synthesis of *exo-nido*-metallacarborane complexes was the synthesis of *closo*-monothioether- (1-SR′-2-R-1,2-$C_2B_{10}H_{10}$) and *closo*-monophosphino-*o*-carborane (1-PR′$_2$-2-R-$C_2B_{10}H_{10}$ ligands), which was achieved by the monolithiation of C-substituted *o*-carborane clusters as illustrated in Scheme 1 *(60–65)*. The synthesis of monothiols 1-SH-2-R-*o*-carborane with almost a 100% yield was done by the reaction of the monolithiated *o*-carborane with sulfur followed by the addition of HCl in Et_2O. The synthesis of monothioethers 1-SR′-2-R-*o*-carborane was accomplished by the addition of R'X and KOH in EtOH after deprotonation of the thiol group (Scheme 1A). Besides, the synthesis of *closo*-monophosphino-*o*-carboranes was performed with *n*BuLi and the corresponding ClPR′$_2$ in EtO_2 or toluene

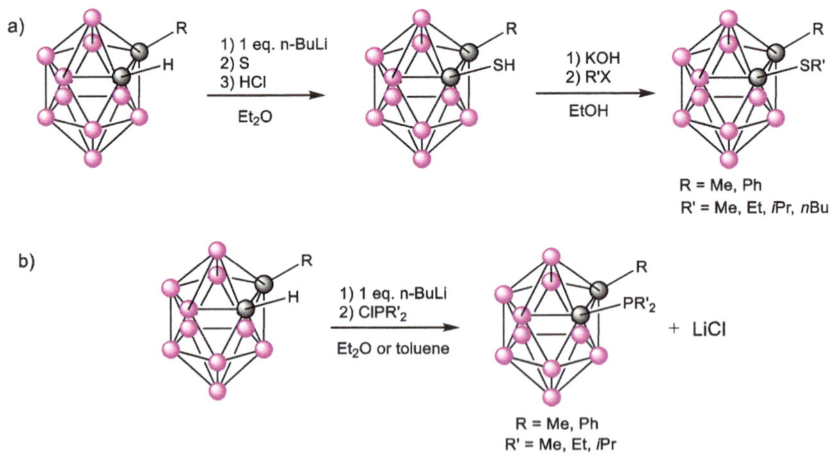

Scheme 1 Synthesis of *closo*-monothioether- and monophosphino-*o*-carboranes.

at reflux (Scheme 1B). In the particular case of the preparation of monosubstituted 1-SR′-*o*-carborane and 1-PR′$_2$-*o*-carborane derivatives, glyme was used as solvent to obtain the target compounds in high yield *(66)*. In a second step, the targeted anionic *nido*-monothioether-*o*-carborane of formula $[7\text{-}SR'\text{-}8\text{-}R\text{-}7,8\text{-}C_2B_9H_{10}]^-$ and *nido*-monophosphino-*o*-carborane of formulae $[7\text{-}PR'_2\text{-}8\text{-}R\text{-}7,8\text{-}C_2B_9H_{10}]^-$ ligands were achieved (Scheme 2). These anionic ligands were obtained as a racemic mixture (not separated) following the optimized conditions for the partial degradation or deboronation reaction of their respective *closo* derivatives $1\text{-}SR'\text{-}2\text{-}R\text{-}1,2\text{-}C_2B_{10}H_{10}$ and $1\text{-}PR'_2\text{-}2\text{-}R\text{-}C_2B_{10}H_{10}$ *(67–69)*.

Once the ligands were isolated, purified and characterized, the next step was the synthesis of their transition metal complexes. Complexes were obtained by the reaction of cesium or tetramethylammonium salts of the anionic *nido*-carborane ligands with Ru(II) and Rh(I) halocomplexes containing bulky or labile ancillary ligands *(70)*. The tetramethylammonium

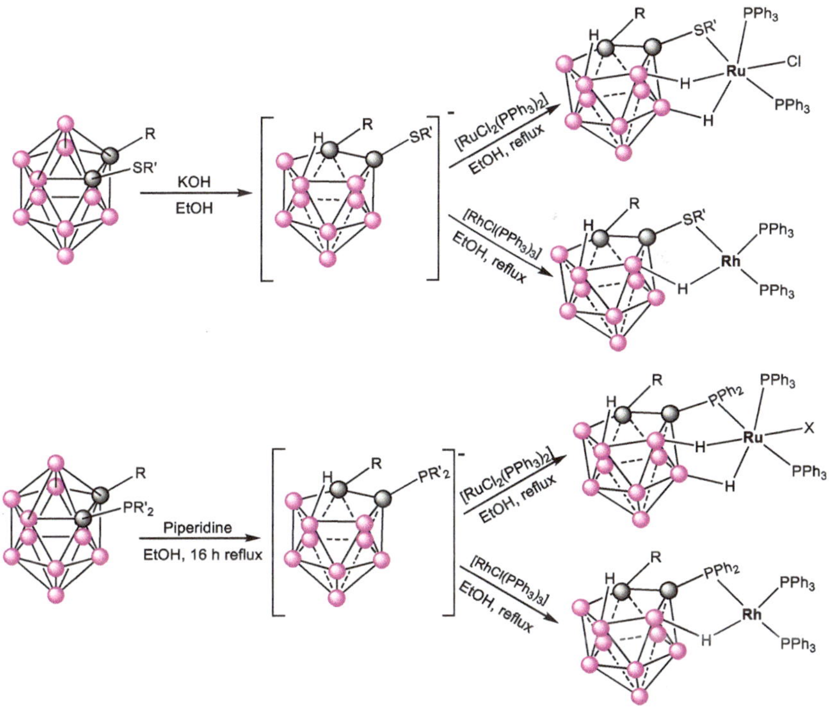

Scheme 2 General degradation reactions of *closo*-monothioether- and *closo*-monophosphino-*o*-carborane derivatives to produce the corresponding *nido* species and their further complexation with Rh and Ru metals.

salts of *nido*-monophosphino-*o*-carborane, $[NMe_4][7\text{-}PR'_2\text{-}8\text{-}R\text{-}7,8\text{-}C_2B_9H_{10}]$, were able to react in EtOH with $[RuCl_2(PPh_3)_3]$ or $[RuH(AcO)(PPh_3)_3]$ in a 1:1 ratio, to afford *exo-nido* complexes with the stoichiometry $[RuX(7\text{-}PR'_2\text{-}8\text{-}R\text{-}7,8\text{-}C_2B_9H_{10})(PPh_3)_2]$ (X = Cl, H; R = H, Me; R′ = Ph, Et, iPr) (**1–5** in Scheme 2 and Fig. 1) *(71,72)*. The ^1H and ^{31}P NMR analysis suggested the formation of complexes with octahedral geometry, where the *nido*-carboranyl behaves as a tridentate ligand coordinating the Ru(II) through the C-PPh_2 and two B-H → Ru agostic bonds from the carborane cage. The X-ray diffraction analysis of the crystal structure of $[RuCl(7\text{-}PPh_2\text{-}8\text{-}Me\text{-}7,8\text{-}C_2B_9H_{10})(PPh_3)_2]$ (**2**) corroborated the suggested geometry (Fig. 2). In this case, the B(2)-H (lower belt) and B(11)-H (upper belt) vertexes participate in the coordination to Ru, in addition a chloride or hydride anion and two PPh_3 fulfill the coordination sphere of the metal. In the same way, the reaction of $[RuCl_2(PPh_3)_3]$ with the cesium or tetramethylammonium salts of $[7\text{-}SPh\text{-}8\text{-}R\text{-}7,8\text{-}C_2B_9H_{10}]^-$ (R = Me, Ph) in EtOH at reflux for 2 h yielded complexes compatible with

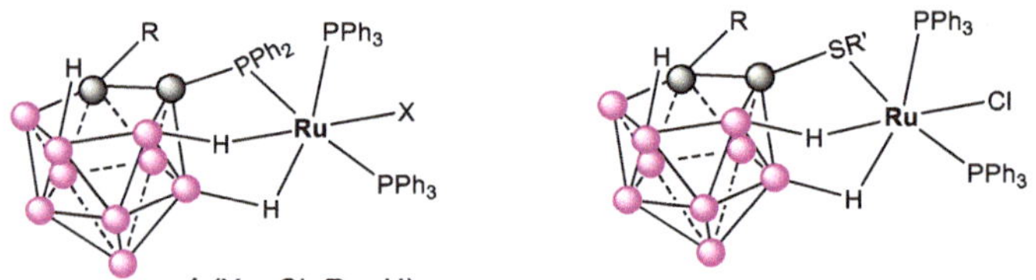

1 (X = Cl, R = H)
2 (X = Cl, R = Me)
3 (X = H, R =H
4 (X = H, R = Me)
5 (X = H, R =Ph)

6 (R = Me, R′ = Ph)
7 (R = Ph, R′= Ph)
8 (R = Me, R' = Et)
9 (R = SPh)

Fig. 1 Structure of *exo-nido*-monothioether and *exo-nido*-monophosphinoruthenacarborane complexes.

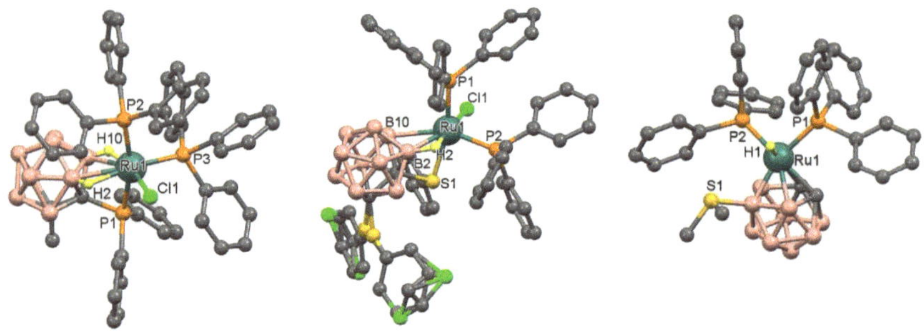

Fig. 2 Crystal structure of ruthenacarborane complexes: **2** (left), $[RuCl\text{-}(7,8\text{-}(SPh)_2\text{–}7,8\text{-}C_2B_9H_{10})(PPh_3)_2]\cdot CH_2Cl_2$ (middle), **22a** (right).

the stoichiometry [RuCl-(7-SR′-8-R-7,8-$C_2B_9H_{10}$)(PPh_3)$_2$] (R=Me, Ph; R′=Ph, Et) (**6–8**) *(73)*. Remarkably, the crystal structure of [RuCl-(7,8-(SPh)$_2$-7,8-$C_2B_9H_{10}$)(PPh_3)$_2$]·CH_2Cl_2 (Fig. 2), revealed that the *nido*-dithioether-*o*-carborane ligand only coordinates the metal through one SPh group and two B—H vertexes giving rise to structures similar to the previous *nido*-monoposphino-*o*-carboranes (**9** in Fig. 1) *(73)*.

Moreover, the reaction of Cs[7-SR′-8-R-7,8-$C_2B_9H_{10}$] with the Wilkinson's catalyst [RhCl(PPh_3)$_3$] in EtOH in a 1:1 ratio gave complexes with the stoichiometry [Rh(7-SR′-8-R-7,8-$C_2B_9H_{10}$)(PPh_3)$_2$] (R=Me, Ph; R′=Ph, Et) (**10 – 12**), Scheme 2, Fig. 3) *(74)*, that was identical when the reaction was performed with [NMe_4][7-PR_2-8-R′-7,8-$C_2B_9H_{10}$] to yield [Rh(7-PR_2-8-R′-7,8-$C_2B_9H_{10}$)(PPh_3)$_2$] (R=Ph, Et, iPr; R′=H, Me, Ph) (**13–16**) (Scheme 2, Fig. 3) *(75)*. A complete spectroscopic analysis (FT-IR and ^{1}H, ^{31}P, ^{11}B and ^{13}C NMR) for all these complexes indicated the formation of the targeted *exo-nido* complexes, which was confirmed by the X-ray analysis of [Rh(7-SPh-8-Me-7,8-$C_2B_9H_{10}$)(PPh_3)$_2$] (**10**) and [Rh(7-PPh_2-7,8-$C_2B_9H_{11}$)(PPh_3)$_2$] (**13**) (Fig. 4). Notably, both complexes showed similar *exo-nido* structures, where the Rh(I) has a square-planar

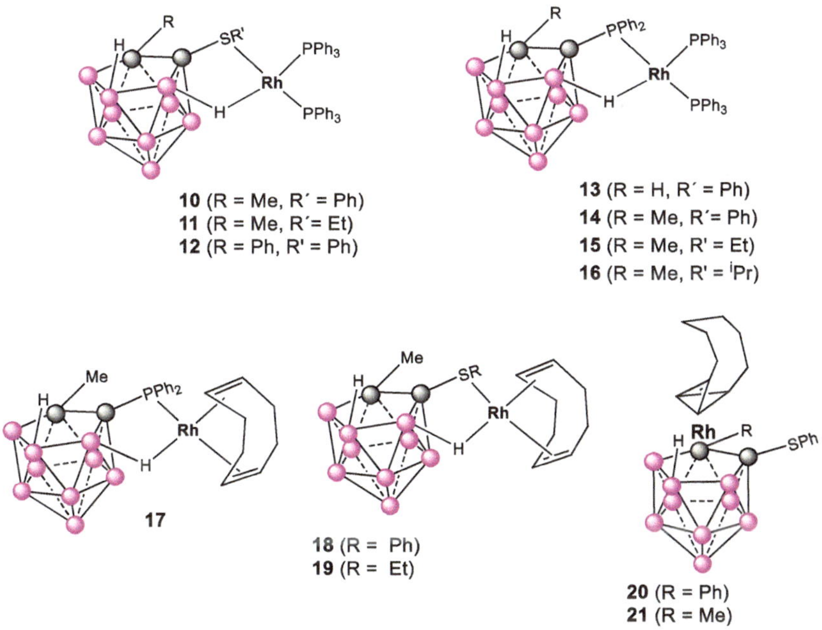

Fig. 3 Structures of the different monothioether and monophosphinorhodacarborane complexes.

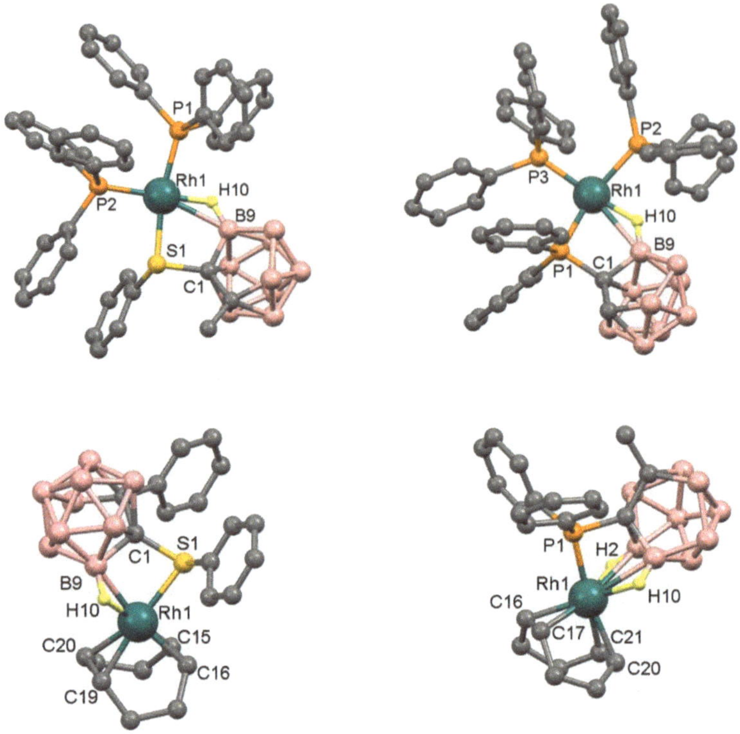

Fig. 4 Crystal structures of monothioether and monophosphinorhodacarborane complexes.

geometry and the *nido*-carboranyl ligands act as bidentate coordinating to the metal through S—Rh or P—Rh bonds and a B(11)–H–Rh agostic bond, as expected. In the latter, the B(11)–H vertex that participated in the coordination sphere of Rh(I) belongs to the C_2B_3 open face (Fig. 4). The two remaining positions of the complex are occupied by two PPh_3 ligands.

In addition to the above, another type of Rh complexes were obtained from the reaction of the *nido*-carborane ligands with [Rh(acac)(cod)] (cod = cyclooctadiene) to afford complexes with the *exo-nido* structure [Rh(7-PPh_2-8-Me-7,8-$C_2B_9H_{10}$)(cod)] (**17**) and [Rh(7-SR′-8-Me-7, 8-$C_2B_9H_{10}$)cod] (R′ = Ph, Et) (**18–19**), in which the carborane cage is coordinated to a $[Rd(cod)]^+$ fragment through a S—Rh and a B(11)–H–Rh bonds *(76, 77)*. The difference with the previous complexes is the presence of a cod ligand instead of two PPh_3 groups. It was also demonstrated that the *exo-nido*-cyclooctadiene complexes **18** and **19** rearranged in solution to give the corresponding *closo*-complexes of stoichiometry [*closo*-3-(C_8H_{13})-1-SPh-2-R-3,2,1-$RhC_2B_9H_9$] (R = Ph, Me) (**18–21**) (Fig. 3). The X-ray analysis of the crystal structures of **17** and an analogous complex to **18** with a Ph group in the C8 atom, confirmed the formation of these compounds (Fig. 4).

Likewise, a set of Ru(III) half-sandwich complexes of the type [RuX(7-R-10-SR′R″-7,8-$C_2B_9H_{10}$)(PPh_3)$_2$] (R = H, Me, R′ = R″ = Me, Et; R′R″ = (CH_2)$_2$, (CH_2)$_4$) (**22a-g**, Fig. 5) were synthesized by the one-pot reaction of the zwitterions 7-R-10-SR′R″-*nido*-7,8-$C_2B_9H_{10}$ with K [t-BuO] and [$RuCl_2(PPh_3)_3$] in 1.5/1.5/1 ratio in EtOH, at 50 °C for 1 h *(78)*. In the reaction, the K[t-BuO] in EtOH removes the open face proton to form the partially charge-compensated monoanionic carborane ligand [7-R-10-SR′R″-*nido*-7,8-$C_2B_9H_{10}$]$^-$ that reacts with [$RuCl_2(PPh_3)_3$]. The spectroscopic data, elemental analyses and finally the X-ray diffraction analysis of the crystal structure of compound **22a** (Fig. 2) confirmed the geometry of these complexes.

In the 90s, we synthesized three 14-electron ruthenium(II) complexes with one or two *nido*-diphosphinocarborane anions (L = [7-(PR_2)$_2$-8-R′-7,8-$C_2B_9H_{10}$]$^-$, R = Ph, *i*-Pr; R′ = H, Me, Ph, and [7,8-(PPh_2)$_2$-7,8-$C_2B_9H_{10}$]$^-$, respectively). Complex **23** (Fig. 6) was obtained from the reaction of [NMe_4][7,8-(PPh_2)$_2$-7,8-$C_2B_9H_{10}$] with [$RuCl_2(PMePh_2)_4$] in 1:1 ratio in refluxing methanol, whereas complexes **24** and **25** (Fig. 6) were prepared by the reaction of [NMe_4][7,8-(PPh_2)$_2$-7,8-$C_2B_9H_{10}$] with

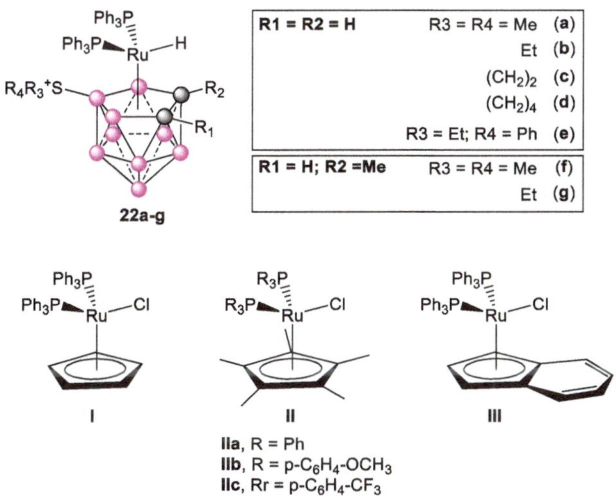

Fig. 5 Ru(III) half-sandwich complexes and Ru(II) arene complexes.

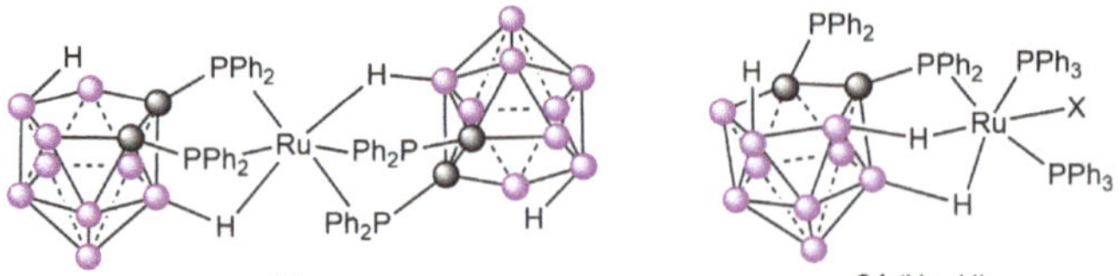

23

24 (X = H)
25 (X = Cl)

Fig. 6 Set of ruthenium complexes containing *nido*-diphosphinocarborane ligands used in Kharasch addition reaction.

$[RuCl_2(PPh_3)_3]$ both in a 1:1 ratio or 1:2 ratio, under the same conditions. According to the $^{1}H\{^{11}B\}$ and ^{31}P NMR data all complexes afford B–H→Ru agostic bonds, which led us to further study their catalytic activities in the Kharasch addition of carbon tetrachloride to representative olefins *(79)*.

3. Hydrogenation of terminal and internal alkenes with rhoda and ruthenacarboranes

The breakthrough that metallic complexes could be used as catalyst precursors in homogeneous hydrogenation and isomerization reactions led to important developments in organometallic and coordination chemistry *(80,81)*. Several decades of research and a large number of advances on this area culminated in the awarding of the Nobel Prize in Chemistry to William S. Knowles and Ryoji Noyori for their work on chiral catalyzed hydrogenation reactions in 2001 *(82)*. Since then around 3817 papers (167 reviews) searching by title and 17,255 (1442 reviews) searching by topics on asymmetric hydrogenation have been published according to Web of Science (10th March 2022) The development of transition metal complexes for their use in catalysis has also been of interest in the chemistry of Boron Clusters.

As mentioned previously, the main target for designing these *nido*-carborane-based complexes was their application as catalyst precursors in the hydrogenation of alkenes. Two sets of ruthenacarboranes containing *exo-nido*-monophosphino and *exo-nido*-monothioethercarborane, $[RuX(7\text{-}PR'_2\text{-}8\text{-}R\text{-}7,8\text{-}C_2B_9H_{10})(PPh_3)_2]$ (**1–5**) and $[RuCl(7\text{-}SR'\text{-}8\text{-}R\text{-}7,8\text{-}C_2B_9H_{10})(PPh_3)_2]$ (**6–9**), were tested in the catalytic hydrogenation of 1-hexene to be converted into *n*-hexane and 2-hexenes. It is important to highlight, that all $[RuX(7\text{-}PR'_2\text{-}8\text{-}R\text{-}7,8\text{-}C_2B_9H_{10})(PPh_3)_2]$ complexes were active catalysts in the hydrogenation of 1-hexene with similar selectivity (Table 1). The catalytic precursor $[RuH(7\text{-}PPh_2\text{-}7,8\text{-}C_2B_9H_{11})(PPh_3)_2]$

Table 1 % Conversion of 1-hexene to *n*-hexane and 2-hexenes using Rhoda and ruthenacarboranes as catalysts.

Complex	T (°C)	P (bar)	10^4 [Cat] (M)	% *n*-Hexane	% 2-Hexenes
1	66	45	$5.2^{b,c}$	73	14
2	66	45	$5.2^{b,c}$	30	3
3	66	45	$5.2^{b,c}$	82	5
4	66	45	$5.2^{b,c}$	70	5
5	66	45	$5.2^{b,c}$	28	4
6	66	45	$5.2^{b,c}$	–	–
10	25	1	$7.4^{a,c}$	47	9
11	25	1	$7.4^{a,c}$	14	5.3
12	25	1	$7.4^{a,c}$	3.6	12
10	25	45	5.2^{b}	98	0.2
13	25	45	5.2^{b}	12	0.7
14	25	45	5.2^{b}	6	0.3
15	25	45	5.2^{b}	0.2	0.1
16	25	45	5.2^{b}	2.4	0.1
10	66	45	5.2^{b}	85	11
13	66	45	5.2^{b}	90	9.3
14	66	45	5.2^{b}	61	20
15	66	45	5.2^{b}	82	11
16	66	45	5.2^{b}	83	14

[a][Hexane] = 1.1.
[b][Hexane] = 3.9.
[c]Toluene was used as solvent and for the rest of experiments THF was used.

(**3**) was the more active giving a 87% of conversion at T = 66 °C and P = 45 bar with low alkene isomerization (5%). [RuCl(7-SPh-8-Me-7, 8-$C_2B_9H_{10}$)(PPh_3)$_2$] (**6**) catalyzes the hydrogenation of 1-hexene, exhibiting a TON of 115 ([1-hexene] = 3.9 M, [**6**] = 5.21×10^{-4} M, 66 °C, P = 45 bar, 1 h) *(72)*. The addition of a base such as NEt_3 which in some ruthenium catalytic systems was shown to favor the heterolytic activation of hydrogen *(83)* did not result in an increase in activity.

The following set of Rh complexes: [Rh(7-SR-8-R′-7,8-$C_2B_9H_{10}$)(PPh_3)$_2$] (**10–12**), [Rh(7-PR_2-8-R′-7,8-$C_2B_9H_{10}$)(PPh_3)$_2$] (**13–16**), [Rh(7-PPh_2-8-Me-7,8-$C_2B_9H_{10}$)(cod)] (**17**), [Rh(7-SR-8-R′-7,8-$C_2B_9H_{10}$)(cod)] (**18, 19**), [*closo*-3-(C_8H_{13})-1-SR-2-R-3,2,1-Rh$C_2B_9H_9$] (**20, 21**), were also tested in the hydrogenation reaction of 1-hexene (Table 1). A comparative study was performed by using similar conditions with the aim to establish the different catalytic activities between complexes containing *nido*-monothioether- or *nido*-monophosphino-*o*-carborane. *exo-nido*-Monothioetherhodacarborane complexes of formulae [Rh(7-SR-8-R′-7,8-$C_2B_9H_{10}$)(PPh_3)$_2$] showed noticeable activity at $T=25$ °C, leading to high yields of *n*-hexane, whereas the homologous *nido*-monophosphino-*o*-carboranes [Rh(7-PR_2-8-R′-7,8-$C_2B_9H_{10}$)(PPh_3)$_2$] required higher pressures and temperatures. On the other hand, similar activity and selectivity were obtained for both type of rhodacarboranes at $T=66$ °C, where a ratio of alkene isomerization to hydrogenation increased for both of them *(75)*. In general, *exo-nido*-monothioetherhodacarboranes exhibited higher catalytic activities than their homologous ruthenacarboranes. Remarkably, the [Rh(7-SPh-8-Me-7,8-$C_2B_9H_{10}$)(PPh_3)$_2$] (**10**) complex exhibited much higher conversion to *n*-hexane (around 8 times higher) than the Wilkinson's catalyst [RhCl(PPh_3)$_3$] at $T=25$ °C and $P=1$ atm *(74)*. Likewise, in an attempt to study the mechanism of the reaction, it was proved that catalytic activity was inhibited by the addition of PPh_3, that suggested that reversible dissociation of PPh_3 is a key point before the activation of hydrogen in the catalytic cycle. The high activity for rhodacarboranes, notably for compound **10**, was attributed to the unique stereochemistry of the complexes. It was suggested that the B-H groups from the open face were more prone to undergone B-Rh(III)-H oxidative addition during the catalytic cycle, giving rise to higher rates. However, the significant differences between the activity of *nido*-monothioether versus *nido*-monophosphinorhodacarboranes seems to indicate that other factors are also playing an important role. Regarding the temperature and pressure, they play a crucial role in the hydrogenation, as higher temperature and pressure lead to a larger % of conversion for almost all the systems.

The high catalytic activity of some *exo-nido*-monophosphinorhoda and *exo-nido*-monophosphinoruthenacarboranes in the hydrogenation of 1-hexane led us to test them in the stereoselective hydrogenation of methacycline that usually gives two diastereomers: doxycycline, which is a potent antibiotic from the family of tetracyclines used in chemotherapy and *epi*-doxycycline *(84)*. For this purpose, the following conditions were used: [methacycline] $= 2.2 \times 10^{-2}$ M, [catalyst] $= 8.7 \times 10^{-4}$ M, $T = 75$ °C, $P = 45$ atm and $t = 7$ h. Noticeably, rhodacarboranes **13** and **14** exhibited high activity and selectivity to give conversions of 85 and 99.7% to doxycycline *(75)*. Very low activity was observed for ruthenacarboranes **1–5** with very low conversions (~2%).

All the previous Rh and Ru complexes exhibited very low activities in the hydrogenation of internal alkenes. With the purpose to enhance the catalytic activity, the *exo-nido*-monothioether complex [Rh(7-SR-8-R′-7,8-$C_2B_9H_{10}$)(cod)] was synthesized and tested in the hydrogenation of cyclohexene *(76)*. In parallel, its *closo* homologous [*closo*-3-(C_8H_{13})-1-SR-2-R′-3,2,1-Rh$C_2B_9H_9$] was also studied for comparative purposes. Notably, *closo* complexes **20** and **21** exhibited an outstanding catalytic activity when compared with their *exo-nido* homologous **18** and **19**, which was still higher when 1 eq. of PPh_3 was added to the reaction (Table 2). On the other hand, no differences were observed between diphosphine complexes [Rh(7-SR-8-R′-7,8-$C_2B_9H_{10}$)(PPh_3)$_2$)] and the monophosphine complex generated in situ from [Rh(7-SR-8-R′-7,8-$C_2B_9H_{10}$)(cod)] and 1 eq. of PPh_3.

Table 2 Conversion of cyclohexene to cyclohexane using carboranyl-base complexes as catalysts.

Complex	% Hydrogenation[a]
10	32
11	8
18[b]	30
19[b]	12
20	76
21	82
21[b]	98

[a]Experimental conditions: [complex] $= 7 \times 10^{-3}$ M; [cyclohexene] $=$ 1.5 M; $P = 20$ atm; $T = 25$ °C; $t = 1$ h.
[b]Experiment performed with 1 equivalent of PPh_3.

4. Cyclopropanation reactions using rhoda and ruthenacarboranes

The exceptional activity of rhodacarboranes in hydrogenation reactions and their ability to liberate a phosphine ligand were the triggers to test these complexes for olefin cyclopropanation reactions. One of the most common procedures to produce cyclopropane derivatives is the well-known cyclopropanation of olefins with diazoacetates catalyzed by transition metal complexes (Scheme 3). At the end of the 90s our group envisaged for the first time the use of the Rh(I) complexes [Rh(7-PPh_2-8-R-7,8-$C_2B_9H_{10}$)$(PPh_3)_2$] (R=H (**13**), Me (**14**)) and [Rh(7-PPh_2-8-Me-7,8-$C_2B_9H_{10}$)(cod)] (**17**) as catalysts in the cyclopropanation of different olefins (styrene, 4-methylstyrene, 4-chlorostyrene and cyclooctene) *(85)*. At that time, Rh(II) and Rh(III) catalyst, usually dirhodium(II) carboxylates, dirhodium(II) carboxamidates, as well as copper complexes modified by nitrogen-containing ligands were the most efficient ones, whereas Rh(I) complexes were less known. The % of cyclopropanation yield was determined from the ethyl diazoacetate by GCL. In general, the three complexes **13**, **14** and **17** showed a good efficiency giving similar high yields ranging between 80% and 95% and stereoselectivities (*cis/tans* or *endo/exo)* in the range 0.52–0.91, suggesting negligible influence of the nearby ligands. Nevertheless, these complexes have two main drawbacks: first they are active at temperatures higher than 60 °C, secondly, their selectivities are not better compared to rhodium(II) carboxylates or copper-based catalysts. Complex **13** was proven as catalyst for the cyclopropanation reaction of different type of olefins (Table 3). The results indicated that activated olefins showed a higher reactivity than not-activated ones, as an example, styrene is 10 times more reactive than cyclooctene or 1-octene. *n*-Butyl vinyl ether is also an activated olefin and was 6 times less reactive than styrene in presence of complex **13**. To be noted was the very low cyclopropane yield obtained with norbornene, where the 1,3-dipolar addition of ethyl diazoacetate was the preferred pathway.

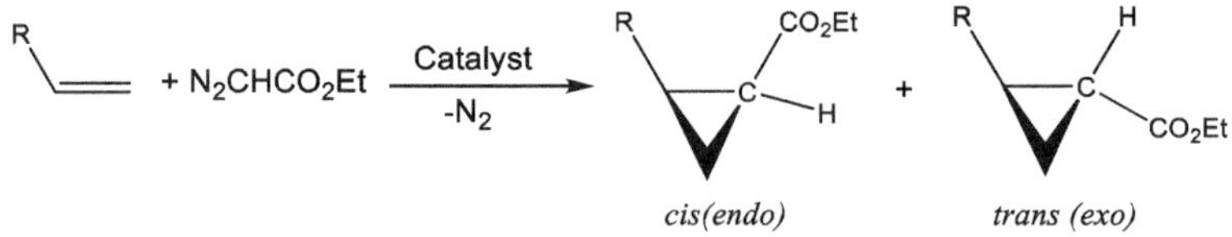

Scheme 3 Cyclopropanation of olefins with ethyl diazoacetate.

Table 3 Cyclopropanation of different olefins with ethyl diazoacetate using **13** as catalyst.

Olefin	Cyclopropanation yield[a]	% *cis/trans* or *endo/exo*[b]
Styrene	91	0.88
4-Methylstyrene	90	0.68
4-t-Butylstyrene	93	0.54
4-Methoxystyrene	87	0.60
4-Fluorostyrene	93	0.62
4-Chlorostyrene	89	0.51
4-Bromostyrene	90	0.47
1,2-Dihydronaphtalene	91	0.69
α-Methylstyrene	94	0.89[c]
n-Butyl vinyl ether (90 °C)	88	0.69
Cyclooctene	86	0.69
1-Octene	71	0.56
1-Dodecene	73	0.66
Norbornene (80 °C)	9	0.69[d]

[a]Reaction conditions: catalyst, 0.0075 mmol; olefin, 20 mmol; ethyl diazoacetate, 1 mmol diluted by the olefin up to 1 mL; addition time, 4 h; temperature, 100 °C.
[b]Based on ethyl diazoacetate and determined by GLC analysis.
[c]Z/E ratio.
[d]**16** (80 °C): yield, 6%; *endo/exo* ratio, 0.71.

In the cyclopropanation reaction, the carbene addition occurred with *trans(exo)* diastereoselectivity, favoring the most thermodynamically stable isomer. In this work, the effect of the size of the olefin substituents was studied using various 4-X-styrene derivatives. In general, a decrease of the *cis/trans* ratio was observed when increasing the steric bulk of X [*cis/trans* ratio = 0.88 (styrene), 0.68 (4-methylstyrene), and 0.54 (4-t-butylstyrene)] (Table 3). This could be improved further by using bulkier diazo compounds, such as t-butyl diazoacetate that decreased *cis/trans* stereoselectivity to 0.57 (styrene) and 0.52 (4-t-butylstyrene).

As it was mentioned before, rhodium complexes were the most efficient catalysts for cyclopropanation reactions, whereas ruthenium complexes always displayed lower activity than their rhodium homologous *(86)*.

The use of ruthenacarboranes as cyclopropanation catalysts were first reported by Demonceau et al. *(87)* It was demonstrated on a yield and selectivity basis that a high catalytic activity could be attributed to *closo*-ruthenacarboranes, whereas a poor catalytic activity was associated to their *exo-nido* congeners. Later, two *exo-nido* Ru(II) complexes containing monophosphinocarboranyl ligands $[RuH(7\text{-}PPh_2\text{-}8\text{-}R\text{-}C_2B_9H_{10})(PPh_3)_2]$ (R=H (**3**), Me (**4**)) were proven by our group as efficient catalysts in the cyclopropanation of a set of representative olefins with ethyl diazoacetate (Table 4) *(88)*. Remarkably, complexes **3** and **4** demonstrated excellent activities for activated olefins such as styrene and styrene derivatives with cyclopropanation yields ranging from 74% to 98%. Nevertheless, cyclopropane yields were lower with cyclic olefins or terminal linear mono olefins than with activated double bonds.

Furthermore, a set of Ru(III) half-sandwich complexes of the type $[RuX(7\text{-}R\text{-}10\text{-}SR'R''\text{-}7,8\text{-}C_2B_9H_{10})(PPh_3)_2]$ (R=H, Me, R′=R″=Me, R′R″=$(CH_2)_2$) (**22a, d, f**, Fig. 5) were used as catalysts for cyclopropanation reactions *(89)*. Their activities were compared to Ru(II)

Table 4 Cyclopropanation reaction of different olefins with ethyl diazoacetate using **3** and **4** as catalysts.

	Cyclopropanation yield,[a] %[b] (*cis/trans* or *endo/exo* ratio)	
	Catalyst:	
Olefin	3	4
Styrene (60 °C)	74 (0.65)[c]	78 (0.62)[d]
Styrene (100 °C)	97 (0.63)	96 (0.64)
4-Methylstyrene	96 (0.52)	96 (0.54)
4-tert-Butylstyrene	93 (0.50)	91 (0.48)
4-Methoxystyrene	90 (0.61)	89 (0.56)
4-Chlorostyrene	94 (0.50)	93 (0.48)
α-Methylstyrene	98 (0.95)	97 (1.02)
Cyclooctene	51 (0.86)	65 (1.08)
1-Octene	61 (0.71)	58 (0.62)
1-Dodecene	59 (0.73)	61 (0.73)

[a]Reaction conditions: 0.0075 mmol of catalyst; olefin, 20 mmol; ethyl diazoacetate, 1 mmol diluted by the olefin up to 1 mL; addition time, 4 h; temperature, 100 °C.
[b]Based on ethyl diazoacetate and determined by GLC analysis.
[c]Polymers were formed in the cyclopropanation of styrene and its X-derivatives, i.e., styrene at 60 °C, reaction time 12 h: polymer yield 15%.
[d]Polymer yield 17%.

complexes containing cyclopentadienyl derivatives (Cp or Cp*) or indenyl as ligands (**I–III**, Fig. 5). The main aim was to determine how changes in electronic and steric properties of the different ligands affect the efficiency and stereoselectivity of the reaction: Cp* has a higher electron donating capacity than Cp; the $[C_2B_{10}H_{11}]^{2-}$ ligand has the unique ability to stabilize higher metal oxidation states than common arenes; Cp* is bulkier than Cp and carboranyl ligands, and indenyl is able to produce a metal ring slippage from η^5 to η^3-coordination, giving rise to the formation of a vacant coordination site on the metal that facilitate the hosting of a substrate. In Table 5 we have summarized the yields (%) and *cis/trans* ratio from the cyclopropanation reaction of styrene with ethyl diazoacetate using this set of catalysts under standard conditions. It is important to notice that complex **I** containing Cp as ligand has shown higher activity (85%) than other similar catalysts with more hindered Cp* ligands and has given the highest *cis/trans* ratio (1.85). When we moved to study the ruthenacarborane complexes, we observed a similar or higher activity than **I** (~86%) with a reverse selectivity giving rise to the *trans* isomer as the major one. In case of complexes **IIa** and **III** the reaction of ethyl diazoacetate with styrene proceeded smoothly showing the lowest catalytic activities (around 60% cyclopropanation yield); on the other hand, carbene insertion into the vinylic C—H bonds was also observed. It was determined that in addition to forming the expected cyclopropanes other byproducts due to C—H insertion were obtained.

Later on, we used complexes **I** and **22a, d, f** as catalysts in the cyclopropanation of a set of styrene derivatives and other olefins (1-octene), using

Table 5 Cyclopropanation reaction of styrene with ethyl diazoacetate with different ruthenium complexes as catalysts.[a]

Complex	Cyclopropanation %	*cis/trans*[b]
21a	85	0.48
21d	86	0.52
21f	86	0.52
I	85	1.85
IIa	57	0.17
IIb	59	0.25
IIc	61	0.30
III	68	0.48

[a]*Reaction conditions*: complex, 0.005 mmol; styrene, 2 mL; ethyl diazoacetate, 1 mmol diluted in 1 mL of styrene; addition time, 4 h; 40 °C.
[b]Based on ethyl diazoacetate.

ethyl diazoacetate at 40 °C. In general, these Ru complexes have shown good yields in the formation of cyclopropane, with predominantly *cis* stereoselectivity for **I** and *trans* stereoselectivity for **22**. On the other hand, non-activated olefins (1-octene and cyclooctene) were much less reactive, mainly producing dimethyl maleate. Apparently, electron-rich olefins show a higher reactivity toward the intermediate carbene species, and thus carbene dimerization can be suppressed.

5. Kharasch addition (ATRA) and radical polymerization (ATRP)

Control of radical polymerization has been one of the most challenging frontiers in polymerization chemistry. This section presents the discovery of phosphine ruthenacarboranes as catalysts for Karasch addition (ATRA) and atom transfer living radical polymerization reactions (ATRP).

The beginnings of the Kharasch reaction date back to 1945 when was first reported the direct addition of carbon tetrachloride to olefinic double bonds catalyzed by peroxides as radical initiators. A major breakthrough occurred by the mid-1950s with the discovery that a number of transition metal complexes were able to catalyze the reaction *(90–93)*. After two decades of moderate interest, the Kharasch addition got a major renaissance with Sawamoto's *(94)* and Matyjaszewski's *(95)* independent findings demonstrating that the Kharasch reaction could be successfully extended to olefin polymerization. In the presence of an excess of olefin (monomer) relative to the halogenated compound (initiator), highly controlled polymers in terms of molecular weights (Mn) and molecular weight distributions (Mw/Mn) could be synthesized provided that favorable equilibria between the active species (–C$^{\bullet}$) and the dormant species (–C–X) were established (Scheme 4) *(96–98)*. This is the so-called "atom transfer radical polymerization" (ATRP) which was coined from "atom transfer radical addition" (ATRA)—the Kharasch addition.

Ruthenium phosphine complexes have played a prominent role in Kharasch chemistry since the early 70s of the 20th century *(99)*. Formation of unsaturated 14-electron ruthenium species (through phosphine dissociation) is assumed to take place prior to halocarbon activation *(100)*. Thus, [$RuCl_2(PPh_3)_3$] *(98)*, [$RuCl_2$(arene)(PR_3)] *(101)* or [$RuCl_2$(=CHPh)$(PR_3)_2$] *(102,103)* complexes were first introduced. Key features such as the lability of the arene and bulky tertiary phosphine (PR_3) ligands were searched at the time. The latter provides steric protection to the empty coordination sites at the metal and to generate stable well-defined

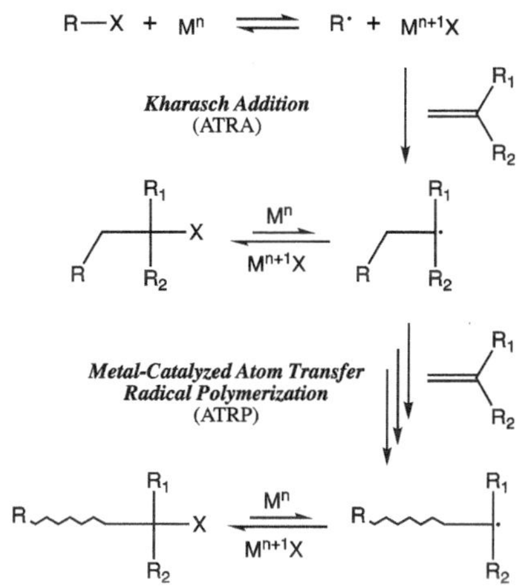

Scheme 4 Evolution of the Kharasch addition reaction into metal-catalyzed living radical polymerization. X=Cl, Br, and M^{n+} is a transition metal in its oxidation state n.

14-electron complexes that could give a direct access to the catalytic species *(104)*. The steric crowding forces the substituents of the ligands in close proximity to the metal center to induce an agostic interaction between a C—H bond in the substituents and the metal atom, thus preventing dimerization or solvent coordination. Similarly we reported B–H→Ru agostic bonds in ruthenium complexes of the dianionic η^5-$[C_2B_9]^{2-}$ thioether/s and phosphine/s carboranyl ligands in the late 90s *(71)*. Based on these considerations, a new set of ruthenium complexes of the monoanionic η^5-$[8\text{-}R\text{-}C_2B_9]^-$ ($R=SR_2$, PR_3) ligands (Fig. 5) were synthesized to discover these new families of ruthenacarborane catalytic systems *(105–108)*.

The catalytic activity of three 14-electron ruthenium(II) complexes with one or two *nido*-diphosphinocarborane anions ($L=[7\text{-}(PR_2)_2\text{-}8\text{-}R'\text{-}7,8\text{-}C_2B_9H_{10}]^-$, R=Ph, *i*-Pr; R′=H, Me, Ph, and $[7,8\text{-}(PPh_2)_2\text{-}7,8\text{-}C_2B_9H_{10}]^-$, respectively, **23–25** in Fig. 6) was studied in the Kharasch addition of carbon tetrachloride to four representative olefins, using standardized reaction conditions (Table 6, Scheme 5) *(79)*.

Nearly quantitative conversion of (meth)acrylates was observed with all three complexes **23–25**. However, bis-carborane complex **23** showed much less activity than the monomeric complexes **24**, **25** for the conversion of 1-Decene or Styrene (Table 6). In terms of catalyst activity, loading, and reaction conditions, these ruthenacarborane complexes compare to $RuCl_2(PPh_3)_3$, which was the most active Ru complex at the time *(109,110)*. The more efficient Karasch addition of CCl_4 was observed for styrene and hydride complex **24** (78%), while other complexes gave lower yields. However, **24** showed no $CHCl_3$ addition to Styrene. The difference

Table 6 Kharasch addition of carbon tetrachloride and chloroform to representative olefins catalyzed by ruthenium complexes **23–25**.[a]

			Substrate conversion (Kharasch addition) (%)[b]		
Substrate	**Alkyl halide**	**Temperature**	23	24	25
1-Decene	CCl_4	60	8 (7)	64 (44)	52 (42)
Styrene	CCl_4	60	8 (3)	90 (78)	56 (46)
n-Butyl acrylate	CCl_4	85	99 (40)	99 (23)	99 (17)
Methyl methacrylate	CCl_4	85	96 (69)	91 (51)	97 (40)
Styrene	$CHCl_3$	60	0 (0)	8 (0)	18 (0)

[a]Reaction conditions: Prior to use, the reagents, the solvent (toluene) and the internal standard (dodecane) were dried using well established procedures, distilled and kept under nitrogen at −20 °C. The catalyst (0.03 mmol) was dissolved in toluene (1 mL) and subsequently added through a septum to the solution of alkene (9 mmol), CCl4 or CHCl3 (13 mmol), dodecane (0.25 mL) in toluene (3 mL). The reaction mixture was heated at 60 or 85 °C for 24 h.
[b]Conversion and yields, based on GLC using dodecane as internal standard.

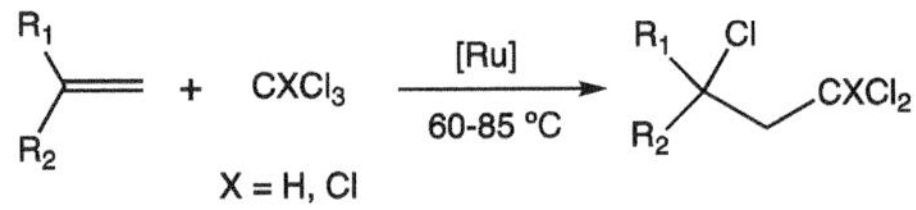

Scheme 5 Kharasch addition of carbon tetrachloride to olefins.

Table 7 Polymerization of methyl methacrylate and styrene catalyzed by ruthenium complexes **23-25**[a].

Monomer	Catalyst	Polymer yield (%)	M_n[b]	M_w/M_n[b]
Methyl methacrylate, 85 °C	**23**	8	4700	2.3
	24	84	28,000	1.25
	25	72	13,000	1.95
Styrene, 110 °C	**23**	52	30,000	1.75
	24	61	28,000	2.25
	25	60	23,000	1.9

[a]Reaction conditions: catalyst, 0.0117 mmol; initiator (CCl_4), 0.0234 mmol solution 0.1 M in toluene, 0.25 mL); monomer, 1 mL; polymerization time, 16 h.
[b]Size exclusion chromatography calibrated with PMMA and PS standards, respectively.

between conversion and Karasch additions shown in Table 6 was explained by the formation of oligomers or telomers, as metathesis did not occur *(107)*.

Quite interestingly, polymerization of methyl methacrylate and styrene took place by **23–25** in the presence of excess of olefin (Table 7). Remarkably, polymerization of methyl methacrylate occurred in high yield and controlled way (MW distributions narrower than 1.25) by the hydride complex **24**. The latter also works in the absence of cocatalyst, so that ruthenacarborane complex **24** is a promising catalyst for atom transfer radical addition (ATRP) of methyl methacrylates.

Soon after the first report of *exo-nido*-ruthenacarboranes **23–25** (Ru is not coordinated to the C_2B_3 pentagonal face of the *nido*-carboranes; Fig. 6), we moved to the half-sandwich ruthenium complexes **22a-g** (top of Fig. 5) *(53,105,106)*. In these complexes, the metal "sits" on the C_2B_3 pentagonal face of the *nido*-carboranes, with a similar coordination mode to that of the C5 cyclopentadienyl (Cp) ligand (bottom of Fig. 5). Thus, we decided to study the possible influence of the *nido*-carboranes on the ATRA and ATRP reactions and in comparison with the isolobal Cp-type ruthenium complexes **I-III** (Fig. 5) *(101,111–115)*. The particular size and electronic properties of *nido*-carboranes were expected to have a significant impact in the ATRP. Indeed, ruthenacarboranes (top of Fig. 5) proved highly efficient and versatile catalyst for the Kharasch addition to olefins and controlled polymerization *(71,105,106)*.

Studies of the radical polymerization of styrene and *n*-butyl acrylate showed that Cp-ruthenium complex **II** was the most efficient catalyst precursor (high yield (91%) and high degree of control (Mw/Mn = 1.2)) for the controlled radical polymerization among ruthenium complexes and ruthenacarboranes in Fig. 5. Complexes **I**, **III** and **22a-d** and **22d-g** were found inefficient and all afforded polymers in an uncontrolled way *(106)*. It was found, however, that ruthenacarboranes **22** polymerized *n*-butyl acrylate close to room temperature (Table 8). This was a quite relevant result as radical polymerizations by **II** were done at 110 °C *(106)*.

Table 8 Polymerization of *n*-butyl acrylate catalyzed by complex **21d** and initiated by ethyl 2-bromopropionate[a].

Temperature (°C)	Polymer yield (%)	M_n[b]	M_w/M_n[b]
30	87	835,000	2.65
45	96	250,000	5.15
60	97	71,000	6.1
75	98	23,100	6.7
85	97	22,500	7.3

[a]Reaction conditions. $[styrene]_0:[initiator]_0:[Ru]_0 = 750:2:1$ (initiator, (1-bromoethyl)benzene; temperature, 110 °C; reaction time, 16 h). $[n\text{-butyl acrylate}]_0:[initiator]_0:[Ru]0 = 1200:4:1$ (**24d**) (initiator, ethyl 2-bromopropionate; reaction time, 16 h).

[b]Determined by size-exclusion chromatography (SEC) with polystyrene and PMMA calibration, respectively.

The low temperature ATRP reaction of the ruthenacarboranes (**21**) prompted us to study the initial step of the reaction, that is, the Karasch addition (ATRA; Scheme 4) *(106)*. As summarized in Table 9, all half-sandwich ruthenacarboranes (**22a-f**) showed the highest and fastest activities for the Karasch addition of CCl_4 to methyl methacrylate (MMA) and Styrene, surpassing that for the related Cp-ruthenium complexes (bottom of Fig. 5). In particular, **22e** proved the most efficient ATRA catalyst reported so far, including a highly active pincer nickel catalyst *(116)*. Turnover numbers (TTN) of 4200 and 9000, and initial turnover frequencies (TOF) of 1880 and 1500 h^{-1} were obtained at 40 °C, respectively, for MMA and Styrene with **22e** *(105)*.

The high activity of these half-sandwich ruthenacarboranes for the Karasch addition is related in reaching an optimum electrical potential that is more easily accessible for the ruthenacarboranes than for Cp-ruthenium complexes **I–III** (Table 9).

Ruthenacarboranes **22** were also found to be efficient catalyst precursors for the addition of CCl_4 to *n*-butyl acrylate (89%–95% Karasch addition), but are not so efficient for 1-Decene (41%–66% Karasch addition) *(105)*.

Table 9 Cyclic voltammetry[a] data and Kharasch addition of carbon tetrachloride to styrene and methyl methacrylate[b].

			Conversion[c] (%)/yield[c] (%)	
ΔE (mV)	E (mV)	Catalyst	Styrene	MMA
86	+133	**I**	24/10	62/5
94	−83	**II**	86/69	79/79
94	−10	**III**	74/69	83/83
76	−266	**22a**	100/99	98/94
82	−266	**22b**	100/99	98/94
98	−287	**22e**	100/99	100/100
80	−266	**22c**	100/99	99/99
80	−366	**22f**	100/99	92/84
88	−368	**22 g**	100/99	92/84

[a]Sample, 1 mM; Bu_4NPF_6 (0.1 M) in CH_2Cl_2; $\nu = 50$ mV s^{-1}; potentials are reported in millivolt versus ferrocene as an internal standard.
[b]Reaction conditions: olefin (9 mmol), carbon tetrachloride (13 mmol), catalysts (0.03 mmol), toluene (4 mL), dodecane (0.25 mL), under nitrogen atmosphere. The reactions were carried out at 40 °C and stopped after 6 h.
[c]Conversions and yields are based on the olefin, and determined by GC using dodecane as internal standard. The Kharasch adducts were characterized by comparison with literature data.

The reaction was even less efficient when $CHCl_3$ was used, giving yields of addition below 33% and 7% for styrene and MMA, respectively. Mechanistic studies confirmed the radical nature of the reactions and the dissociation of one of the phosphine ligands to provide 16-electrons ruthenacarborane species. The latter can react with CCl_4 to provide $^{\bullet}CCl_3$ radical that can add to the olefins and therefore initiating the catalytic cycle.

A comparative study for the ATRP polymerization of *n*-butyl acrylate or methyl methacrylate by the ruthenium semi-sandwich complexes **I-III** (Fig. 5) that parallel the corresponding sandwich complexes or ruthenacarboranes **22**, clearly demonstrated that the expected ring-slippage of the indenyl moieties in the complexes has no positive effect in the polymerization reactions *(117)*.

6. Suzuki coupling reaction using *o*-carboranyl NBN based pincer Pd complexes as catalysts

Pincer complexes represent an exciting platform for homogeneous catalysis *(118–120)*. Pincer ligands feature a rigid and tridentate binding pocket, usually a coordinating bridgehead position and two donor arms. Boron was surprisingly absent in this picture until work reported by Teixidor *(57)* in 1991 and later by Nozaki *(121)* and Mirkin *(122)* in 2009. Teixidor and co-workers reported the first dithio-7,8-*nido*-carboranyl based SBS pincer rhodium complex. In 2014, we obtained a first family of o-carboranyl NBN based pincer palladium complexes and extended the synthesis to the corresponding *m*-carboranyl NBN analogues (Fig. 7).

Catalytic applications of **(*o*CB-L1)Pd** and **(*m*CB-L1)Pd** have shown that the complexes are good catalyst precursors in Suzuki coupling reactions in water, and with remarkably low amounts of catalyst loadings (0.0001 mol%) and with good functional group tolerance for the substrates (Table 10).

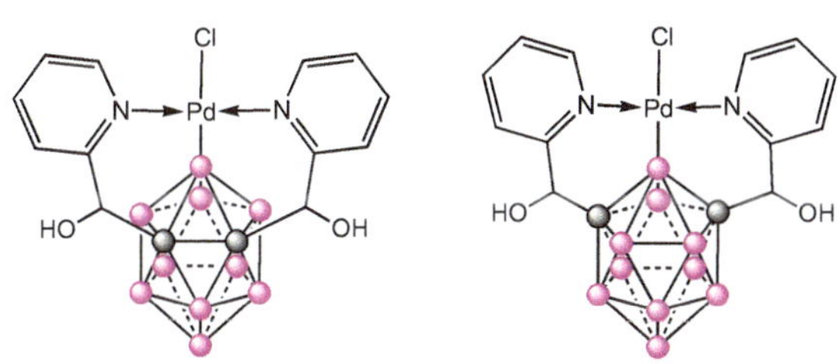

(oCB-L1)Pd **(*m*CB-L1)Pd**

Fig. 7 Pd complexes containing *m*-carboranyl NBN analogues as ligands.

Table 10 Suzuki coupling in the presence of carborane-based NBN pincer complexes.

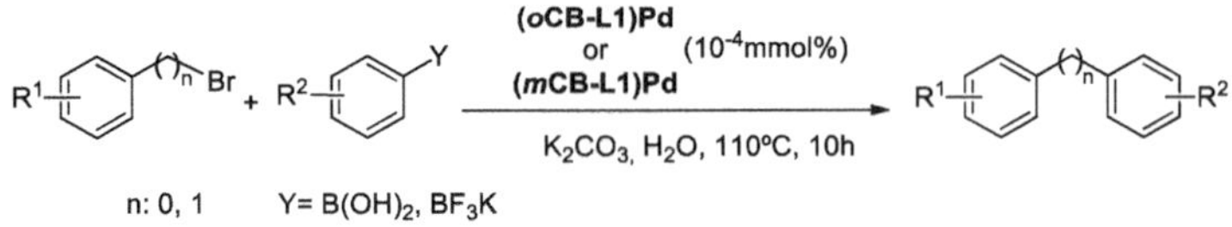

Entry	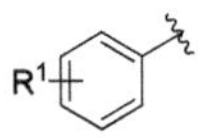	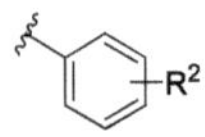	n	Y	(*o*CB-L1)Pd (%)[a]	(*m*CB-L1)Pd (%)[a]
1	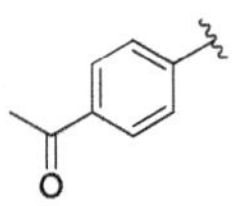	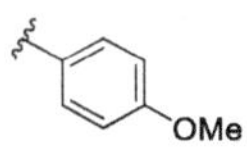	0	$B(OH)_2$	99	60
2	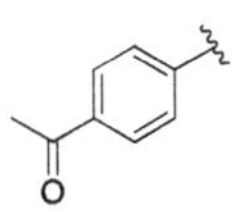	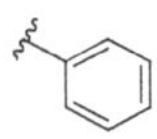	0	$B(OH)_2$	99	85
3[b]	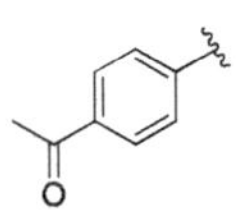	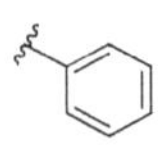	0	BF_3K	99	93
4	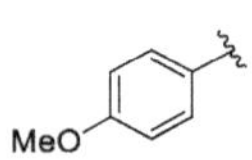	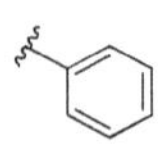	0	$B(OH)_2$	84	90
5[b]	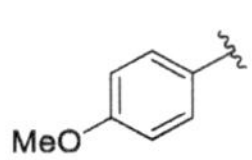	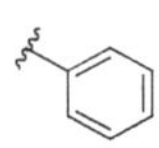	0	BF_3K	92	60
6	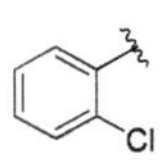	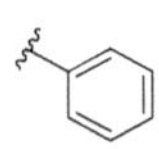	0	$B(OH)_2$	91	5
7[b]	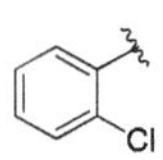	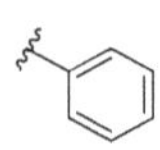	0	BF_3K	97	69
8	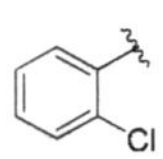	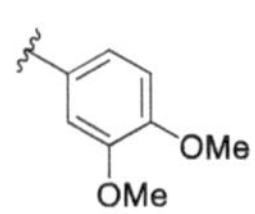	0	$B(OH)_2$	77	20
9	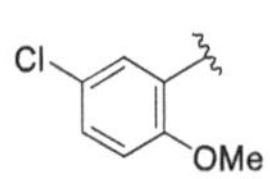	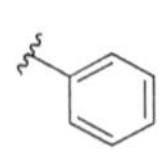	0	$B(OH)_2$	66	56

Table 10 Suzuki coupling in the presence of carborane-based NBN pincer complexes.—cont'd

Entry		n	Y	(*o*CB-L1)Pd (%)[a]	(*m*CB-L1)Pd (%)[a]
10[b]	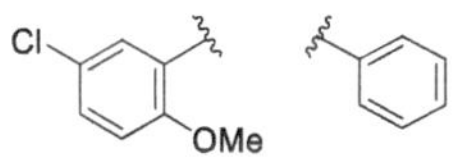	0	BF_3K	61	54
11	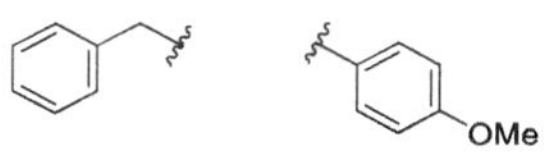	1	$B(OH)_2$	82	99
12	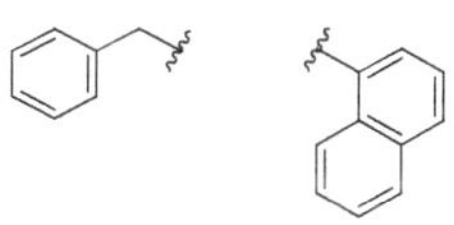	1	$B(OH)_2$	99	99
13	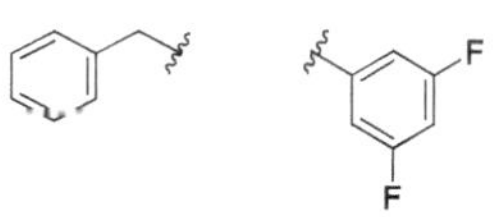	1	$B(OH)_2$	87	95
14	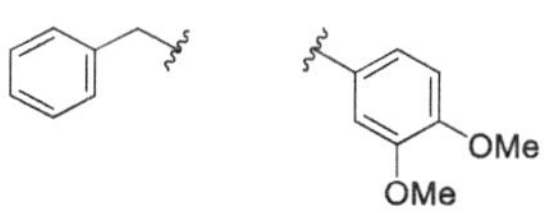	1	$B(OH)_2$	99	99

[a]Aryl or arylmethyl bromide, arylboronic acid, or aryl trifluoroborate (1.5 equiv), (oCB-L1)Pd or (oCB-L1)Pd (10^{-4} mol %), K_2CO_3 (2.0 equiv), H_2O (1 mL per mmol of ArBr or ArCH2Br), 110 °C, 10 h. Isolated yields.
[b]2.5 equiv of K_2CO_3 were used.

Complex **(*o*CB-L1)Pd** shows a better catalytic profile than **(*o*CB-L1)Pd** and with excellent conversions and TON values ranging from 770,000 to 990,000, thus showing a very high catalytic activity which rivals previous reports on Suzuki coupling performed by very low amounts of palladium catalysts, even with other pincer complexes *(123)*.

7. Metallabisdicarbollides as efficient light-driven oxidation catalysts

7.1 Background

In this section, we will focus on photoredox catalysis for oxidation processes. Before we will define some concepts such as *photochemistry*. This is a discipline of chemistry, different from thermal chemistry, that alters the nature of substrates induced by light. In photochemistry a reaction takes place by absorbing light. In this reaction, the activation energy necessary to change a substrate is provided by light. Normally in a photochemical reaction, light generates free radicals in molecules. Radical reactivity is complementary to polar or two-electron processes. Sometimes, a catalyst is required to achieve or accelerate a chemical reaction; the combination of photochemistry with catalysis is what is known as *photocatalysis*. In photocatalysis the reactions are accelerated by light, but maintaining the same course as the thermal reactions. *Photoredox catalysis* can be defined as the methodology where light is used to alter the redox properties of compounds then accelerating chemical reactions by electron transfer between a photocatalyst and a substrate *(124)*. Photoredox catalysis is a branch of photochemistry that employs small quantities of a light-sensitive compound that, when excited by light, trigger single-electron transfer (SET) processes with substrates upon radiation.

The development of the term photocatalysis responds to the evolution and research of alternative energy sources, as hydrogen by water splitting *(125)*; also to the cleaning of water and air avoiding the addition of chemicals *(126)* and to the chemical synthesis *(127)*. These applications fit into the field of green and sustainable chemistry. From a perspective of sustainability and respect for the environment, we must transform the current economic development paradigm to turn it into one that takes us on the path of sustainable development, inclusive and with a long-term vision. Mimicking processes that occur in nature is also one step forward in the development of sustainable chemistry. In this sense, photosynthesis is a source of inspiration for scientists; in this process, plants convert light energy into chemical energy *(128,129)*. Photoredox catalysis is the right way to mimicking nature in efficiency and sustainability *(130)*. Photoredox catalysts can generally be complexes of transition metals, organic dyes, and semiconductors. Soluble transition metal complexes are more commonly used today.

When we use the photoredox catalysis in synthesis using molecules of complexes as homogeneous photocatalysts, the interaction between an electronically excited photocatalyst and the substrate can result in the generation of diverse reactive intermediates, which involve one-electron oxidation or one-electron reduction of a substrate by the photocatalyst, occurring an outer-sphere reaction on the photoexcited metal center.

In this photoinduced electron transfer mechanism (PET), the photoinduced promotion of an electron from the HOMO to the LUMO orbital results in the formation of an excited state that can formally be conceptualized as a charge-separated electron-hole pair. The excited state can undergo relaxation back to the ground state via radiative or non-radiative processes. The redox properties of the excited state can lead to a fast electron transfer to an electron-deficient acceptor or oxidative quencher (A) or from an electron-rich donor or reductive quencher (D) compound. Regeneration of the photochemically active state requires a second electron transfer process from another donor (reducing) or acceptor (oxidizing) species (Fig. 8).

Electron transfer processes (ET) can be classified as outer or inner sphere. In the former, the photocatalyst remains intact before, during and after the ET, and no new bonds are broken or formed. In the latter, the ET occurs via a bridging ligand and bonds are broken and formed.

Photocatalysis can be systemized into three categories: (A) photocatalysis that enables the transformation to occur either via redox- or energy transfer process; in this case, the photocatalyst is not involved in the bond-forming / breaking event. (B) Cooperative/dual photocatalysis, which synergistically combines the first strategy with a transition metal catalyst. (C) The photocatalyst is a transition metal that enables the bond forming/breaking event.

In this work, we highlight recent studies mostly in our group in the development of new green photocatalysts for the oxidation of organic substrates where a traditional (A) or a cooperative photocatalysis (B) take place (Scheme 6).

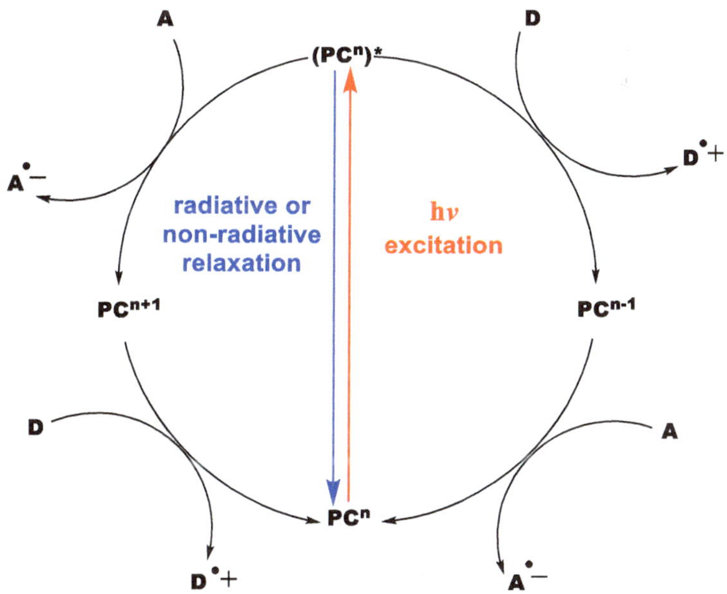

Fig. 8 Photoinduced electron transfer mechanism (PET).

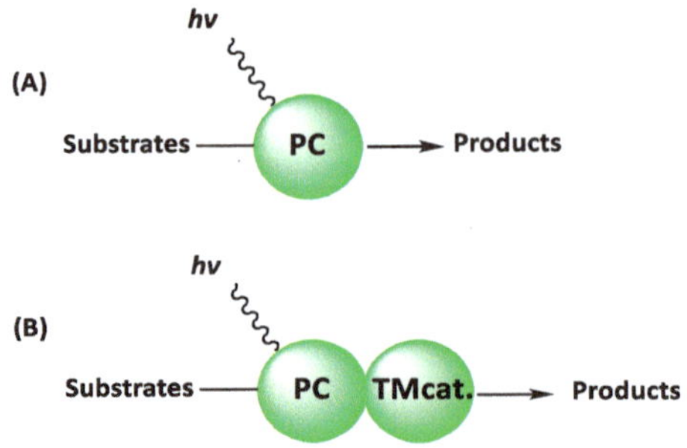

Scheme 6 Photocatalytic approaches.

Electron-transfer processes are initiated by the absorption of a photon by a chromophore, (i.e., the photocatalyst). These systems take the advantage of the enhanced redox reactivity of the chromophore in its excited state to facilitate a reaction that would not proceed otherwise; in this direction, light provides the extra energy needed to make the reaction thermodynamically viable *(131)*. The photoredox catalyst can transform the light into chemical energy and perform the selective molecule activation, simultaneously. The most common studied photocatalysts are based on transition metal complexes: polypyridyl ruthenium *(132, 133)* and iridium *(134)* complexes are the typical ones, being the tris (2,2′-bipyridine)ruthenium (II) ($[Ru(bpy)_3]^{2+}$) complex, the most studied (Fig. 9).

Upon absorption of a photon in the visible region by $[Ru(bpy)_3]^{2+}$, an electron is excited from one of the metal-centered *d*-orbitals to a ligand-centered π* orbital *(135)*, taking place a metal to ligand charge transfer (MLCT). During this process it exists a reductive and oxidative quenching cycle in which are produced $[Ru(bpy)_3]^{+}$, a strong reducing agent −1.76 V vs $Fc^{+/0}$ in CH_3CN, and $[Ru(bpy)_3]^{3+}$, a strong oxidant (+0.85 V vs $Fc^{+/0}$ in CH_3CN), respectively *(136)*. The complex participates in electron transfer through an outer-sphere mechanism, which implies that no major structural modification occurs on ET *(137)*.

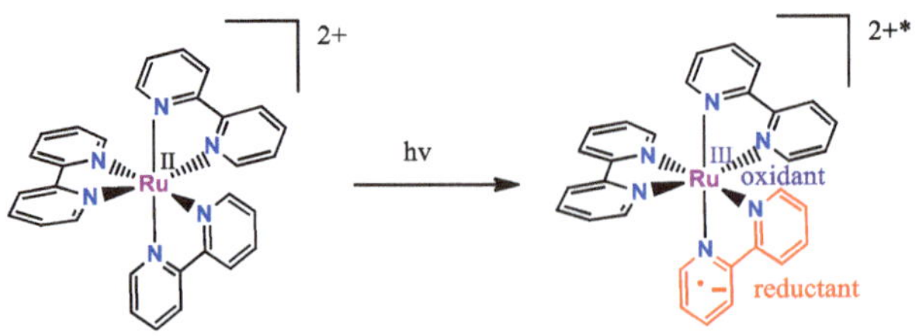

Fig. 9 Representation of MLCT in tris (2,2′-bipyridine)ruthenium (II), $[Ru(bpy)_3]^{2+}$.

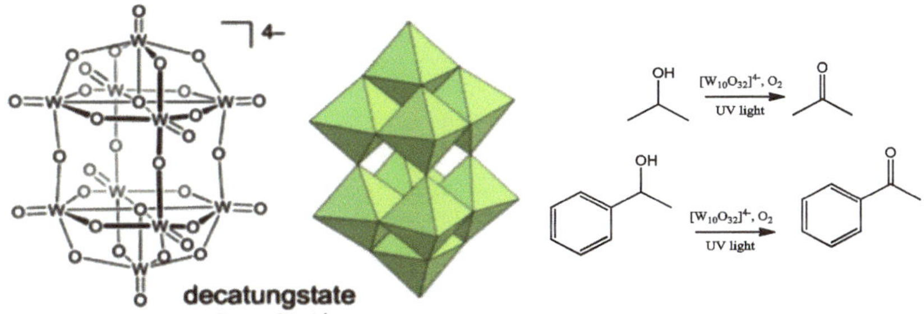

Fig. 10 Anionic structure of decatungstate POMs and Photooxidation of aliphatic and aromatic alcohols under aerobic conditions, using POMs as photocatalysts.

Besides the use of noble metal ions, some progresses are made with respect to the use of complexes made from earth-abundant elements in photoredox oxidation catalysis such as tungsten *(138)*, iron *(139)*, or copper *(140)*. In most cases organic ligands, organic solvents and high catalyst loads are used in the above mentioned photocatalytic systems.

Polyoxometalates (POMs) are stable metal oxide nanoclusters that can adopt a wide variety of structures *(141)*. POMs are attractive materials for application as photocatalysts. Upon irradiation with UV and/or near UV light, it takes place an intramolecular charge transfer from the O^{2-} based highest occupied molecular orbital (HOMO) to the W^{6+}-based lowest unoccupied molecular orbital (LUMO), leading to the formation of photoexcited states that are highly reactive, as oxidizing agents, allowing them to be used for photocatalytic reactions. The reaction of the excited catalyst may occur by hydrogen atom transfer (HAT) and/or single-electron transfer (SET). Photocatalytic oxidation of organic compounds such as alcohols can be carried out under aerobic conditions using UV-driven decatungstate photocatalysis *(138)*, as is shown in Fig. 10.

7.2 Metallabis(dicarbollide) compounds as homogeneous photooxidation catalysts

Cobaltabisdicarbollide compounds are highly compact Greek letter θ shaped molecules with interesting and suitable properties, thanks to which, cobaltabis(dicarbollide) compounds have proven to be efficient catalysts in the photooxidation of alcohols and alkenes in water, upon UV light irradiation. Among these properties are their high solubility in water *(142)*, their lack of photoluminescence, and high oxidizing power for the couple $Co^{4+/3+}$ *(143)*. Compound $[3,3'\text{-}Co(1,2\text{-}C_2B_9H_{11})_2]^-$, **[1]**$^-$ can undergo stepwise substitutions in B that sequentially modify their redox potentials and keep them reversible *(9)*. Other relevant properties of cobaltabis(dicarbollides) for their application and effectiveness in photoredox catalysis are their non-localized negative charge spread all over the molecule *(1)*, their many possibilities to produce hydrogen bonds, e.g., C_c-H···O and dihydrogen bonding C_c-H···H-B, that participate in their self-assembling *(36,144)*, micelle and vesicle formation *(35)*. Moreover, these interactions via a supramolecular approach seem to be significant in the electron transfer processes *(145)*.

Cobaltabis(dicarbollide), $[3,3'\text{-Co}(1,2\text{-}C_2B_9H_{11})_2]^-$, **[1]$^-$** and its chloro derivatives, dichloro $[3,3'\text{-Co}(8\text{-Cl-}1,2\text{-}C_2B_9H_{10})_2]^-$ **[Cl$_2$–1]$^-$** and hexachloro $[3,3'\text{-Co}(8,9,12\text{-}Cl_3\text{-}1,2\text{-}C_2B_9H_8)_2]^-$, **[Cl$_6$–1]$^-$** (Chart 2) are efficient tunable photoredox catalysts for the oxidation of alcohols in water through SET processes *(146)*, (Scheme 7). These catalysts show high performance in the oxidation of alcohols in neutral media, using a low catalyst load of 0.01 mol% and $Na_2S_2O_8$ or air as an oxidizing agent (Table 11). The studied systems have displayed quantitative conversion of reactants into products, in some cases over 99% yield, exhibiting the catalyst high efficiency to reach TON = 10,000. These results seem to indicate that lower concentrations of catalyst are more effective probably because at low concentrations, the metallacarborane is present in the form of free molecules

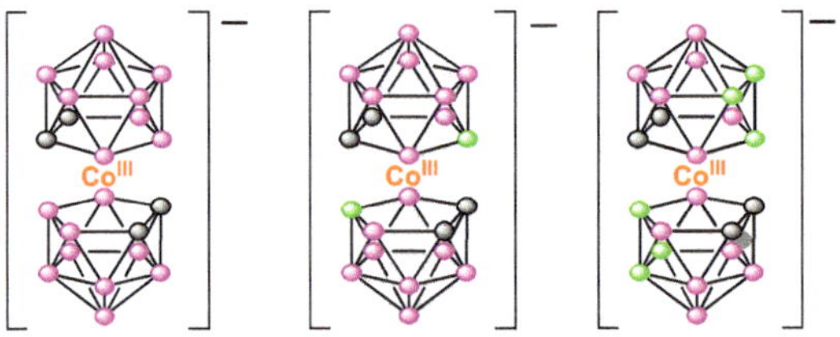

Chart 2 Cobaltabis(dicarbollide) and its chloro derivatives.

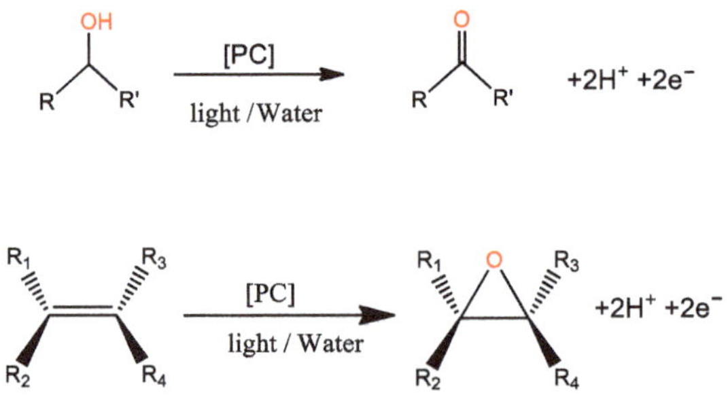

Scheme 7 Photooxidation reactions studied using metallacarboranes as photoredoxcatalysts.

Table 11 Product set obtained from photooxidation of alcohols by **Na[1]**.[a]

	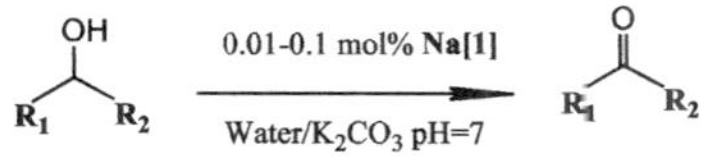	
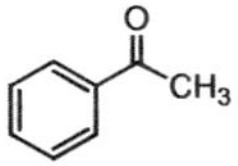	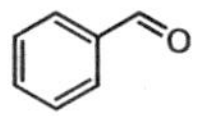	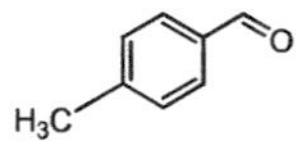
1	**2**	**3**
>99%, TON=10,000 (0.01 mol%)	>96%, TON=960 (0.1 mol%)	>97%, TON=970 (0.1 mol%)
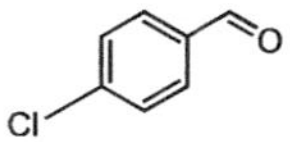	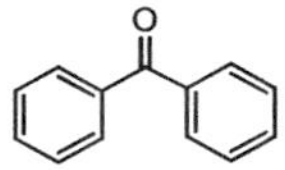	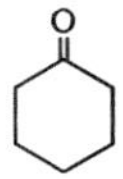
4	**5**	**6**
>94%, TON=9400 (0.01 mol%)	>99%, TON=10,000 (0.01 mol%)	93%, TON=9300 (0.01 mol%)
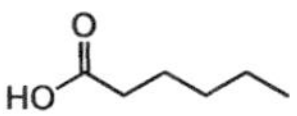	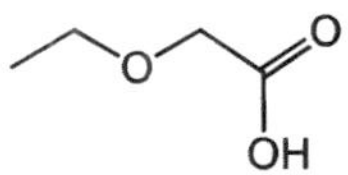	
7	**8**	
91%, TON=9100 (0.01 mol%)	97%, TON=9700 (0.01 mol%)	

[a]Conditions: **Na[1]** (0.002 mM), substrate (20 mM), $Na_2S_2O_8$ (40 mM), 5 mL potassium carbonate solution at pH=7; light irradiation 8 h using a lamp with λ=300 nm; yields and TONs values obtained are shown together with loading of **Na[1]** in parenthesis.

and aggregates (vesicles and micelles) while at higher concentrations it has a greater tendency to aggregate, thus reducing the number of active centers available to carry out the oxidation process *(35)*.

The substitution of one B—H by one B-Cl in cobaltabis(dicarbollide) **[1]**$^-$ produces a shift of the redox potential near 0.1 V, moving toward more anodic voltages and longer wavelengths concerning the non-substituted compound **[1]**$^-$. This behavior allows a shift of the absorption maximum of the UV–visible spectra toward the visible region.

The photocatalytic oxidation of alcohols performed using the chlorinated compounds, **[Cl$_2$-1]**$^-$ and **[Cl$_6$-1]**$^-$ as catalysts, shows higher yields at lower times (6 h) when irradiated with wavelength lamps close to the absorption maximum of the UV spectra, ($\lambda \sim 352$ nm) observed for these derivatives. As expected, the performance of the catalysts is lower when the irradiation of the samples was done with a lamp of wavelength $\lambda \sim 253$ nm.

The cobaltabis(dicarbollide) **Na[1]** is also an effective photoredox catalyst for alkene oxidation to epoxides in water (Scheme 7) *(147)*. The styrene photooxidation leads to the formation of different oxidation products after 30 min of catalysis. The yield in epoxide is higher at short reaction times (5 min), however, the amount of diol increases after 15 min, observing also the formation of benzaldehyde and benzoic acid. Using a catalyst load of 0.1 mol%, **Na[1]** epoxidates aromatic alkenes whose conversion values are enhanced by the presence of electron-donating substituents on the double bond and lowered by the presence of electron-withdrawing substituents. For aliphatic alkenes, the 1-octene epoxidation is less effective than for cyclooctene and 4-vinyl-1-cyclohexene. The photooxidation of this last substrate occurs on the external double bond, indicating that this reaction is regiospecific. For all the studied substrates, it can be observed that yield and selectivity values in epoxide decrease with increasing catalysis time as a consequence of the ring-opening of epoxides in aqueous media. The process leads to the formation of 1,2-diols, an important functional group to produce pharmaceutical, surfactants, or their intermediates *(148)*.The formation of the corresponding diols in the short times of reaction studied happens if **Na[1]** is present, concluding that the compound acts as a photocatalyst in both processes, in the epoxidation of alkenes and their hydroxylation in an aqueous medium. With a catalyst load of 0.01 mol%, the selectivity for the corresponding epoxides is high, being the only compound formed after 15 min of reaction (selectivity >99%) for

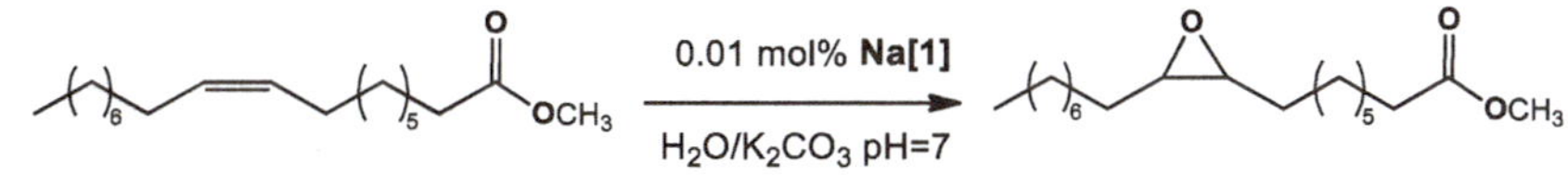

Scheme 8 Photooxidation of methyl oleate by **Na[1]**.

most of the alkenes studied. These results show a higher rate for epoxidation than for the hydroxylation process.

Cobaltabisdicarbollide, **Na[1]** epoxidizes selectively the double bond of methyl oleate as substrate (Scheme 8) which opens a promising starting point for the development of practical methods for the processing of unsaturated fatty acids.

7.3 Cobaltabis(dicarbollide) compound as heterogeneous photooxidation catalyst

From the perspective of sustainability and industrial applicability, the immobilization of homogeneous catalysts on a solid support is a good approach since homogeneous catalysts are often expensive, create additional waste and their regeneration is difficult. In this context, magnetic nanoparticles (MNPs) *(149)* coated with silica have been widely used due to their inert properties and because they help in improving the stability and the physicochemical properties of the MNPs, preventing their aggregation in solution.

The metallacarborane H[3,3′-Co(1,2-$C_2B_9H_{11})_2$] (**H[1]**) is supported on silica-coated magnetic nanoparticles in a non-covalent manner, producing hybrid materials linked by ionic interactions between **H[1]** and the functionalized magnetic nanoparticles with NH_2 groups, Fe_3O_4@SiO_2-NH_2 **(MSNP-NH$_2$)** *(150)* (Scheme 9).

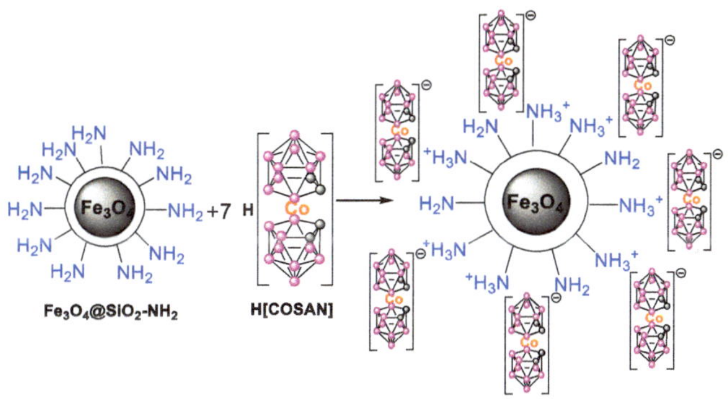

Scheme 9 Immobilization of H[COSAN], **H[1]** on MSNP-NH$_2$, MSNP-NH$_2$@**H[1]**.

The activity and selectivity of the homogeneous catalyst **[1]**$^-$ in the photoxidation of alcohols is fully preserved after heterogenization, reaching yield and selectivity values close to 100%. These are achieved after 8 h of catalysis previous sonication after 4 h since a certain degree of aggregation in the nanoparticles is observed. Back into a colloidal stable phase, the photocatalyst displays high efficiency in the photooxidation of aromatic and aliphatic alcohols, using molar ratios of 1:1000:2000 of catalyst loaded on the MNP, to the substrate, to the oxidizing agent and also of 1:10000:20000.

The addition of cetyl trimethyl ammonium chloride as surfactant decreases the aggregation of the nanoparticles during the time of the catalysis and does not block the active sites of the supported catalyst; on the other hand it provides adequate stability without hampering a quick magnetic separation. A total conversion of 1-phenylethanol was achieved in 4 h of reaction, even after 2 h a high yield of 95% was obtained; the selectivity values were >99% in all cases.

The recyclability of the catalytic system MSNP-NH_2@**H[1]** was demonstrated throughout 12 runs for the photooxidation of 1-phenylethanol, (Fig. 11) preserving the high yields and selectivity.

Overall, the cobaltabisdicarbollide catalyst supported on silica-coated magnetite nanoparticles has proven to be a robust, efficient and easily reusable system for the photooxidation of alcohols in water, resulting in a green and sustainable catalytic system.

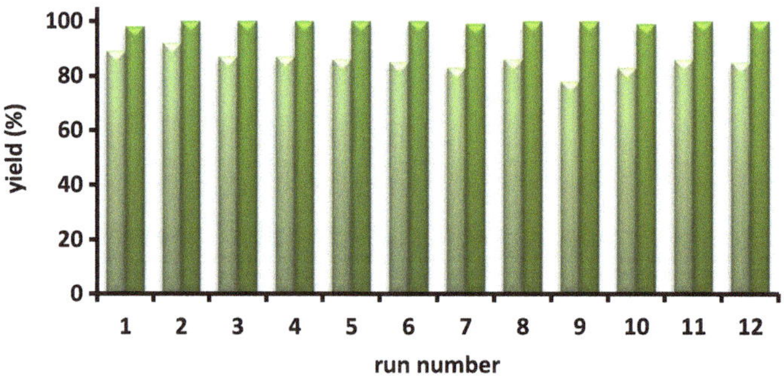

Fig. 11 Yield values obtained throughout a number of successive reuses of catalyst MSNP-NH_2@**H[1]** in the photooxidation of diphenylmethanol (light green) and 1-phenylethanol (dark green); Conditions: MNSP-NH_2@**H[1]** (0.1 μmol), substrate (0.1 mmol), $Na_2S_2O_8$ (0.2 mmol), CTAC (2 μmol), potassium carbonate solution (5 mL) at pH = 7; light irradiation using a lamp with $\lambda = 300$ nm, 4 h per run

7.4 Non-covalent cooperative coupling in a ruthenium cobaltabis(dicarbollide) system as photoredox oxidation catalyst

Cooperative photoredox catalysis means new progress in photoredox catalysis and refers to a first catalyst that is photochemically active, while the second catalyst is redox-active in the absence of light *(151)*. This strategy allows multielectron photoredox catalysis with the help of a second redox catalyst, avoiding the single-electron transfer (SET) imposed by the photoredox catalyst. Different cooperative systems can be found in the literature, based on (i) two separate complexes (photoredox catalyst and transition metal catalyst); (ii) one complex that merges both parts the photoredox and catalyst connected with an organic linker, where both parts can operate independently in their common roles; and, (iii) the transition-metal and the photoredox catalysts share a common bridging ligand or an organic linker or simply are directly connected.

Cooperativity in photoredox catalysis in two separate complexes for the oxidation of organic substrates uses a ruthenium or iridium photoredox catalyst and transition-metal catalysts, such as Mn, Fe, or Ru among others and in most cases, the substrates are oxidized to the corresponding products, through proton-coupled electron transfer (PCET) processes.

The merging of two complexes in a single complex could result in higher efficiency and may lead to new selective light-induced organic transformations. However, this union involves the formation of covalent bonds. Currently, systems focused on the light-driven oxidation of substrates are based on chromophore-catalyst dyads of ruthenium polypyridyl compounds *(152,153)*. In some cases where the photosensitizer and catalyst are linked by a connecting fragment, the electron transferred may be trapped in this connector, resulting in reduced catalytic activity. On the other hand, the construction of these systems can be laborious and expensive.

An air-stable cooperative ion pair photoredox system $[Ru^{II}(trpy)(bpy)(H_2O)][(3,3'-Co(1,2-C_2B_9H_{11})_2]_2$, **Ru-Co,** (Fig. 12) is formed by **[1]**$^-$, as photoredox catalyst, and the aqua ruthenium complex, $[Ru^{II}(trpy)(bpy)(OH_2)]^{2+}$, as oxidation catalyst in absence of light *(154)*. Both, are linked by non-covalent interactions, that are quite prevalent even after water dissolution, avoiding costly covalent bonding. **Ru-Co,** displays excellent photoredox catalyst properties in water, through a Proton coupled electron transfer process (PCET), using low loads of catalyst in relatively short reaction times (4 h). The electrochemical studies show a significant electronic coupling between the two units in the non-covalent ion pair

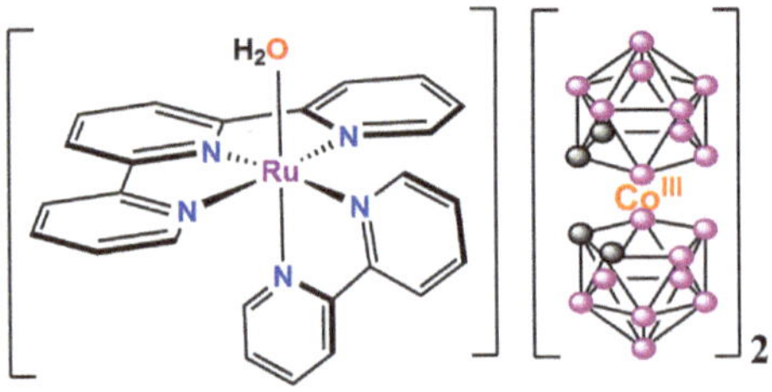

Fig. 12 Structure of $[Ru^{II}(trpy)(bpy)(H_2O)][(3,3'\text{-}Co(1,2\text{-}C_2B_9H_{11})_2]_2$, **Ru-Co**.

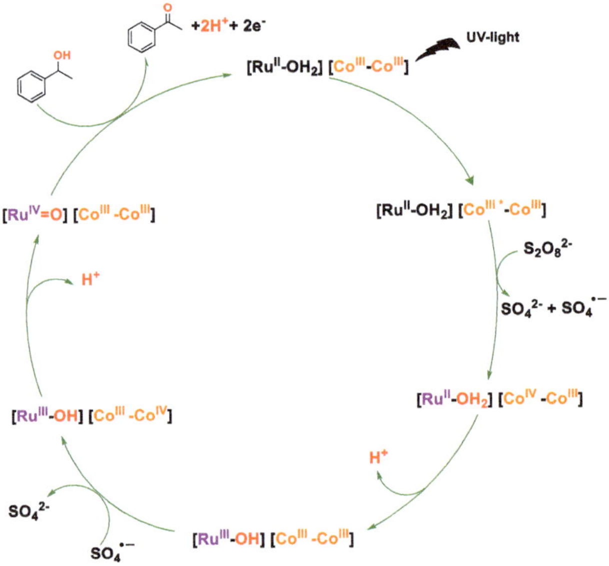

Fig. 13 Postulated pathway for the alcohol oxidation with **Ru—Co**.

ruthenium-cobaltabis(dicarbollide) compound and that the photogenerated Co^{IV} can easily oxidize $[Ru^{III}\text{-}OH]^{2+}$ to $[Ru^{IV}{=}O]^{2+}$ in water via PCET processes. Thus, the metallacarborane can act as a good photosensitizer for photocatalytic oxidation of organic substrates in this cooperative system. The possible pathway proposed is displayed in (Fig. 13).

8. Summary and outlook

Over the years, our group has devoted its efforts to the development of synthetic methodologies to efficiently and specifically functionalize the vertices of icosahedral boron clusters, thus obtaining an enormous number of ligands and complexes, which have been used for specific purposes. We have also demonstrated the versatility of these compounds and the ability to modulate their structural, electronic, physicochemical or thermal properties by simply making certain changes in the cluster. The functionalization of the cluster and the preparation of its derivatives has always been carried out in search of certain properties and applications, among them the use in catalysis, as has been demonstrated in this chapter.

The design and subsequent development of monothioether and monophosphino Rhoda- and ruthenacarboranes containing unique B–H–M type agostic bonds in their structure has allowed their use as catalytic precursors in the hydrogenation of different alkenes. This is particularly demonstrated for rhodacarboranes, which show an outstanding catalytic activity, in some cases even higher than other Rh complexes with arene ligands. Moreover, some of these Rhoda and ruthenacarborane complexes have also shown moderate to good catalytic activities and selectivities in the cyclopropanation of olefins. Remarkably, half-sandwich ruthenacarboranes showed excellent catalytic properties in the Kharash reaction of CCl_4 addition to double bonds. This high activity was found to be related to the fine-tuning to get an optimal potential in the ruthenacarboranes that is not so easy with Cp-ruthenium complexes. We cannot leave behind the relevant role of Pd complexes containing carborane-Based NBN Pincer ligands as catalysts in the Suzuki coupling reaction in water with remarkably low amounts of catalyst.

In the last section, it has been highlighted the high potential of cobaltabis(dicarbollides) as efficient, sustainable and tunable photoredox catalysts in oxidation processes, both as homogeneous or supported complexes. To this stage they have been applied in the oxidation of alcohols to aldehydes or ketones and in the oxidation of alkenes to epoxides or diols. In these reactions the metal in the catalyst is an earth-abundant transition metal, the ligand is inorganic, the medium of catalysis is water, uses an inorganic salt as a base, the oxidizing agent is inorganic but also has been demonstrated with air, the excitation source is light, and the amount of catalyst to date is of the order of 10^{-3} vs substrate. Cooperative systems with non-covalent bonding have not been described previously, and represent a step forward in getting cooperativity.

All these results have demonstrated the effectiveness of these systems in different catalytic reactions, so we are convinced that their use would help to solve problems as topical as the energy and the environment.

References

1. Grimes, R. N., Ed. *Carboranes*; 3rd ed.; Elsevier: Oxford, UK, 2016.
2. Scholz, M.; Hey-Hawkins, E. *Chem. Rev.* **2011**, *111*, 7035–7062.
3. Teixidor, F.; Barberà, G.; Vaca, A.; Kivekäs, R.; Sillanpää, R.; Oliva, J.; Viñas, C. *J. Am. Chem. Soc.* **2005**, *127*, 10158.
4. Poater, J.; Sola, M.; Viñas, C.; Teixidor, F. *Angew. Chem., Int. Ed.* **2014**, *53*, 12191–12195.
5. Poater, J.; Viñas, C.; Bennour, I.; Escayola, S.; Solà, M.; Teixidor, F. *J. Am. Chem. Soc.* **2020**, *142*, 9396.
6. Cabrera-González, J.; Ferrer-Ugalde, A.; Bhattacharyya, S.; Chaari, M.; Teixidor, F.; Gierschner, J.; Núñez, R. *J. Mater. Chem. C* **2017**, *5*, 10211–10219.
7. Sivaev, I. B.; Bregadze, V. I. *Boron in Organometallic Chemistry*, Vol. 1; 2019; pp. 147–203.
8. Mukherjee, S.; Thilagar, P. *Chem. Commun.* **2016**, *52*, 1070–1093.
9. Núñez, R.; Tarrés, M.; Ferrer-Ugalde, A.; de Biani, F. F.; Teixidor, F. *Chem. Rev.* **2016**, *116*, 14307–14378.
10. Fisher, S. P.; Tomich, A. W.; Lovera, S. O.; Kleinsasser, J. F.; Guo, J.; Asay, M. J.; Nelson, H. M.; Lavallo, V. *Chem. Rev.* **2019**, *119*, 8262–8290.
11. Ochi, J.; Tanaka, K.; Chujo, Y. *Angew. Chem. Int. Ed. Engl.* **2020**, *59* (25), 9841–9855.
12. Hosmane, N. S.; Maguire, J. A. In *Comprehensive Organometallic Chemistry III*; Crabtree, R. H., Mingos, D. M. P., Eds.; Vol. 3; Elsevier: Oxford, 2007 (Chapter 5).
13. Zhao, D.; Xie, Z. *Coord. Chem. Rev.* **2016**, *314*, 14–33.
14. (a) Tan, F.; Lopez-Periago, A.; Light, M. E.; Cirera, J.; Ruiz, E.; Borras, A.; Teixidor, F.; Viñas, C.; Domingo, C.; Planas, J. G. *Adv. Mater.* **2018**, *30*, e1800726.
15. Rodriguez-Hermida, S.; Tsang, M. Y.; Vignatti, C.; Stylianou, K. C.; Guillerm, V.; Perez-Carvajal, J.; Teixidor, F.; Vinas, C.; Choquesillo-Lazarte, D.; Verdugo-Escamilla, C.; Peral, I.; Juanhuix, J.; Verdaguer, A.; Imaz, I.; Maspoch, D.; Giner Planas, J. *Angew. Chem., Int. Ed.* **2016**, *55*, 16049–16053.
16. Núñez, R.; Romero, I.; Teixidor, F.; Viñas, C. *Chem. Soc. Rev.* **2016**, *45* (19), 5147–5173.
17. Valliant, J. F.; Guenther, K. J.; King, A. S.; Morel, P.; Schaffer, P.; Sogbein, O. O.; Stephenson, K. A. *Coord. Chem. Rev.* **2002**, *232*, 173–230.
18. Sivaev, I. B.; Bregadze, V. V. *Eur. J. Inorg. Chem.* **2009**, *2009*, 1433–1450.
19. Issa, F.; Kassiou, M.; Rendina, L. M. *Chem. Rev.* **2011**, *111*, 5701–5722.
20. Viñas, C. *Future Med. Chem.* **2013**, *5*, 617–619.
21. Stockmann, P.; Gozzi, M.; Kuhnert, R.; Sárosi, M. B.; Hey-Hawkins, E. *Chem. Soc. Rev.* **2019**, *48*, 3497–3512.
22. Lee, H.-S.; Bae, J.-Y.; Ko, J.; Kang, Y. S.; Kim, H. S.; Kim, S.-J.; Chung, J.-H.; Kang, S. O. *J. Organomet. Chem.* **2000**, *614–615*, 83–91.
23. Chizhevsky, I. T. *Coord. Chem. Rev.* **2007**, *251*, 1590–1619.
24. Lyubimov, S. E.; Davankov, V. A.; Petrovskii, P. V.; Hey-Hawkins, E.; Tyutyunov, A. A.; Rys, E. G.; Kalinin, V. N. *J. Organomet. Chem.* **2008**, *693*, 3689–3691.
25. Lee, J.-D.; Co, T. T.; Kim, T.-J.; Kang, S. O. *Synlett* **2009**, *5*, 771–774.
26. Yao, Z. J.; Deng, W. *Coord. Chem. Rev.* **2016**, *309*, 21–35.
27. Hosmane, N. S.; Eagling, R. *Boron in Catalysis*, Vol. 2; World Scientific Publishing Europe Ltd, 2018.
28. Gozzi, M.; Schwarze, B.; Hey-Hawkins, E. *Boron in Catalysis*, Vol. 2; 2019; pp. 27–80.
29. Zhao, J.; Huang, P. C.; Zhan, M. S. *Prog. Chem.* **2012**, *24*, 556–567.
30. Pepiol, A.; Teixidor, F.; Sillanpää, R.; Lupu, M.; Viñas, C. *Angew. Chem., Int. Ed.* **2011**, *50*, 12491–12495.

31. Tarrès, M.; Arderiu, V. S.; Zaulet, A.; Viñas, C.; Fabrizi de Biani, F.; Teixidor, F. *Dalton Trans.* **2015**, *44*, 11690–11695.
32. Housecroft, C. E. *Encyclopedia of Inorganic and Bioinorganic Chemistry*; John Wiley & Sons: New York, 2011.
33. Cabrera-González, J.; Sánchez-Arderiu, V.; Viñas, C.; Parella, T.; Teixidor, F.; Núñez, R. *Inorg. Chem.* **2016**, *55*, 11630–11634.
34. Masalles, C.; Llop, J.; Viñas, C.; Teixidor, F. *Adv. Mater.* **2002**, *14*, 826–829.
35. Bauduin, P.; Prevost, S.; Farràs, P.; Teixidor, F.; Diat, O.; Zemb, T. *Angew. Chem., Int. Ed.* **2011**, *50*, 5298–5300.
36. Brusselle, D.; Bauduin, P.; Girard, L.; Zaulet, A.; Viñas, C.; Teixidor, F.; Ly, I.; Diat, O. *Angew. Chem., Int. Ed.* **2013**, *52*, 12114–12118.
37. Verdiá-Báguena, C.; Alcaraz, A.; Aguilella, V. M.; Cioran, A. M.; Tachikawa, S.; Nakamura, H.; Teixidor, F.; Viñas, C. *Chem. Commun.* **2014**, *50*, 6700–6703.
38. Gassin, P.-M.; Girard, L.; Martin-Gassin, G.; Brusselle, D.; Jonchère, A.; Diat, O.; Viñas, C.; Teixidor, F.; Bauduin, P. *Langmuir* **2015**, *31*, 2297–2303.
39. Uchman, M.; Ďorďovič, V.; Tošner, Z.; Matějíček, P. *Angew. Chem. Int. Ed.* **2015**, *54*, 14113–14117.
40. Fernández-Alvarez, R.; Ďorďovič, V.; Uchman, M.; Matějíček, P. *Langmuir* **2018**, *34*, 3541–3554.
41. Malaspina, D. C.; Viñas, C.; Teixidor, F.; Faraudo, J. *Angew. Chem. Int. Ed.* **2020**, *59*, 3088–3092.
42. Parejo, L.; Chaari, M.; Santiago, S.; Guirado, G.; Teixidor, F.; Núñez, R.; Hernando, J. *Chem. A Eur. J.* **2021**, *27*, 270–280.
43. García-Mendiola, T.; Bayon-Pizarro, V.; Zaulet, A.; Fuentes, I.; Pariente, F.; Teixidor, F.; Viñas, C.; Lorenzo, E. *Chem. Sci.* **2016**, *7*, 5786–5797.
44. Leśnikowski, Z. J. *J. Med. Chem.* **2016**, *59*, 7738–7758.
45. Jung, C. W.; Hawthorne, M. F. *J. Am. Chem. Soc.* **1980**, *102*, 3024.
46. Baker, T.; Delaney, M. S.; King, R. E., III; Knobler, C. B.; Long, J. A.; Marder, T. B.; Paxson, T. E.; Teller, R. G.; Hawthorne, M. F. *J. Am. Chem. Soc.* **1984**, *106*, 2965–2978.
47. Long, J. A.; Marder, T. B.; Behncken, P. E.; Hawthorne, M. F. *J. Am. Chem. Soc.* **1984**, *106*, 2979–2989.
48. Knobler, C. B.; Marder, T. B.; Mizusawa, E. A.; Teller, R. G.; Long, J. A.; Behnken, P. E.; Hawthorne, M. F. *J. Am. Chem. Soc.* **1984**, *106*, 2990–3004.
49. Long, J. A.; Marder, T. B.; Hawthorne, M. F. *J. Am. Chem. Soc.* **1984**, *106*, 3004–3010.
50. Hewes, J. D.; Kreimendahl, C. W.; Marder, T. B.; Hawthorne, M. F. *J. Am. Chem. Soc.* **1984**, *106*, 5757–5759.
51. Behnken, P. E.; Busby, D. C.; Delaney, M. S.; King, R. E., III; Kreimendahl, C. W.; Marder, T. B.; Wilczynski, J. J.; Hawthorne, M. F. *J. Am. Chem. Soc.* **1984**, *106*, 7444–7450.
52. Teixidor, F.; Núñez, R.; Flores, M. A.; Demonceau, A.; Viñas, C. *J. Organomet. Chem.* **2000**, *614–615*, 48–56.
53. Teixidor, F.; Núñez, R.; Viñas, C. *Science* **2000**, *1* (4), 435–449.
54. Popescu, A. R.; Teixidor, F.; Viñas, C. *Coord. Chem. Rev.* **2014**, *269*, 54–84.
55. Teixidor, F.; Rius, J.; Miravitlles, C.; Viñas, C.; Escriche, L.; Sánchez, E.; Casabó, J. *Inorg. Chim. Acta* **1990**, *176*, 61–65.
56. Teixidor, F.; Casabó, J.; Viñas, C.; Sánchez, E.; Escriche, L.; Kivekäs, R. *Inorg. Chem.* **1991**, *30*, 3053–3058.
57. Teixidor, F.; Romerosa, A. M.; Viñas, C.; Rius, J.; Miravitlles, C.; Casabó, J. *J. Chem. Soc., Chem. Commun.* **1991**, 192–193.

58. Teixidor, F.; Ayllón, J. A.; Viñas, C.; Kivekäs, R.; Sillanpää, R.; Casabó, J. *J. Chem. Soc., Chem. Commun.* **1992**, 1281–1282.
59. Teixidor, F.; Ayllón, J. A.; Viñas, C.; Kivekäs, R.; Sillanpää, R.; Casabó, J. *Organometallics* **1994**, *13*, 2751–2760.
60. Kivekäs, R.; Sillanpää, R.; Teixidor, F.; Viñas, C.; Núñez, R. *Acta Cryst.* **1994**, *C50*, 2027–2030.
61. Kivekäs, R.; Sillanpää, R.; Teixidor, F.; Viñas, C.; Núñez, R.; Abad, M. M. *Acta Cryst.* **1995**, *C51*, 1864–1968.
62. Kivekäs, R.; Sillanpää, R.; Teixidor, F.; Viñas, C.; Núñez, R. *Acta Cryst.* **1995**, *C51*, 1968–1970.
63. Sillanpää, R.; Kivekäs, R.; Teixidor, F.; Viñas, C.; Núñez, R. *Acta Cryst.* **1996**, *C52*, 2223–2225.
64. McWhannell, M. A.; Rosair, G. M.; Welch, A. J.; Teixidor, F.; Viñas, C. *Acta Cryst.* **1996**, *C52*, 3135–3138.
65. Kivekäs, R.; Benakki, R.; Viñas, C.; Sillanpää, R. *Acta Cryst.* **1999**, *C55*, 1581.
66. Viñas, C.; Benakki, R.; Teixidor, F.; Casabó, J. *Inorg. Chem.* **1995**, *34*, 3844–3845.
67. Wiesboeck, R. A.; Hawthorne, M. F. *J. Am. Chem. Soc.* **1964**, *86*, 1642–1643.
68. Hawthorne, M. F.; Young, D. C.; Garret, P. M.; Owen, D. A.; Schwerin, S. G.; Tebbe, F. N.; Wegner, P. M. *J. Am. Chem. Soc.* **1968**, *90*, 862.
69. Teixidor, F.; Viñas, C.; Abad, M. M.; Núñez, R.; Kivekäs, R.; Sillanpää, R. *J. Organomet. Chem.* **1995**, *503*, 193–203.
70. Teixidor, F.; Viñas, C.; Casabó, J.; Romerosa, A. M.; Rius, J.; Miravitlles, C. *Organometallics* **1994**, *13*, 914–919.
71. Viñas, C.; Núñez, R.; Flores, M. A.; Teixidor, F.; Kivekäs, R.; Sillanpää, R. *Organometallics* **1995**, *14*, 3952–3957.
72. Viñas, C.; Núñez, R.; Teixidor, F.; Kivekäs, R.; Sillanpää, R. *Organometallics* **1996**, *15*, 3850–3858.
73. Teixidor, F.; Flores, M. A.; Viñas, C.; Kivekäs, R.; Sillanpää, R. *Organometallics* **1998**, *17*, 4675–4679.
74. Teixidor, F.; Flores, M. A.; Viñas, C.; Kivekäs, R.; Sillanpää, R. *Angew. Chem., Int. Ed. Engl.* **1996**, *35* (19), 2251–2253.
75. Viñas, C.; Flores, M. A.; Núñez, R.; Teixidor, F.; Kivekäs, R.; Sillanpää, R. *Organometallics* **1998**, *17*, 2278–2289.
76. Viñas, C.; Núñez, R.; Teixidor, F.; Kivekäs, R.; Sillanpää, R. *Organometallics* **1998**, *17*, 2376–2378.
77. Teixidor, F.; Flores, M. A.; Viñas, C.; Sillanpää, R.; Kivekäs, R. *J. Am. Chem. Soc.* **2000**, *122*, 1963–1973.
78. Tutusaus, O.; Núñez, R.; Viñas, C.; Teixidor, F.; Mata, I.; Molins, E. *Inorg. Chem.* **2004**, *43*, 6067–6074.
79. Núñez, R.; Viñas, C.; Teixidor, F.; Abad, M. M. *Appl. Organometal. Chem.* **2003**, *17*, 509–517.
80. Dubois, M. R. *Chem. Rev.* **1989**, *89*, 1–9.
81. Simpson, M. C.; Cole-Hamilton, D. J. *Coord. Chem. Rev.* **1996**, *155*, 163.
82. http://nobelprize.org/nobel_prizes/chemistry/laureates/2001/index.html.
83. Cotton, F. A.; Wilkinson, G. *Advanced Inorganic Chemistry*, 5th ed.; Wiley-Interscience: New York, 1988; p. 1248.
84. Pirottea, B.; Felekidis, A.; Fontaine, M.; Demonceau, A.; Noels, A. F.; Delargea, J.; Chizhevsky, I. T.; Zinevich, T. V.; Pisareva, I. V.; Bregadze, V. I. *Tetrahedron Lett.* **1993**, *34*, 1471–1474.
85. Demonceau, A.; Simal, F.; Noels, A. F.; Viñas, C.; Nuñez, R.; Teixidor, F. *Tetrahedron Lett.* **1997**, *38*, 7879–7882.

86. Demonceau, A.; Abreu Dias, E.; Lemoine, C. A.; Stumpf, A. W.; No, A. F.; Pietraszuk, C.; Gulinski, J.; Marciniec, B. *Tetrahedron Lett.* **1995**, *36*, 3519–3522.
87. Demonceau, A.; Saivea, E.; de Froidmont, Y.; Noels, A. F.; Hubert, A. J.; Chizhevsky, I. T.; Lobanova, I. A.; Bregadze, V. I. *Tetrahedron Lett.* **1992**, *33*, 2009–2012.
88. Demonceau, A.; Simal, F.; Noels, A. F.; Viñas, C.; Nuñez, R.; Teixidor, F. *Tetrahedron Lett.* **1997**, *38*, 4079–4082.
89. Tutusaus, O.; Delfosse, S.; Demonceau, A.; Noels, A. F.; Núñez, R.; Viñas, C.; Teixidor, F. *Tetrahedron Lett.* **2002**, *43* (6), 983–987.
90. Kharasch, M. S.; Jensen, E. V.; Urry, W. H. *Science (80-.)* **1945**, *102* (2640), 128.
91. Minisci, F. *Acc. Chem. Res.* **1975**, *8* (5), 165–171.
92. Iqbal, J.; Bhatia, B.; Nayyar, N. K. *Chem. Rev.* **1994**, *94* (2), 519–564.
93. Gossage, R. A.; van de Kuil, L. A.; van Koten, G. *Acc. Chem. Res.* **1998**, *31* (7), 423–431.
94. Kato, M.; Kamigaito, M.; Sawamoto, M.; Higashimura, T. *Macromolecules* **1995**, *28* (5), 1721–1723.
95. Wang, J.-S.; Matyjaszewski, K. *J. Am. Chem. Soc.* **1995**, *117* (20), 5614–5615.
96. Sawamoto, M.; Kamigaito, M. *Chemtech* **1999**, *29* (6), 30–38.
97. Matyjaszewski, K. *Chem. A Eur. J.* **1999**, *5* (11), 3095–3102.
98. Matyjaszewski, K. In *Education in Advanced Chemistry, Mechanistic Aspects of Molecular Catalysis*; Marciniec, B., Ed.; Vol. 6; Wydawnictwo Poznanskie: Poznan-Wroclaw, 1999; pp. 95–105.
99. Matsumoto, H.; Nakano, T.; Nagai, Y. *Tetrahedron Lett.* **1973**, *14* (51), 5147–5150.
100. Bland, W. J.; Davis, R.; Durrant, J. L. A. *J. Organomet. Chem.* **1984**, *267* (3), c45–c48.
101. Simal, F.; Demonceau, A.; Noels, A. F. *Angew. Chem. Int. Ed.* **1999**, *38* (4), 538–540.
102. Simal, F.; Demonceau, A.; Noels, A. F. *Tetrahedron Lett.* **1999**, *40* (31), 5689–5693.
103. Tallarico, J. A.; Malnick, L. M.; Snapper, M. L. *J. Org. Chem.* **1999**, *64* (2), 344–345.
104. Baratta, W.; Herdtweck, E.; Rigo, P. *Angew. Chem. Int. Ed.* **1999**, *38* (11), 1629–1631.
105. Tutusaus, O.; Viñas, C.; Núñez, R.; Teixidor, F.; Demonceau, A.; Delfosse, S.; Noels, A. F.; Mata, I.; Molins, E. *J. Am. Chem. Soc.* **2003**, *125* (39), 11830–11831.
106. Tutusaus, O.; Delfosse, S.; Demonceau, A.; Noels, A. F.; Viñas, C.; Teixidor, F. *Tetrahedron Lett.* **2003**, *44* (46), 8421–8425.
107. Tutusaus, O.; Delfosse, S.; Simal, F.; Demonceau, A.; Noels, A. F.; Núñez, R.; Viñas, C.; Teixidor, F. *Inorg. Chem. Commun.* **2002**, *5* (11), 941–945.
108. Simal, F.; Sebille, S.; Demonceau, A.; Noels, A. F.; Nuñez, R.; Abad, M.; Teixidor, F.; Viñas, C. *Tetrahedron Lett.* **2000**, *41* (28), 5347–5351.
109. Matsumoto, H.; Nakano, T.; Nagai, Y. *Tetrahedron Lett.* **1973**, *14* (51), 5147–5150.
110. Simal, F.; Wlodarczak, L.; Demonceau, A.; Noels, A. F. *Eur. J. Org. Chem.* **2001**, *2001* (14), 2689–2695.
111. Simal, F.; Delfosse, S.; Demonceau, A.; Noels, A. F.; Denk, K.; Kohl, F. J.; Weskamp, T.; Herrmann, W. A. *Chem. A Eur. J.* **2002**, *8* (13), 3047–3052.
112. Takahashi, H.; Ando, T.; Kamigaito, M.; Sawamoto, M. *Macromolecules* **1999**, *32* (11), 3820–3823.
113. Simal, F.; Wlodarczak, L.; Demonceau, A.; Noels, A. *Tetrahedron Lett.* **2000**, *41*, 6071–6074.
114. Simal, F.; Wlodarczak, L.; Demonceau, A.; Noels, A. F. *Eur. J. Org. Chem.* **2001**, *2001* (14), 2689–2695.
115. Watanabe, Y.; Ando, T.; Kamigaito, M.; Sawamoto, M. *Macromolecules* **2001**, *34* (13), 4370–4374.
116. van de Kuil, L. A.; Grove, D. M.; Gossage, R. A.; Zwikker, J. W.; Jenneskens, L. W.; Drenth, W.; van Koten, G. *Organometallics* **1997**, *16* (23), 4985–4994.

117. Delfosse, S.; Richel, A.; Simal, F.; Demonceau, A.; Noels, A. F.; Tutusaus, O.; Núñez, R.; Viñas, C.; Teixidor, F. *Advances in Controlled/Living Radical Polymerization; ACS Symposium Series*, Vol. 854; American Chemical Society, 2003; pp. 116–129.
118. Vogt, M.; Langer, R. *Eur. J. Inorg. Chem.* **2020**, *2020* (41), 3885–3898.
119. Peris, E.; Crabtree, R. H. *Chem. Soc. Rev.* **2018**, *47* (6), 1959–1968.
120. Lawrence, M. A. W.; Green, K.-A.; Nelson, P. N.; Lorraine, S. C. *Polyhedron* **2018**, *143*, 11–27.
121. Segawa, Y.; Yamashita, M.; Nozaki, K. *J. Am. Chem. Soc.* **2009**, *131* (26), 9201–9203.
122. Spokoyny, A. M.; Reuter, M. G.; Stern, C. L.; Ratner, M. A.; Seideman, T.; Mirkin, C. A. *J. Am. Chem. Soc.* **2009**, *131* (27), 9482–9483.
123. Tsang, M. Y.; Viñas, C.; Teixidor, F.; Planas, J. G.; Conde, N.; SanMartin, R.; Herrero, M. T.; Dominguez, E.; Lledos, A.; Vidossich, P.; Choquesillo-Lazarte, D. *Inorg. Chem.* **2014**, *53* (17), 9284–9295.
124. Kiwi, J.; Kalyanasundaram, K.; Grätzel, M. *Solar Energy Materials*; Springer Berlin Heidelberg: Berlin, Heidelberg, 1982; pp. 37–125.
125. Turner, J. A. *Science* **2004**, *305*, 972–974.
126. Serpone, N.; Pelizzetti, E. *Photocatalysis: Fundamentals and Applications*; Wiley-Interscience: New York, 1989.
127. Ravelli, D.; Dondi, D.; Fagnoni, M.; Albini, A. *Chem. Soc. Rev.* **2009**, *38*, 1999–2011.
128. Crisenza, G. E. M.; Melchiorre, P. *Nat. Commum.* **2020**, *11*, 803.
129. Renger, G. In *The Photosystems: Structure, Function and Molecular Biology*; Barber, J., Ed.; Elsevier Science Publishers: Amsterdam, The Netherlands, 1992; pp. 45–99.
130. Shaw, M. H.; Twilton, J.; MacMillan, D. W. C. *J. Org. Chem.* **2016**, *81*, 6898–6926.
131. Arias-Rotondo, D. M.; McCusker, J. K. *Chem. Soc. Rev.* **2016**, *45*, 5803–5820.
132. Yoon, T. P.; Ischay, M. A.; Du, J. *Nat. Chem.* **2010**, *2*, 527–532.
133. Lin, S.; Ischay, M. A.; Fry, C. G.; Yoon, T. P. J. *J. Am. Chem. Soc.* **2011**, *133*, 19350–19353.
134. Huo, H.; Shen, X.; Wang, C.; Zhang, L.; Röse, P.; Chen, L. A.; Harms, K.; Marsch, M.; Hilt, G.; Meggers, E. *Nature* **2014**, *515*, 100–103.
135. Campagna, S.; Puntoriero, F.; Nastasi, F.; Bergamini, G.; Balzani, V. *To. Curr. Chem.* **2007**, *280*, 117–214.
136. Hohloch, S.; Schweinfurth, D.; Sommer, M. G.; Weisser, F.; Deibel, N.; Ehret, F.; Sarkar, B. *Dalton Trans.* **2014**, *43*, 4437–4450.
137. Prier, C. K.; Rankic, D. A.; MacMillan, D. W. C. *Chem. Rev.* **2013**, *113*, 5322–5363.
138. Suzuki, K.; Mizuno, N.; Yamaguchi, K. *ACS Catal.* **2018**, *8*, 10809–10825.
139. Chen, J.; Draksharapu, A.; Harvey, E.; Rasheed, W.; Que, L.; Browne, W. R. *Chem. Commun.* **2017**, *53*, 12357–12360.
140. Zhou, J.; Huang-Fu, X.; Huang, Y.-Y.; Cao, C.-N.; Han, J.; Zhao, X.-L.; Chen, X.-D. *Inorg. Chem.* **2020**, *59*, 254–263.
141. Vila-Nadal, L.; Cronin, L. *Nat. Rev.* **2017**, *2*, 17054.
142. Tarres, M.; Viñas, C.; Gonzalez-Cardoso, P.; Hänninen, M. M.; Sillanpää, R.; Dordovic, V.; Uchman, M.; Teixidor, F.; Matejicek, P. *Chem. Eur. J.* **2014**, *20*, 6786–6794.
143. Corsini, M.; Fabrizi de Biani, F.; Zanello, P. *Coord. Chem. Rev.* **2006**, *250*, 1351–1372.
144. Fox, M. A.; Huges, A. K. *Coord. Chem. Rev.* **2004**, *248* (5–6), 457–476.
145. Li, H.; Li, F.; Zhang, B.; Zhou, X.; Yu, F.; Sun, L. *J. Am. Chem. Soc.* **2015**, *137*, 4332–4335.
146. Guerrero, I.; Kelemen, Z.; Viñas, C.; Romero, I.; Teixidor, F. *Chem. Eur. J.* **2020**, *26*, 5027–5036.
147. Guerrero, I.; Viñas, C.; Romero, I.; Teixidor, F. *Green Chem.* **2021**, *23*, 10123–10131.
148. Danov, S. M.; Kazantsev, O. A.; Esipovich, A. L.; Belousov, A. S.; Rogozhin, A. E.; Kanakov, E. A. *Catal. Sci. Technol.* **2017**, *7*, 3659–3675.

149. Agotegaray, M. A.; Lassalle, V. L. *Silica-Coated Magnetic Nanoparticles*, 1st ed.; Springer Publications: Cham, Switzerland, 2017.
150. Guerrero, I.; Saha, A.; Xavier, J.; Viñas, C.; Romero, I.; Teixidor, F. *ACS Appl. Mater. Interfaces* **2020**, *12* (50), 56372–56384.
151. Lang, X.; Zhao, J.; Chen, X. *Chem. Soc. Rev.* **2016**, *45*, 3026–3038.
152. Chen, W.; Rein, F. N.; Rocha, R. C. *Angew. Chem., Int. Ed.* **2009**, *48*, 9672–9675.
153. Farràs, P.; Maji, S.; Benet-Buchholz, J.; Llobet, A. *Chem. Eur. J.* **2013**, *19*, 7162–7172.
154. Guerrero, I.; Viñas, C.; Fontrodona, X.; Romero, I.; Teixidor, F. *Inorg. Chem.* **2021**, *60*, 8898–8907.

About the authors

Francesc Teixidor graduated in Chemistry from the Universitat Autònoma de Barcelona (UAB). He got his PhD at the UAB under the supervision of Prof. Dr. Heribert Barrera, working with thiol-containing ligands. Then moved to the University of Michigan (USA) for a Postdoctoral position under the supervision of Prof. Ralph Rudolph to work on Boron Clusters and Naked Metal Clusters. In 1982 got an Assistant Professor position at the UAB and began his independent career focusing on macrocycles with thioether binding units. He then moved to the CSIC at the ICMAB with an Associate Professor position and combined thioether chemistry and membranes to finally work with boron clusters, which had always been his passion. He combined chemical synthesis of Boron clusters, their electronic interpretation, the activation of the B—H bond for catalysis; he produced outer and inner cluster metal complexes. In 1999, he became Research Professor at the CSIC. His interest to apply boron clusters moved him to focus on highly stable θ-shaped Co and Fe metallacarboranes, and explore its electrochemical and photophysical properties that lead to produce excellent biosensors. However, perhaps the most attractive applications of these metallacarboranes comes from their interactions with light and response to voltage. He envisions a brilliant future for these materials in aqueous catalysis. The self-organization of these metallacarboranes in water demonstrated by him and others widens the applications field of these unexplored but highly stable species.

Clara Viñas graduated in Chemistry at the Universitat Autònoma de Barcelona and in Pharmacy at the Universitat de Barcelona. After earning her degrees, she was a predoctoral student at The University of Michigan (USA) under Prof. Dr. Ralph W. Rudolph supervision. Back to Barcelona, she worked for industry and, later she moved to Laboratori Municipal de Sabadell where she became the director to work on food and environmental control. She got her PhD in Pharmacy at the Universitat de Barcelona. In 1991, she earned a CSIC Tenured Scientist position at ICMAB that followed by promotion to Scientific Researcher in 2002. In 2006, she promoted to CSIC Full Professor where remains active nowadays. Her scientific interests and her main achievements have/are built based on her chemical and pharmaceutical training. She has been active in new synthetic approaches of novel boron cluster compounds to be applied to new bio/materials taking benefit of their self-assembly (catalysts, sensors/biosensors), environment (waste water treatment) and medicine (particularly antineoplastic, antibiotics, antifungal; visualization of drugs by Raman-microscopy techniques and radioimaging in collaboration with experts who own the equipment). In most of these areas, Prof. Viñas pioneered the research.

José Giner Planas received his PhD in 2000 from the Tokyo University of Agriculture and Technology, Japan. After working in the Université Pierre et Marie Courie (2000–2001, France), Universität Erlangen-Nürnber (2001–2003, Germany) and Material Science Institute of Barcelona (ICMAB-CSIC, 2003–2007, Spain) as a Postdoctoral Fellow, he earned a permanent tenure science position at the LMI group (ICMAB-CSIC) in 2008. His research focuses on the synthesis, characterization and applications of inorganic (boron based) molecular and polymeric materials. He has recently directed his research toward new concepts in metal-organic frameworks (MOFs) chemistry.

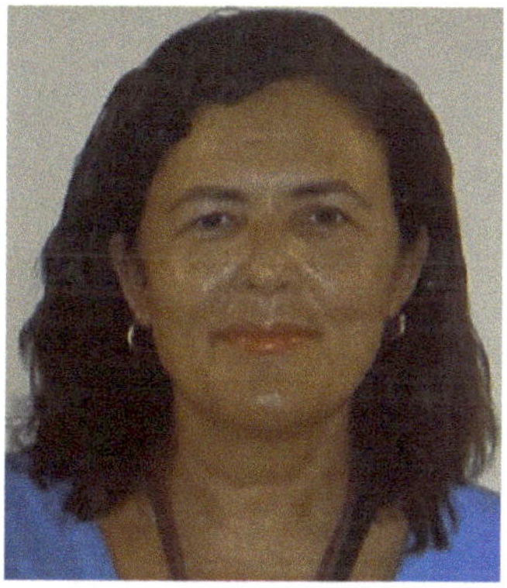

Isabel Romero, graduated in Chemistry at the University of Murcia in 1989; she obtained her PhD in chemistry from the Universitat Autònoma de Barcelona (UAB) in 1995. After postdoctoral studies at the Laboratoire de Electrochimie Organique et Photochimie Redox (CNRS, Université Joseph Fourier Grenoble) under the supervision of Professors Alain Deronzier and M. Noëlle Collomb, she obtained in 2001 a position as lecturer in chemistry at the University of Girona (UdG). Her research interests involve the synthesis and study of transition metal complexes to be used as catalysts and photocatalysts for sustainable and environmentally friendly processes. In the last years, she has been working on the study of (i) complexation of boron clusters to cheap metals and (ii) Metallabis(dicarbollide) compounds as photoredox catalysts, in collaboration with Professors Francesc Teixidor and Clara Viñas.

Rosario Núñez graduated in Chemistry at Universidad de Granada. She performed her thesis at Inorganic Materials & Catalysis Laboratory (LMI) at the Material Science Institute of Barcelona (ICMAB-CSIC) under the supervision of Professor. Clara Viñas and got her PhD from UAB in 1996. After that, she worked as a postdoctoral fellow at Université de Montpellier II under the supervision of Professors Bruno Boury and. Robert Corriu. In 1999, she joined the ICMAB as a fellow staff doctor to work at the LMI led by Professors. Francesc Teixidor and Clara Viñas. From 2001 she has a permanent position as a Tenured Scientist, and in 2018 she was promoted as Research Scientist. Her research interest focuses on the development and application of boron clusters-based materials. She has awareness in the preparation and characterization of boron clusters-containing small molecules, boron rich dendrimers, hybrid materials, photoluminescence dyes and surface functionalized carbon materials for their applications in materials science and biomedicine.

CHAPTER TWO

Transition metal catalyzed synthesis of derivatives of polyhedral boron hydrides with B—N, B—P, B—O and B—S bonds

Igor B. Sivaev*, Sergey A. Anufriev, and Akim V. Shmal'ko
A. N. Nesmeyanov Institute of Organoelement Compounds, Russian Academy of Sciences, Moscow, Russia
*Corresponding author: e-mail address: sivaev@ineos.ac.ru

Contents

Abstract

The use of B—X and B—H activation reactions catalyzed by transition metals in the synthesis of derivatives of polyhedral boron hydrides with boron–nitrogen, boron–phosphorus, boron–oxygen, and boron–sulfur bonds is considered.

Advances in Catalysis, Volume 71
ISSN 0360-0564
https://doi.org/10.1016/bs.acat.2022.04.002

1. Introduction

Polyhedral boron hydrides are inorganic molecules having the structure of polyhedra at the vertices of which there are boron atoms, each of which additionally forms a bond with an *exo*-polyhedral hydrogen atom. The formal replacement of one or two boron atoms in the polyhedron with carbon atoms leads to compounds called carboranes. Due to their extraordinary stability and synthetic availability, 10- and 12-vertex boranes and carboranes are of the greatest practical interest (Fig. 1). Even though these compounds are of artificial origin and have no analogues in nature, polyhedral boron hydrides and their derivatives are of interest for a wide variety of applications from design of new materials *(1–8)* to medicinal chemistry *(9–18)*. Therefore, the development of efficient methods for the functionalization of polyhedral boron hydrides is of great importance. In the case of carboranes, the task is facilitated by the presence of carbon atoms in the carborane cage, which makes them accessible for functionalization using the arsenal of classical organic chemistry *(19)*. On the other hand, the design of new materials and pharmaceuticals based on carboranes and other polyhedral boron hydrides requires the development of methods for their functionalization *via* boron atoms.

Transition metal catalyzed cross-coupling is known as a reliable method to form C—C and C—heteroatom bonds and is widely used in organic synthesis to obtain a variety of organic materials and pharmaceuticals. Moreover, over the past 40 years, new C—C and C–heteroatom cross-coupling protocols are mainly based on transition metal catalyzed reactions *(20–23)*. This is especially true for aromatic and heteroaromatic compounds, the synthesis of various derivatives of which is currently almost impossible

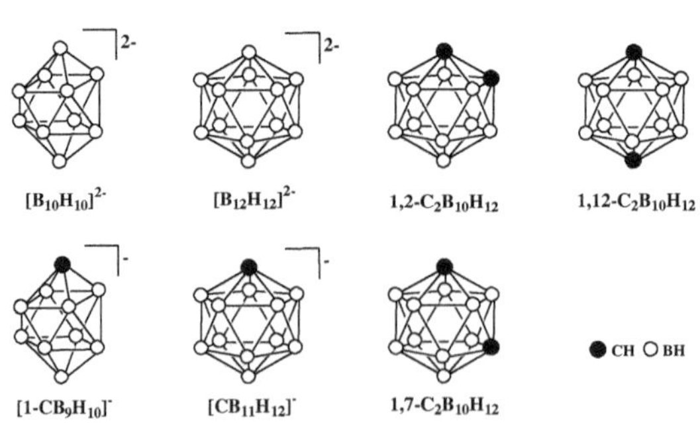

Fig. 1 The most accessible polyhedral boron hydrides (boranes and carboranes).

to imagined without the use of cross-coupling reactions catalyzed by transition metals *(24,25)*. Polyhedral boron hydrides are usually considered as three-dimensional aromatic compounds *(26,27)*. In addition, the most studied substitution reactions at boron atoms in polyhedral boron hydrides are halogenation reactions. Therefore, it is not surprising that transition metal catalyzed cross-coupling reactions have found application for the functionalization of carboranes and other polyhedral boron hydrides.

The first examples of the syntheses of carborane derivatives with B—C bond *via* Pd-catalyzed cross-coupling reactions of their iodo derivatives with Grignard reagents (Kumada cross-coupling) were described 40 years ago *(28)*. This approach was subsequently successfully used to obtain various *B*-alkyl, *B*-alkynyl, and especially *B*-aryl derivatives of carboranes *(29–48)*, however, the number of available substituents in the aromatic ring in these reactions is strictly limited due to the high reactivity of Grignard reagents. The possibility of B-(hetero)aryl-, alkyl- and alkynyl derivatives of carboranes using organozinc reagents (Negishi cross-coupling) has been demonstrated as well *(36,37,42,49,50)*, however the arylzinc reagents were obtained by transmetallation of the corresponding organolithium and organomagnesium reagents that decreases the synthetic utility of this method. Recently, a simple and efficient method has been proposed for the one-pot synthesis of *B*-aryl derivatives of *ortho*-carborane with functional groups sensitive to organolithium and organomagnesium reagents using *in situ* generated organozinc compounds *(51,52)*. Mild *B*-arylation of carboranes *via* Suzuki cross-coupling reactions of their iodo derivatives with boronic acids has been reported, but this approach has found only limited application due to the easy deboronation of the *ortho*-carborane cage in the presence of strong nucleophiles such as F^- or OH^- *(53–55)*. The possibilities of alkenylation and alkynylation of carboranes using the Pd-catalyzed Heck *(56)* and Sonogashira *(37)* reactions have also been demonstrated but have not been widely used. Pd-catalyzed cross-coupling reactions of iodo derivatives of other polyhedral boron hydrides, including *closo*-dodecaborate *(57–60)*, *closo*-decaborate *(61)*, carba-*closo*-dodecaborate *(62–67)* and 1-carba-*closo*-decaborate *(68,69)* anions, to form a B—C bond have also been reported. Transition metal-catalyzed cross-coupling reactions of iodo and bromo derivatives of polyhedral boron hydrides leading to the formation of derivatives with B—C bonds have been recently reviewed *(70,71)*.

Another approach to transition-metal induced functionalization of polyhedral boron hydrides is based on direct BH functionalization.

This approach includes activation of the electron-rich BH vertices with electron-deficient transition metal catalysts, whereas electron-rich transition metals are used for activation of the electron-deficient BH vertices. Another important component of this strategy is the use of diverse directing groups, which have performed well in transition metal catalyzed CH activation reactions *(72–74)*. This approach to the synthesis of derivatives of carboranes and other polyhedral boron hydrides has been especially actively developed during the last 5 years, and its achievements have been reviewed in several recent reviews *(75–77)*.

The purpose of this chapter is to review the current state of functionalization of polyhedral boron hydrides to form B—N, B—P, B—O, and B—S bonds using transition metal-catalyzed BX and BH activation reactions.

2. Reactions with formation of the B—N bonds

Due to the great practical importance of derivatives of polyhedral boron hydrides with a boron–nitrogen bond, the reactions of their formation catalyzed by transition metals have been studied most extensively.

2.1 Icosahedral carboranes

In contrast to the use of transition metal catalysis for the synthesis of carboranes with an *exo*-polyhedral B—C bond, the first example of which was described as early as 40 years ago, the history of the use of transition metal catalysis for the synthesis of carboranes with a B—N bond is much shorter and began only 15 years ago, when it was demonstrated that the Buchwald-Hartwig reaction can be applied to synthesis of various *N*-azole derivatives of *para*-carborane *(78)*.

The reaction of 2-iodo-*para*-carborane with carbazole in the presence of 5 mol% of $Pd(dba)_2$, 5 mol% of BINAP and 1 equiv. of NaH in 1,4-dioxane at 100 °C results in the formation of the *para*-carboranyl carbazole in 90% yield. The similar reactions with benzimidazole, imidazole and indole in the presence of *t*-BuONa as a base give the corresponding *N*-azole derivatives of *para*-carborane in high yields (Scheme 1). The reactions with more acidic azoles, such as triazole and benzotriazole, lead only to trace amounts of the *para*-carborane derivatives with B—N bond *(78)*.

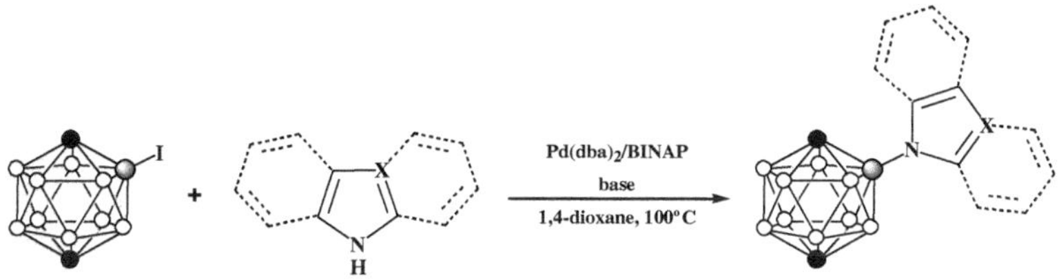

Scheme 1

The reactions of 2-iodo-*para*-carborane with aromatic amines under the same conditions lead to the corresponding carboranyl amines in ~60% yield together with 2-hydroxy-*para*-carborane (~30%) (See below) (Scheme 2) *(78)*.

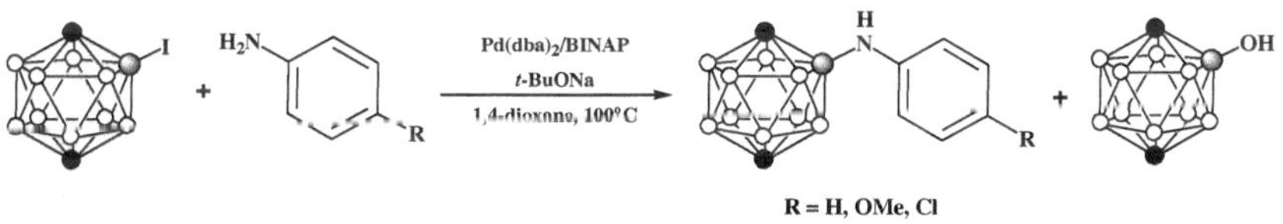

Scheme 2

The reaction with morpholine under the same conditions gives the desired amine in only 30% yield together with 20% of 2-hydroxy-*para*-carborane and the parent *para*-carborane (Scheme 3) *(78)*.

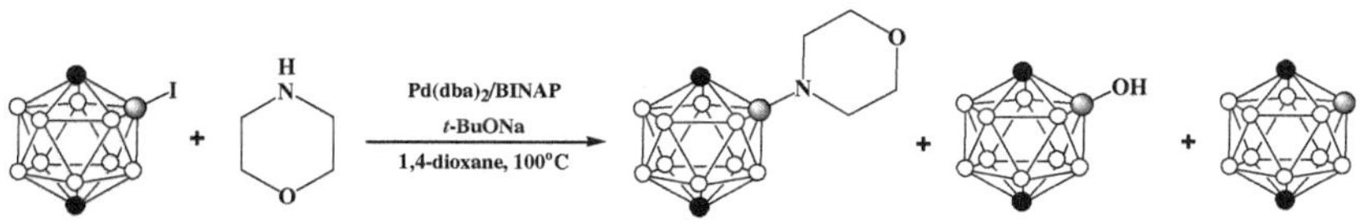

Scheme 3

The reactions of 2-iodo-*para*-carborane with amides in the presence of 5 mol% of $Pd(dba)_2$, 5 mol% of BINAP and 1 equiv. of NaH in 1,4-dioxane at 100 °C or 2.5 mol% of $Pd_2(dba)_3$, 5 mol% of DavePhos and 5 equiv. of K_3PO_4 in toluene at 100 °C result in the corresponding *N*-carboranyl amides in good yields (70–90%) (Scheme 4) *(79–81)*. The base mediated hydrolysis of the carboranyl trifluoroacetate at room temperature leads to 2-amino-*para*-carborane (Scheme 4) *(80)*.

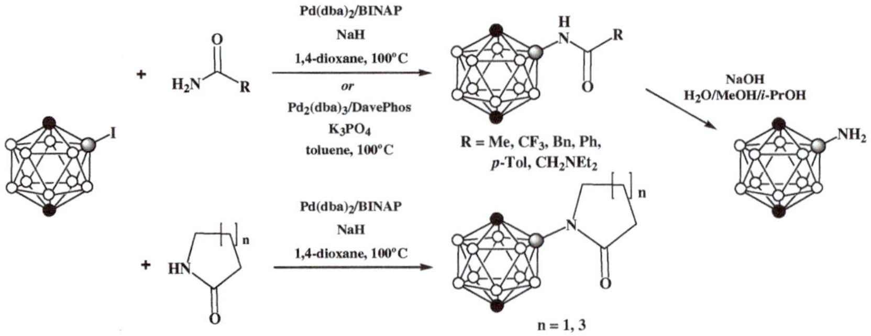

Scheme 4

In a similar way, the reactions of 9-iodo-*meta*-carborane with amides in the presence of 5 mol% of $Pd(dba)_2$, 5 mol% of BINAP and 1 equiv. of NaH in 1,4-dioxane at 100 °C or 2.5 mol% of $Pd_2(dba)_3$, 5 mol% of DavePhos and 5 equiv. of K_3PO_4 in toluene at 100 °C result in the corresponding *N*-carboranyl amides in 60–90% yields (Scheme 5) *(79–81)*. The base mediated hydrolysis of the carboranyl trifluoroacetate at room temperature leads to 9-amino-*meta*-carborane (Scheme 5) *(80)*.

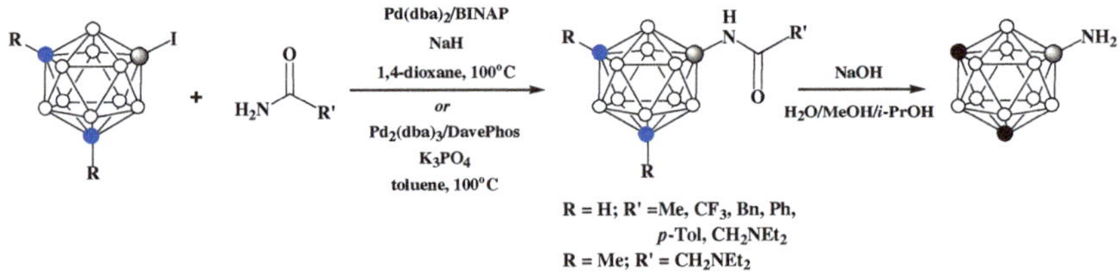

Scheme 5

The reactions of 9,10-diiodo-*meta*-carborane with amides in the presence of 2.5 mol% of $Pd_2(dba)_3$, 5 mol% of DavePhos and 5 equiv. of K_3PO_4 in toluene at 100 °C give the corresponding *N*-carboranyl amides in ~85% yields (Scheme 6) *(80)*.

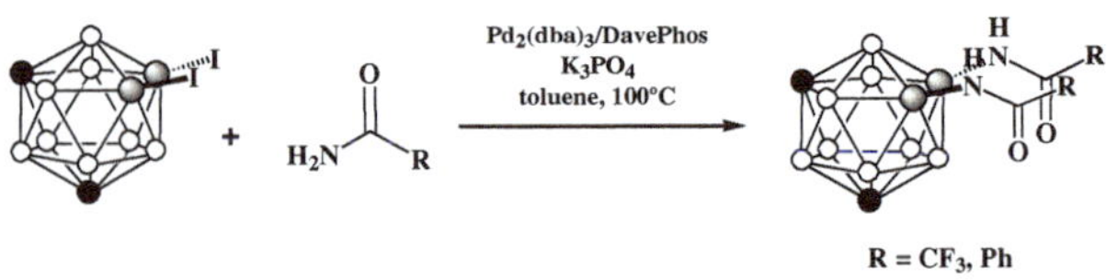

Scheme 6

Bromo derivatives of carboranes, as a rule, do not undergo cross-coupling reactions due to the greater strength of the B-X bonds compared to the aryl C–X bonds, which prevents oxidative addition of bromo carboranes. The difference in the B–X bond strength allows the substitution of iodine for bromine in 9-iodo-*meta*-carborane in the presence of $(Bu_4N)Br$ and $[(Ph_3P)_4Pd]$ *(82)*. Nevertheless, the chemical inertness of the B—Br bond in the cross-coupling reactions can be overcome by using precatalysts based on the 2-aminobiphenyl mesylate palladacycle G3PdL *(83–85)*. The reaction of 9-bromo-*meta*-carborane with trifluoroacetamide in the presence of 5 mol% of [(DavePhos)(ABP)Pd(OMs)], 5 mol% of DavePhos and 3 equiv. of K_3PO_4 in toluene at 80 °C result in the formation of the corresponding carboranyl amide in ~80% yield (Scheme 7) *(86)*. Other types of amides such as sulfonamides and phosphoramidates are also capable of cross-coupling with bromo derivatives of carboranes. The reactions of 9-bromo-*meta*-carborane with phenyl sulfonamide and diethyl phosphoramidate in the presence of 5 mol% of [(SPhos)(ABP)Pd(OMs)], 5 mol% of SPhos and 2 equiv. of K_3PO_4 in 1,2-dimethoxyethane at 80 °C lead to the corresponding carboranyl amides isolated in good (~70%) yields (Scheme 7) *(87)*.

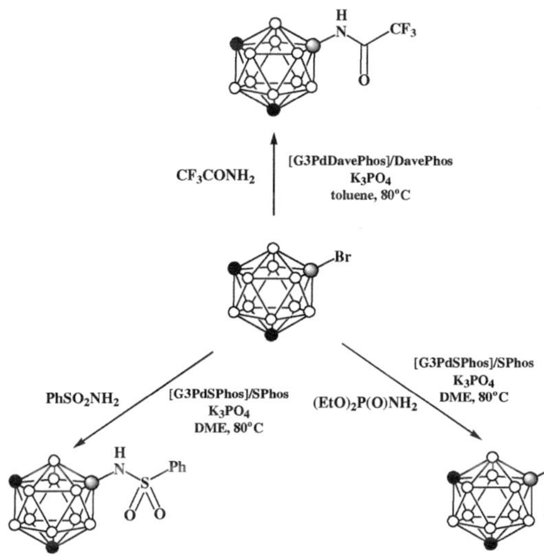

Scheme 7

The reactions of 9-bromo-*meta*-carborane with potassium phthalimide in the presence of 5 mol% of [(DavePhos)(ABP)Pd(OMs)] and 5 mol% of DavePhos or 5 mol% of [(SPhos)(ABP)Pd(OMs)] and 5 mol% of SPhos in 1,4-dioxane at 80 °C give 9-*N*-phthalimido-*meta*-carborane in good yields, whereas a similar reaction in the presence of 5 mol% of [(XPhos)(ABP)Pd(OMs)] and 5 mol% of XPhos results in a mixture of all possible isomers (2-, 4-, 9- and 5-) *(88)*.

The reactions of 9-bromo-*meta*-carborane with ammonia or various amines (cyclohexyl amine, morpholine, 3,5-dimethoxyaniline, 4-methylimidazole) in the presence of 5 mol% of [(SPhos)(ABP)Pd(OMs)], 5 mol% of SPhos and *t*-BuOK or K_3PO_4 in 1,4-dioxane at 80 °C result in the formation of the corresponding amines in near quantitative yields (Scheme 8) *(86)*.

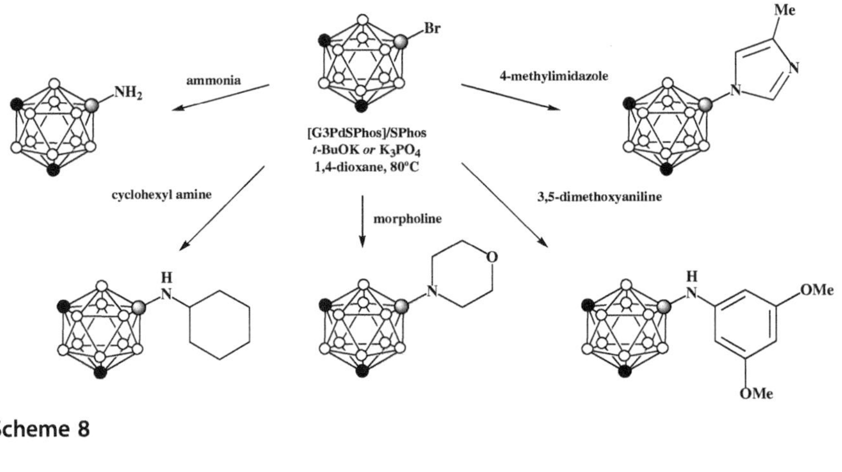

Scheme 8

Pd-catalyzed reactions of 9-bromo-*meta*-carborane with sodium azide and cyanate also could be used for preparation of the corresponding carboranyl azide and isocyanate (Scheme 9) *(87)*.

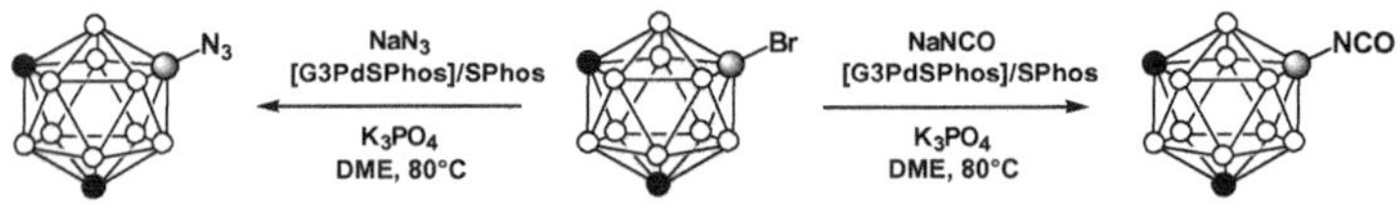

Scheme 9

At the same time, it should be noted that, unlike halogenated aromatics, iodo derivatives of polyhedral boron hydrides, including carboranes, have a greater synthetic availability compared to other halogen derivatives, which, together with the use of more complex and expensive catalytic systems, makes the use of bromo carboranes less attractive than iodo carboranes.

In the case of *ortho*-carborane, there are two most available isomeric iodo derivatives. The first one, 9-iodo-*ortho*-carborane, in which the substituent is located at the position furthest from the carbon atoms, is formed by electrophilic iodination of *ortho*-carborane *(89,90)*. The second one, 3-iodo-*ortho*-carborane, in which the substituent is located at the boron atom directly bonded to both carbon atoms, is obtained by insertion of BI_3 into the open face of *nido*-carborane *(29,91)* or by nucleophilic substitution of the diazo group in the diazonium derivative of *ortho*-carborane *(92,93)*.

A characteristic feature of *ortho*-carborane is its easy deboronation (removal one boron vertex) under the action of strong nucleophiles, which makes the Pd-catalyzed Buchwald-Hartwig amination using aliphatic amines impossible, and also significantly limits the range of bases used in other cross-coupling reactions. Nevertheless, the reactions of 3-iodo-*ortho*-carborane with amides in the presence of 2.5 mol% of $Pd_2(dba)_3$, 5 mol% of DavePhos and 5 equiv. of K_3PO_4 in toluene at 100 °C result in the corresponding *N*-carboranyl amides (Scheme 10) *(80,81)*.

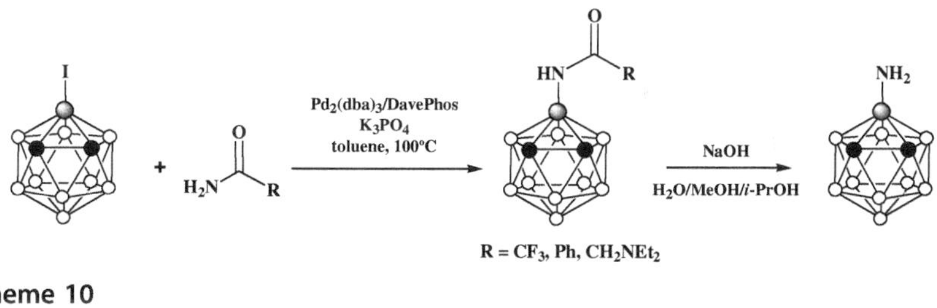

Scheme 10

The base mediated hydrolysis of the carboranyl trifluoroacetate at room temperature leads to 3-amino-*ortho*-carborane (Scheme 10) *(80)*. It should be noted that 3-amino-*ortho*-carborane can be obtained more simply by sequential one-pot reduction of *ortho*-carborane with sodium in liquid ammonia, followed by oxidation of the resulting dianion with $KMnO_4$ or $CuCl_2$ *(94–96)*.

The only example of a Pd-catalyzed Buchwald-Hartwig amination of *ortho*-carborane is the reaction of 3-bromo-*ortho*-carborane with 3,5-dimethoxyaniline in the presence of 5 mol% of [(SPhos)(ABP)Pd(OMs)], 5 mol% of SPhos and 4 equiv. of K_3PO_4 in 1,4-dioxane at 70 °C (Scheme 11) *(86)*.

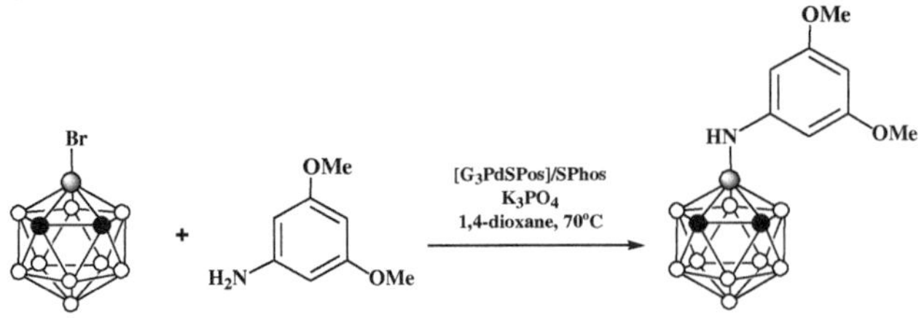

Scheme 11

A series of various *N*-carboranyl amides was prepared by the reactions of 9-iodo-*ortho*-carborane with the corresponding amides in the presence of 2.5 mol% of $Pd_2(dba)_3$, 5 mol% of DavePhos and 5 equiv. of K_3PO_4 in toluene at 100 °C (Scheme 12) *(80,81,97,98)*.

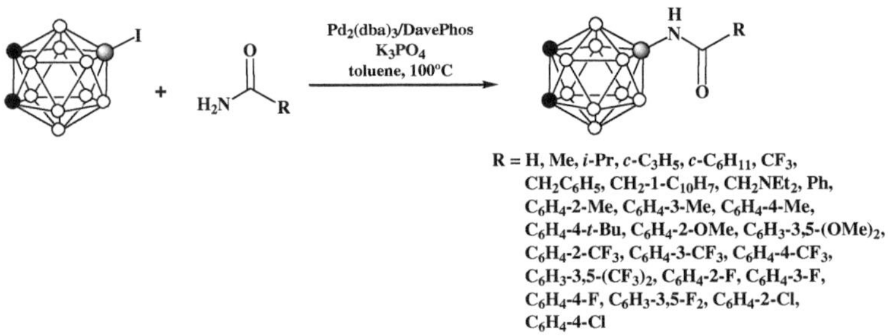

Scheme 12

The reactions of 9-bromo-*ortho*-carborane with phenyl sulfonamide and diethyl phosphoramidate in the presence of 5 mol% of [(SPhos)(ABP)Pd(OMs)], 5 mol% of SPhos and 2 equiv. of K_3PO_4 in 1,2-dimethoxyethane at 80 °C lead to the corresponding carboranyl amides in good yields (Scheme 13) *(87)*.

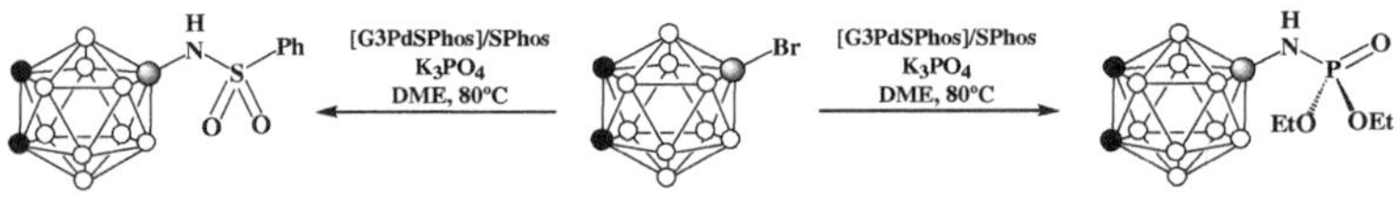

Scheme 13

Pd-catalyzed reactions of 9-bromo-*ortho*-carborane with sodium azide and cyanate lead to the corresponding carboranyl azide and isocyanate (Scheme 14) *(87)*.

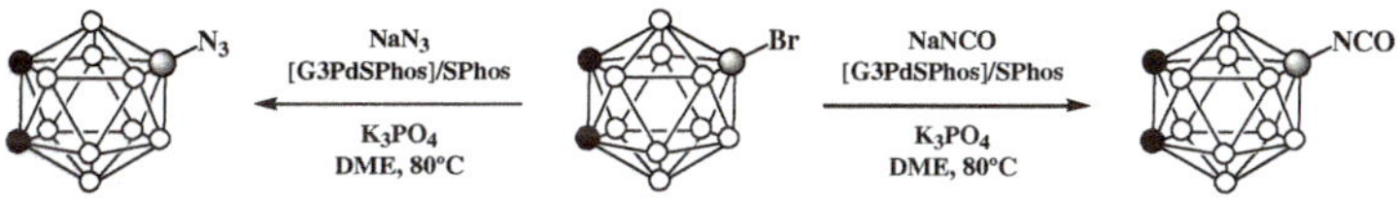

Scheme 14

Another interesting example of transition metal mediated amidation of *ortho*-carborane includes reaction of 9-chloromercuro-*ortho*-carborane with $[(COD)Pd(CH_2SiMe_3)_2]$ and XPhos in tetrahydrofuran resulting in complex **A** with a Pd—Hg bond. Subsequent heating complex **A** with trifluoroacetamide in the presence of XPhos and K_3PO_4 in toluene leads to 9-CF_3CONH-1,2-$C_2B_{10}H_{11}$ (Scheme 15). Complex **A** also can serve as a catalyst in the reaction of 9-iodo-ortho-carborane with trifluoroacetamide *(99)*.

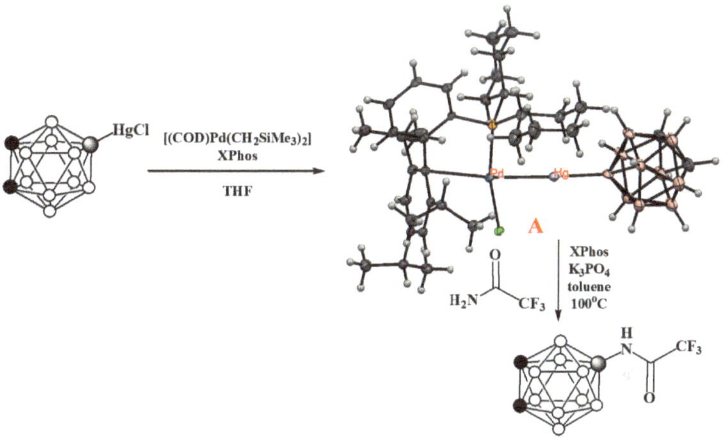

Scheme 15

Transition metal catalyzed direct B—H bond functionalization is considered as a tempting strategy as it occurs with high atom-and-step economy. An example of such direct functionalization is the reaction of *ortho*-carborane with ammonia in the presence of 5 mol% of $[(COD)Ir(OMe)]_2$, 30 mol% of PCy_3 and 3 equiv. of K_2CO_3 in tetrahydrofuran at 110 °C resulting in 3-amino-*ortho*-carborane in 80% yield (Scheme 16) *(100)*. As noted above, 3-amino-*ortho*-carborane can be obtained by one-pot sequential reduction and oxidation reactions of *ortho*-carborane in liquid ammonia *(94–96)*. Therefore, the most important here is the ability to use this approach for the synthesis of 3-amino derivatives of various *C*- and *B*-substituted *ortho*-carboranes (Scheme 16) *(100)*.

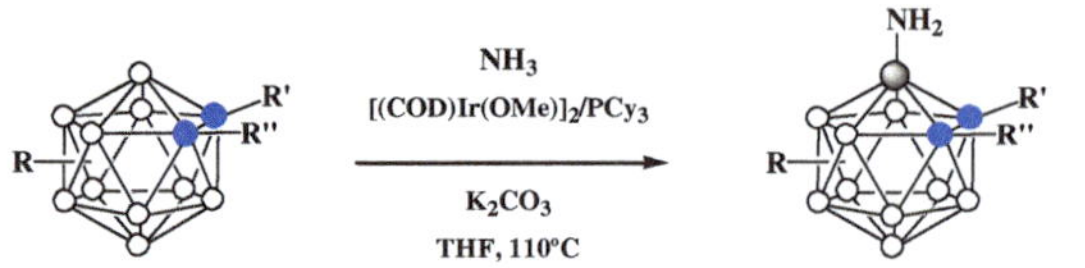

R = R' = R'' = H;
R = 9-Me, 9-*cyclo*-C_5H_9, 9-Bn, 9-Ph, 9-(2'-MeC_6H_4), 9-(4'-MeC_6H_4), 9-(4'-$MeOC_6H_4$), 9-(3',5'-$(MeO)_2C_6H_3$), 9-(4'-FC_6H_4), 9-(4'-ClC_6H_4), 9-(4'-$CF_3C_6H_4$), 9-(4'-MeC_6H_4), 9-(2'-SC_4H_3); R' = R'' = H;
R = 4-Ph; R' = R'' = H;
R = 5-Ph; R' = R'' = H;
R = 6-Ph; R' = R'' = H;
R = R' = H; R'' = Me, *i*-Pr, *n*-Bu, Bn, 4'-MeC_6H_4;
R = H; R' = R'' = Me;
R = 6-Ph, 6-(4'-*t*-BuC_6H_4); R' = H; R'' = Me

Scheme 16

The introduction of substituents at other positions of the *ortho*-carborane cage requires the use of directing groups. The amination of *ortho*-carboranyl carboxylic acid 1-HOOC-1,2-$C_2B_{10}H_{11}$ with *O*-benzoyl-*N*-hydroxylmorpholine in the presence of 10 mol% of $Pd(OAc)_2$ and 1 equiv. of K_2HPO_4 in toluene at 100 °C proceeds with elimination of the directing carboxylic group to form the 4-morpholine derivative of *ortho*-carborane 4-$O(CH_2CH_2)_2N$-1,2-$C_2B_{10}H_{11}$ in 46% yield (Scheme 17) *(101)*. The same product in some higher yield (53%) can be obtained starting from 1-HOOC-2-Me_3Si-1,2-$C_2B_{10}H_{10}$ *(101)*. In a similar way, the reactions of *C*-substituted *ortho*-carboranyl carboxylic acids 1-HOOC-2-R-1,2-$C_2B_{10}H_{10}$ with *O*-benzoyl-*N*-hydroxylmorpholine give the corresponding 7-morpholine derivatives 7-$O(CH_2CH_2)_2N$-1-R-1,2-$C_2B_{10}H_{10}$ *(101)*.

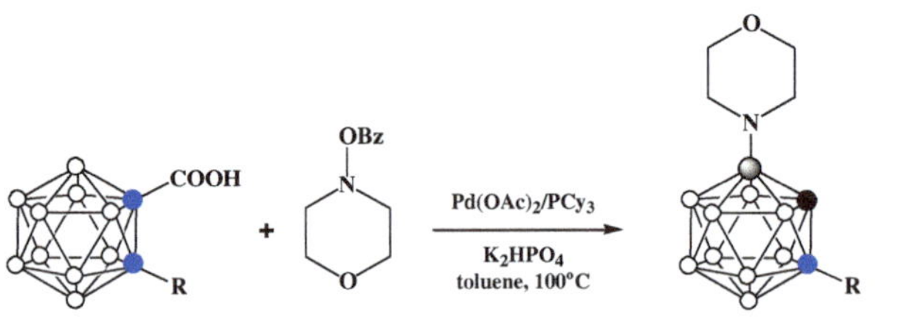

R = H, Me, Et, *i*-Pr, *n*-Bu, $CH_2C_6H_5$, $CH_2C_6H_4$-3-Me, $CH_2C_6H_4$-4-Me, $CH_2C_6H_4$-4-OMe, $CH_2C_6H_4$-4-F, $CH_2C_6H_4$-4-Cl, $CH{=}CHC_6H_5$, $CH{=}CHC_6H_4$-4-Me

Scheme 17

The same approach can be applied to the synthesis of other dialkylamino derivatives (Scheme 18) *(101)*. The Pd-catalyzed reduction of 7-dibenzylamino-1-methyl-*ortho*-carborane with H_2 leads to the corresponding amine 7-H_2N-1-Me-1,2-$C_2B_{10}H_{10}$ (Scheme 18) *(101)*.

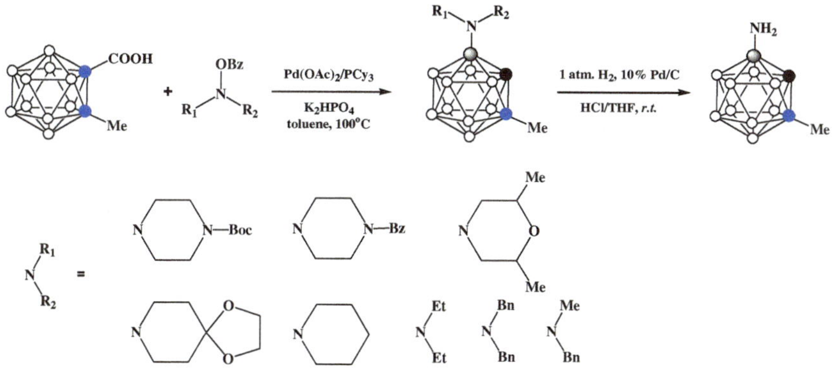

Scheme 18

The reactions of *C*-substituted *ortho*-carboranyl carboxylic acids 1-HOOC-2-R-1,2-$C_2B_{10}H_{10}$ with aryl and alkyl azides R'SO_2N_3 in the presence of 2.5 mol% of [Cp*$IrCl_2$]$_2$, 10 mol/% of $AgNTf_2$ and 1.5 equiv. of NaOAc in 1,2-dichloroethane at 85 °C result in the corresponding monoaryl- and alkylamino derivatives 7-R'SO_2NH-1-R-1,2-$C_2B_{10}H_{10}$ (Scheme 19) *(102)*.

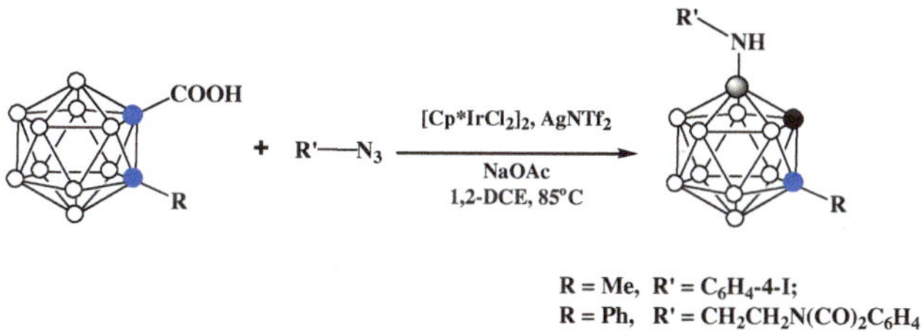

Scheme 19

The reactions of *C*-substituted *ortho*-carboranyl carboxylic acids 1-HOOC-2-R-1,2-$C_2B_{10}H_{10}$ with 3-aryl- and alkyl-1,4,2-dioxazolones in the presence of 2.5 mol% of [Cp*$RhCl_2$]$_2$, 10 mol% of $AgNTf_2$ or 2.5 mol% of [Cp*$IrCl_2$]$_2$, 10 mol% of $AgSbF_6$ and NaOAc in 1,2-dichloroethane at 80 °C lead to the corresponding amides 7-R'(O)CNH-1-R-1,2-$C_2B_{10}H_{10}$ in high yields (Scheme 20) *(103,104)*. Mechanochemical Ir-catalyzed amidation of *C*-substituted *ortho*-carboranyl carboxylic acids 1-HOOC-2-R-1,2-$C_2B_{10}H_{10}$ with dioxazolones proceeds in a similar way, whereas the reactions dioxazolones with 1-HOOC-1,2-$C_2B_{10}H_{11}$ lead to isomeric mixtures of the corresponding 3- and 4-carboranyl amides *(105)*.

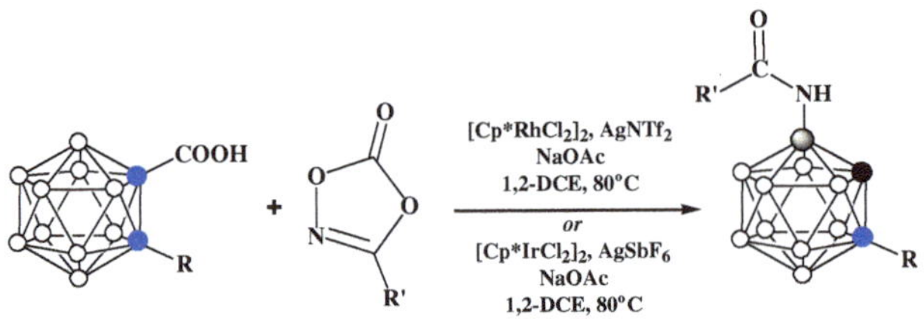

R = Me, R' = Me, *n*-Pr, Bn, Ph, C_6H_4-2-Me, C_6H_4-3-Me, C_6H_4-4-Me, C_6H_4-4-*t*-Bu, C_6H_4-4-OMe, C_6H_4-4-NMe_2, C_6H_4-4-F, C_6H_4-3-Cl, C_6H_4-4-Cl, C_6H_4-4-Br, C_6H_4-4-CN, C_6H_4-4-CF_3, C_6H_4-4-NO_2, 1-Naph, 2-Naph, azylen-2-yl, 2-thienyl, 2-furyl;

R = *n*-Bu, R' = Ph; R = C_6H_4-4-Me, R' = Ph;
R = *c*-Hx, R' = Ph; R = C_6H_4-4-OMe, R' = Ph;
R = Ph, R' = Ph; R = C_6H_4-4-Br, R' = Ph;
R = C_6H_4-3-Me, R' = Ph; R = $_2$-thienyl, R' = Ph;

Scheme 20

The reactions of *C*-substituted *ortho*-carboranyl carboxylic acids 1-HOOC-2-R-1,2-$C_2B_{10}H_{10}$ with sulfonyl azides R'SO_2N_3 in the presence of 2.5 mol% of [(*p*-cymene)$RuCl_2$]$_2$ and 2 equiv. of NaOAc in toluene at 100 °C or in the presence of 2.5 mol% of [Cp*$IrCl_2$]$_2$, 10 mol/% of $AgNTf_2$ and 1.5 equiv. of NaOAc in 1,2-dichloroethane at 85 °C result in the corresponding sulfonamides 7-R'SO_2NH-1-R-1,2-$C_2B_{10}H_{10}$ (Scheme 21) *(101,102)*. The Ir-catalyzed reaction of *ortho*-carboranyl carboxylic acid 1-HOOC-1,2-$C_2B_{10}H_{11}$ with *p*-acetamidophenylsulfonyl azide

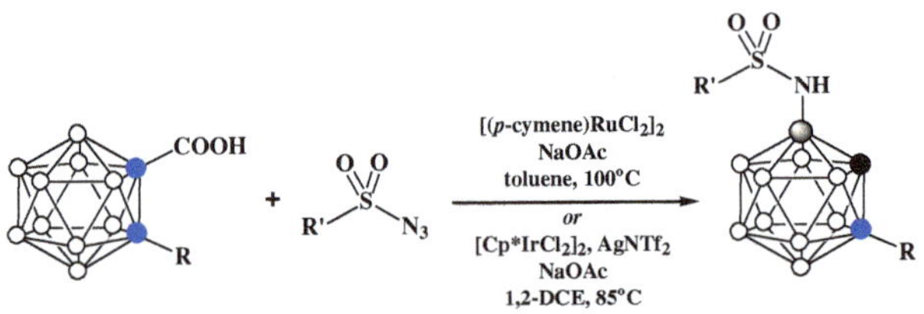

R = Me, R' = *n*-Bu, Bn, Ph, C_6H_4-2-Me, C_6H_4-4-Me, C_6H_4-4-*t*-Bu, C_6H_2-2,4,6-^{i}Pr_3, C_6H_4-4-OMe, C_6H_4-4-NHCOMe, C_6H_4-4-OCF_3, C_6H_4-2-F, C_6H_4-4-F, C_6H_4-4-Cl, C_6H_3-2,6-Cl_2, C_6H_4-2-Br, C_6H_4-3-Br, C_6H_4-4-Br, C_6H_4-4-I, C_6H_4-4-CN, C_6H_4-4-COOMe, C_6H_4-4-COMe, C_6H_4-2-CF_3, C_6H_4-3-CF_3, C_6H_4-4-CF_3, C_6H_4-2-NO_2, C_6H_4-3-NO_2, C_6H_4-4-NO_2, 1-Naph, 2-Naph, 2-SC_4H_3;

R = Et; R' = C_6H_4-4-Me;
R = *i*-Pr; R' = C_6H_4-4-Me, C_6H_4-4-NHCOMe;
R = Bn; R' = C_6H_4-4-Me, C_6H_4-4-NHCOMe;
R = CH_2CH_2OMe; R' = C_6H_4-4-Me;
R = $CH_2C_6H_4$-4-Me; R' = C_6H_4-4-Me;
R = $CH_2C_6H_4$-4-OMe; R' = C_6H_4-4-Me;
R = $CH_2C_6H_4$-4-Cl; R' = C_6H_4-4-Me;
R = Ph; R' = *n*-Pr, C_6H_4-4-NHCOMe;
R = C_6H_4-4-Br; R' = C_6H_4-4-NHCOMe

Scheme 21

results in an inseparable mixture of isomeric sulfonamides 3- and 4-*p*-$MeCONHC_6H_4SO_2NH$-1,2-$C_2B_{10}H_{11}$ *(102)*, whereas the Ru-catalyzed reaction of 1-HOOC-2-Me_3Si-1,2-$C_2B_{10}H_{10}$ with *p*-tolylsulfonyl azide selectively leads to 4-*p*-$MeC_6H_4SO_2NH$-1,2-$C_2B_{10}H_{11}$ *(101)*.

A special case of functional group directed B—H activation are cyclization reactions in which the directing group is part of the resulting heterocycle. Thus, heating 3-(2-amidophenyl)-*ortho*-carborane in the presence of 10 mol% of $[Pd(MeCN)_4](BF_4)_2$, Ag_2CO_3 and $Cu(OPiv)_2$ in toluene at 100 °C results in the formation of *ortho*-carborane fused isoquinolinone 4,3-$NHC(O)C_6H_4$–1,2-$C_2B_{10}H_{10}$ in ~90% yield. The cyclization also proceeds with high yields in the case of substituted benzamides containing electron-releasing groups and with somewhat worse yields in the case of benzamides containing electron-withdrawing substituents (Scheme 22) *(106)*.

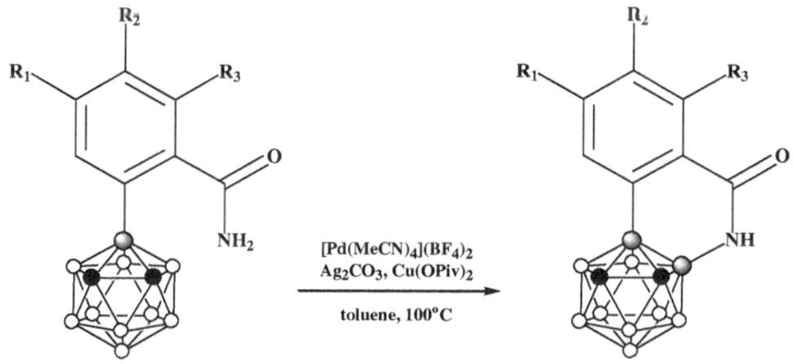

$R_1 = R_2 = R_3 = H$;
$R_1 = R_3 = H$, R_2 = Me, *t*-Bu, OMe, F, Cl, Br, CF_3, NO_2;
R_1 = Me, OMe, CF_3, $R_2 = R_3$ =H;
$R_1 = R_2 = H$, R_3 = F

Scheme 22

2.2 Carba-*closo*-dodecaborate anion $[CB_{11}H_{12}]^-$

In spite of the chemistry of the carba-*closo*-dodecaborate anion has been the focus of many research groups over the past 20 years *(107–110)*, only a few examples of its functionalization through transition metal-catalyzed B—N bond formation have been reported. The microwave-assisted Buchwald-Hartwig amination of 12-iodo- and 7-iodo derivatives of the carba-*closo*-dodecaborate anion with lithium bis(trimethylsilyl)amide $LiN(SiMe_3)_2$ in the presence of 2 mol% of $[Pd_2(dba)_3]$ and 10 mol% of CyJohnPhos in tetrahydrofuran at 55 °C followed by acidic removal of silyl groups leads to the corresponding amino derivatives (Scheme 23) *(111, 112)*.

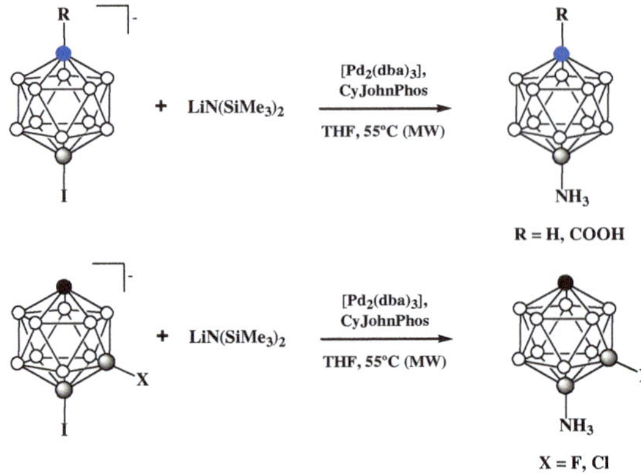

Scheme 23

2.3 *closo*-Dodecaborate anion $[B_{12}H_{12}]^{2-}$

Unlike carboranes, the chemistry of derivatives of the *closo*-dodecaborate anion with a boron-nitrogen bond is well developed *(113)*. Methods for the synthesis of its amino *(58,114)* and alkylamino *(58,115–120)* derivatives, amides *(121,122)*, and other derivatives *(121–125)* are described. Nevertheless, the synthesis of arylamino and azaheterocyclic derivatives of the *closo*-dodecaborate anion remains a problem to be solved.

The microwave-assisted reactions of the iodo derivative of the *closo*-dodecaborate anion $[B_{12}H_{11}I]^{2-}$ with anilines in the presence of 10 mol% of $[Pd_2(dba)_3]$, 20 mol% of DavePhos and 2.5 equiv. of *t*-BuOK in dimethyl sulfoxide at 150 °C lead to the corresponding arylamino derivatives $[B_{12}H_{11}NHAr]^{2-}$ in 50%–85% yields (Scheme 24). At the same time, the reaction with 1-aminonaphthaline results in the corresponding amine only in 23% yield *(126)*.

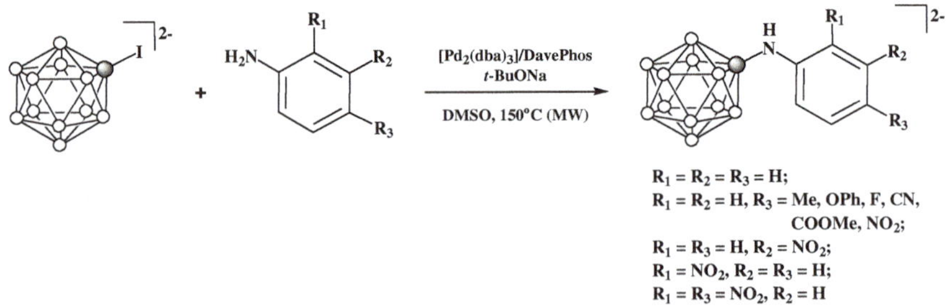

Scheme 24

The microwave-assisted reaction of the iodo derivative of the *closo*-dodecaborate anion $[B_{12}H_{11}I]^{2-}$ with carbazole in the presence of 10 mol% of $[Pd_2(dba)_3]$, 20 mol% of DavePhos and 2.5 equiv. of *t*-BuOK in dimethyl sulfoxide at 150 °C results in the formation of the corresponding carbazole derivative in 83% yield. The similar reactions with pyrrole, indole, pyrazole, and imidazole lead to the corresponding *N*-azole derivatives of the *closo*-dodecaborate anion in 82%–87% yields (Scheme 25) *(126)*.

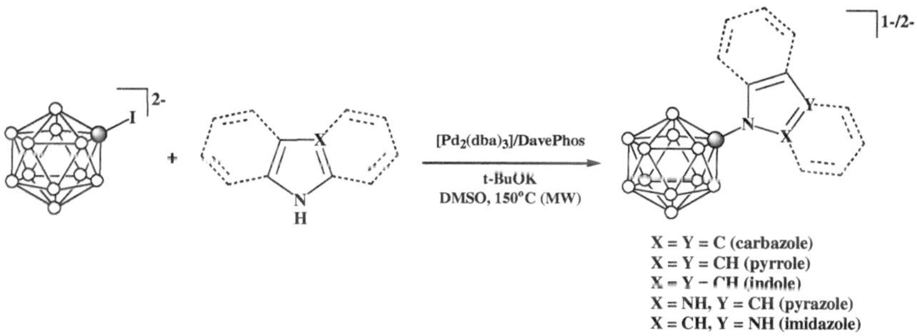

Scheme 25

The microwave-assisted reactions of the tetrabutylammonium salt of the iodo derivative of the *closo*-dodecaborate anion $[B_{12}H_{11}I]^{2-}$ with amides in the presence of 10 mol% of $[Pd_2(dba)_3]$, 20 mol% of DavePhos and 2.5 equiv. of *t*-BuOK in dimethyl sulfoxide at 150 °C lead to the corresponding *N*-borylated amides (Scheme 26) *(126)*.

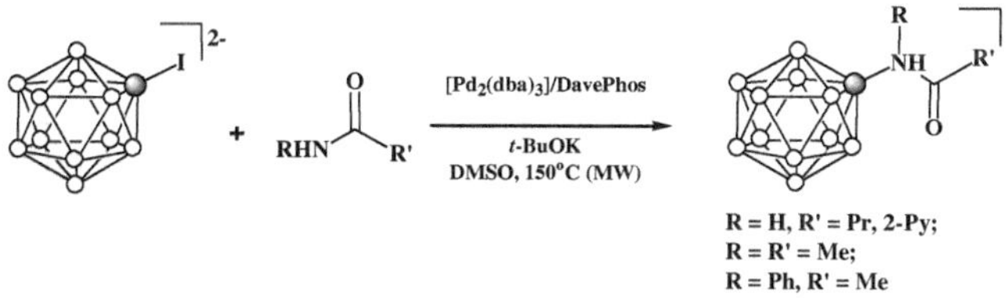

Scheme 26

In a similar way, the microwave-assisted reactions of the iodo derivative of the *closo*-dodecaborate anion $[B_{12}H_{11}I]^{2-}$ with sulfamide, *p*-tolyl sulfonamide and thiourea lead to the corresponding *closo*-dodecaborane derivatives with B—N bond (Scheme 27) *(126)*.

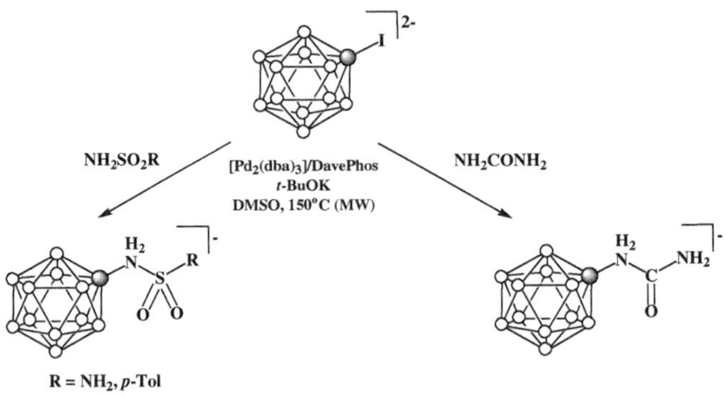

Scheme 27

The reaction of the tetramethylammonium salt of the iodo derivative of the *closo*-dodecaborate anion $[B_{12}H_{11}I]^{2-}$ with 1,8-diazabicyclo(5.4.0) undec-7-ene (DBU) under the same conditions resulted in the corresponding amidine (Scheme 28) *(126)*.

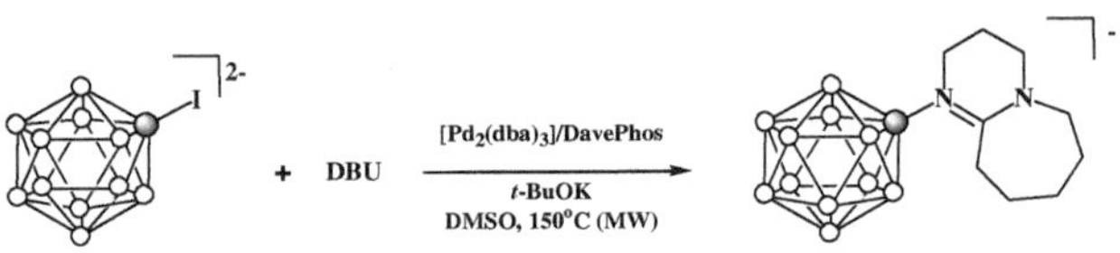

Scheme 28

2.4 1-Carba-*closo*-decaborate anion $[1\text{-}CB_9H_{10}]^-$

The *closo*-decaborate anion is less studied than its 12-vertex analogue *(127)*. Nevertheless, the synthesis of its derivatives with a B—N bond using the Buchwald-Hartwig cross-coupling reaction has received more attention than in the case of the carba-*closo*-dodecaborate anion. The Buchwald-Hartwig amination of 6-iodo- and 10-iodo derivatives of the 1-carba-*closo*-decaborate anion $[1\text{-}HOOC\text{-}6\text{-}I\text{-}1\text{-}CB_9H_8]^-$ and $[1\text{-}HOOC\text{-}10\text{-}I\text{-}1\text{-}CB_9H_8]^-$ with lithium hexamethyldisilazane $LiN(SiMe_3)_2$ in the presence of 2 mol% of $[Pd_2(dba)_3]$ and 8 mol% of CyJohnPhos in refluxing tetrahydrofuran at 55 °C followed by acidic removal of silyl groups leads to the corresponding amino derivatives (Scheme 29) *(128–130)*.

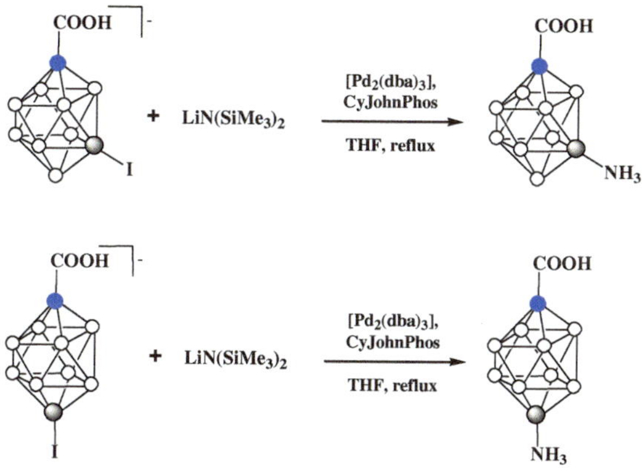

Scheme 29

The microwave-assisted reactions of the tetrabutylammonium salt of the 6-iodo derivative of the 1-carba-*closo*-decaborate anion with anilines in the presence of 3 mol% of $[Pd_2(dba)_3]$, 6 mol% of DavePhos and 2.5 equiv. of *t*-BuOK in dimethyl sulfoxide at 150 °C lead to the corresponding arylamino derivatives $[6\text{-ArNH-1-CB}_9H_9]^-$ in 88%–94% yields (Scheme 30) *(131)*.

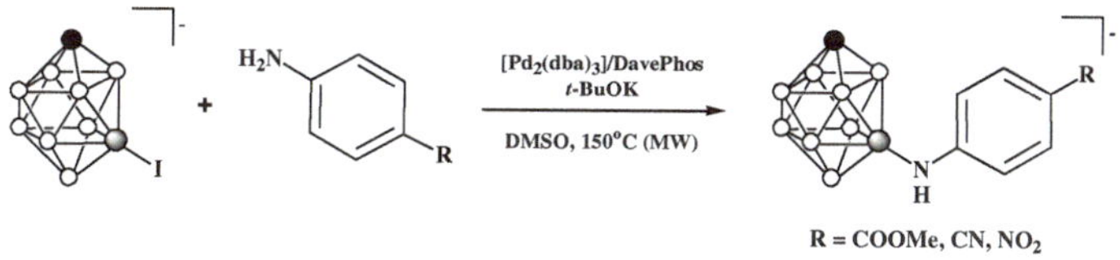

Scheme 30

In a similar way, the reaction of the 6-iodo derivative of the 1-carba-*closo*-decaborate anion with benzylamine in the presence of 4 mol% of $[Pd_2(dba)_3]$, 8 mol% of DavePhos and 2.5 equiv. of *t*-BuOK in dimethyl sulfoxide at 150 °C results in the corresponding benzylamino derivative $[6\text{-BnNH}_2\text{-1-CB}_9H_9]$ in 92% yield (Scheme 31) *(131)*.

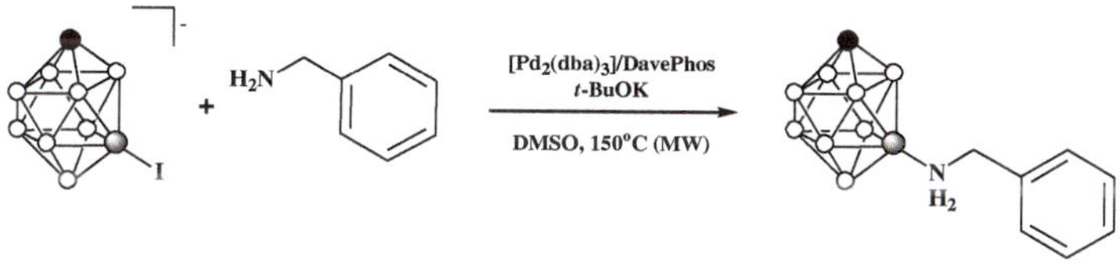

Scheme 31

The microwave-assisted reactions of the 6-iodo derivative of the 1-carba-*closo*-decaborate anion $(Bu_4N)[6\text{-}I\text{-}CB_9H_9]$ with indole, indazole or benzimidazole benzylamine in the presence of 3–4 mol% of $[Pd_2(dba)_3]$, 6–8 mol% of DavePhos and 2.5 equiv. of *t*-BuOK in dimethyl sulfoxide at 150 °C give the corresponding *N*-boronated heteroaromatics in 71%–92% yields (Scheme 32) *(131)*.

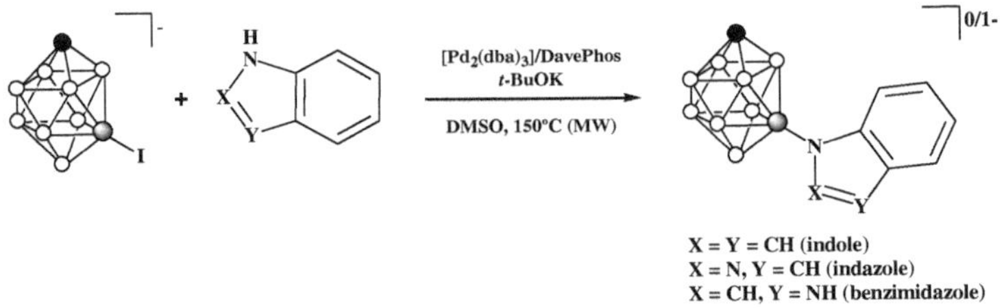

Scheme 32

The microwave-assisted reaction of the 6-iodo derivative of the 1-carba-*closo*-decaborate anion with nicotinamide results in the corresponding carboranyl amide in 77% yield, whereas the reaction with acetanilide gives the expected carboranyl amide only in 33% yield (Scheme 33) *(131)*.

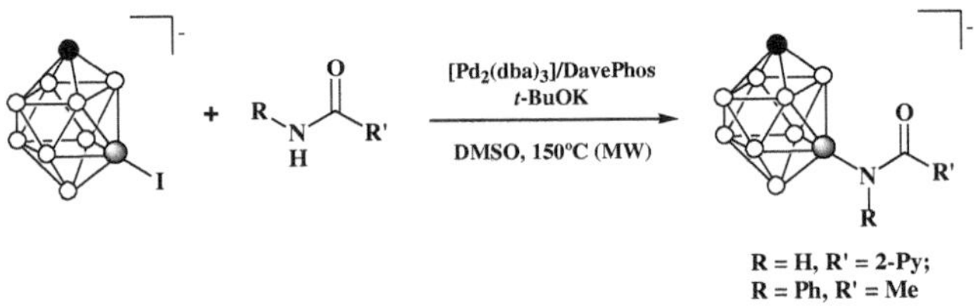

Scheme 33

3. Reactions with formation of the B—P bonds

The formation of derivatives with a B—P bond is characteristic of the reactions of anionic polyhedral boron hydrides with phosphine complexes of palladium.

3.1 Carba-*closo*-dodecaborate anion $[CB_{11}H_{12}]^-$

The microwave-assisted reaction of the cesium salt of the 12-iodo derivative of the carba-*closo*-dodecaborate anion with triphenylphosphine in the presence of 6 mol% of $[(Ph_3P)_2PdCl_2]$ and 1 equiv. of K_2CO_3 in tetrahydrofuran

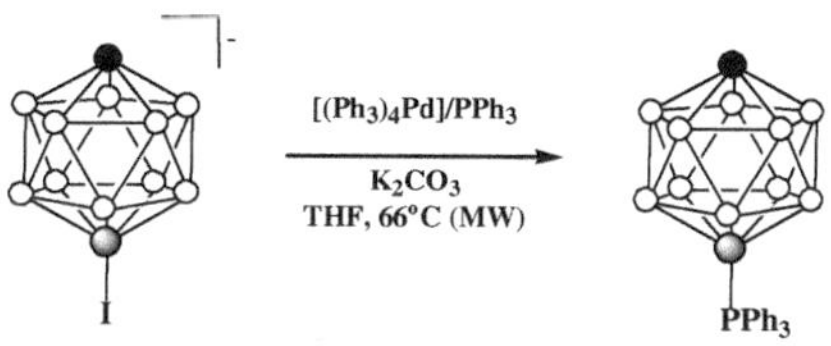

Scheme 34

at 66 °C results in the corresponding triphenylphosphonium derivative [12-Ph_3P-1-$CB_{11}H_{11}$] in 88% yield (Scheme 34) *(64)*.

3.2 *closo*-Dodecaborate anion $[B_{12}H_{12}]^{2-}$

The reaction of the tetrabutylammonium salt of the iodo derivative of the *closo*-dodecaborate anion with [$(Ph_3P)_4Pd$] in the presence of Na_2CO_3 in tetrahydrofuran at 50 °C results in the triphenylphosphonium derivative (Bu_4N)[$B_{12}H_{11}PPh_3$] (Scheme 35) *(132)*.

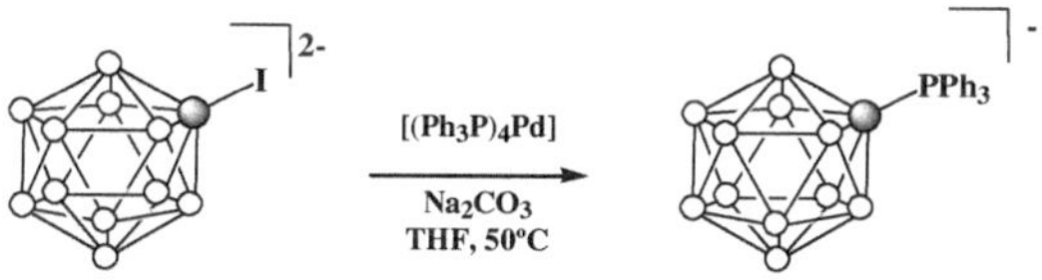

Scheme 35

Phosphonium derivatives of the *closo*-dodecaborate anion can be also obtained using Pd-mediated BH-activation. The reaction of sodium *closo*-dodecaborate $Na_2[B_{12}H_{12}]$ with [$(Me_2PhP)_2PdCl_2$] in tetrahydrofuran at room temperature followed by refluxing with $NaBH_4$ leads to a mixture of phosphonium Na[$B_{12}H_{11}PMe_2Ph$] and isomeric bis(phosphonium) [1,7-$B_{12}H_{10}(PMe_2Ph)_2$] and [1,12-$B_{12}H_{10}(PMe_2Ph)_2$] derivatives (Scheme 36) *(133,134)*.

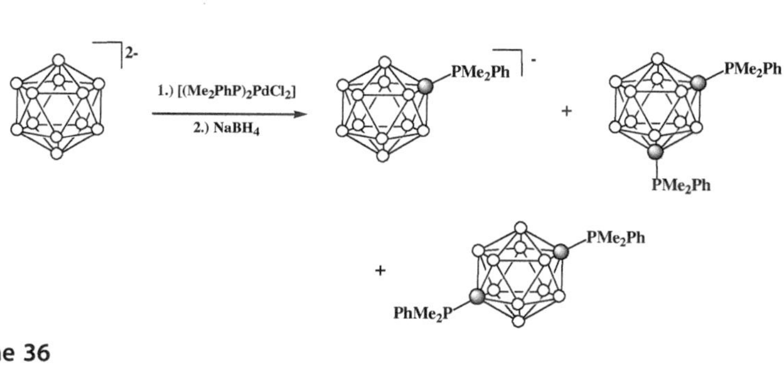

Scheme 36

3.3 *closo*-Decaborate anion $[B_{10}H_{10}]^{2-}$

Similarly to the *closo*-dodecaborate anion, phosphonium derivatives of the *closo*-decaborate anion can be obtained *via* Pd-mediated *BH*-activation methods. The reaction of lithium *closo*-decaborate $Li_2[B_{10}H_{10}]$ with $[(PhMe_2P)_2PdCl_2]$ in dichloromethane at room temperature followed by treatment with $NaBH_4$ in refluxing benzene leads to a mixture of isomeric bis(phosphonium) derivatives $[1,6\text{-}B_{10}H_8(PMe_2Ph)_2]$, $[1,10\text{-}B_{10}H_8(PMe_2Ph)_2]$ and $[2,7\text{-}B_{10}H_8(PMe_2Ph)_2]$ (Scheme 37) *(133)*.

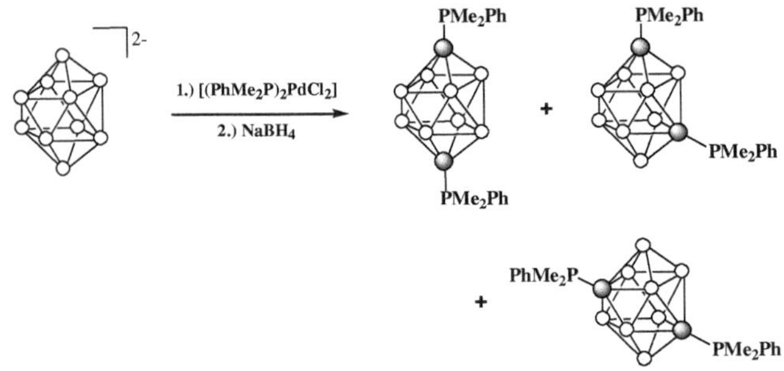

Scheme 37

Bis(phosphonium) *closo*-decaborate derivatives $[1,6\text{-}B_{10}H_8(PPh_3)_2]$ and $[2,4\text{-}B_{10}H_8(PPh_3)_2]$ were obtained by the reaction of tetraethylammonium *closo*-decaborate $(Et_4N)_2[B_{10}H_{10}]$ with $[(Ph_3P)_2PdCl_2]$ in ethanol at room temperature (Scheme 38) *(135,136)*.

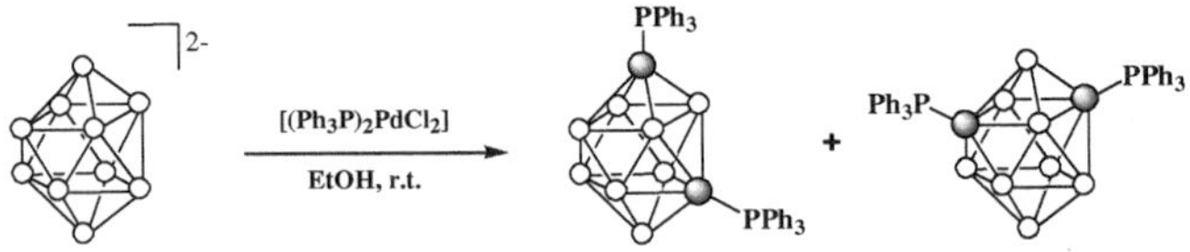

Scheme 38

The reaction of the 1-diazonium derivative of the *closo*-decaborate anion $[1\text{-}B_{10}H_9N_2]^-$ with Ph_2PH in the presence of catalytic amounts of $[(Ph_3P)_2PdCl_2]$ and CuI at 150 °C results in the formation of the corresponding phosphonium derivative $[1\text{-}B_{10}H_9PPh_2H]^-$, which is easily oxidized to $[1\text{-}B_{10}H_9PPh_2OH]^-$, and stable isomeric bis(phosphonium) derivatives $[1,6\text{-}B_{10}H_8(PPh_2H)_2]$ and $[1,10\text{-}B_{10}H_8(PPh_2H)_2]$ (Scheme 39) *(137)*.

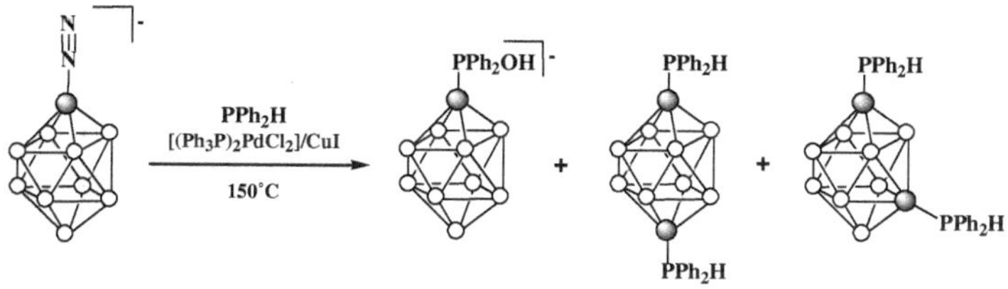

Scheme 39

3.4 7,8-Dicarba-*nido*-undecaborate anion (*nido*-carborane) $[7,8\text{-}C_2B_9H_{12}]^-$

A series of triphenylphosphonium derivatives of *nido*-carborane [X-Ph_3P-7,8-$C_2B_9H_{11}$] (X=1, 3, 5, 9) were synthesized by the reactions of the corresponding iodo derivatives with PPh_3 in the presence of 10 mol% of [$(Ph_3P)_4Pd$] in 1,4-dioxane at 90 °C (Scheme 40) *(138)*.

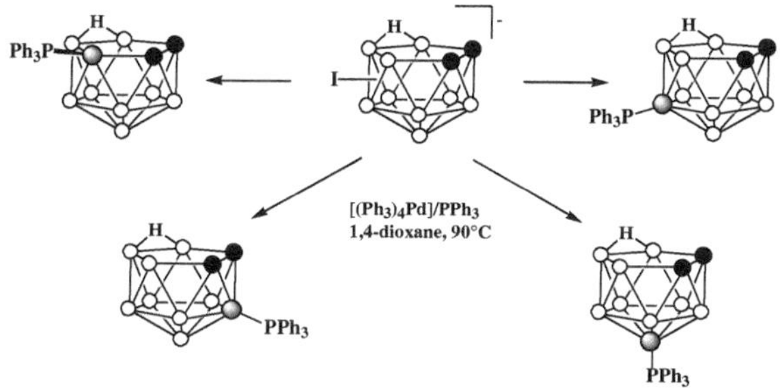

Scheme 40

The symmetrically substituted diiodo derivatives [5,6-I_2–7,8-$C_2B_9H_{10}$]$^-$ and [9,11-I_2–7,8-$C_2B_9H_{10}$]$^-$ under the same conditions give the corresponding monocoupled products [5-I-6-Ph_3P-7,8-$C_2B_9H_{10}$] и [9-I-11-Ph_3P-7,8-$C_2B_9H_{10}$] in good yields, whereas the reaction of the asymmetrical diiodo derivative with substituents at boron atoms in both pentagonal belts [6,9-I_2–7,8-$C_2B_9H_{10}$]$^-$ results in the selective coupling at open pentagonal belt to form [6-I-9-Ph_3P-7,8-$C_2B_9H_{10}$] in a moderate yield (Scheme 41) *(138)*.

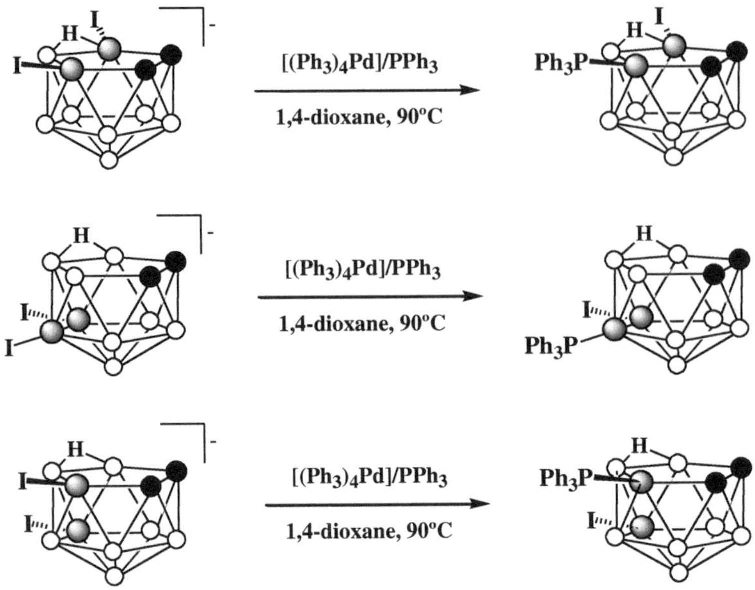

Scheme 41

The bis(phosphonium) derivatives [5,6-$(Ph_3P)_2$-7,8-$C_2B_9H_{10}$] and [9,11-$(Ph_3P)_2$-7,8-$C_2B_9H_{10}$] can be prepared obtained in a similar way with the addition of Cs_2CO_3 as a base to remove the bridging hydrogen (Scheme 42) *(138)*. [9,11-$(Ph_3P)_2$–7,8-$C_2B_9H_{10}$] also can be obtained in two-step way through [9-I-11-Ph_3P-7,8-$C_2B_9H_{10}$] (Scheme 42) *(138)*.

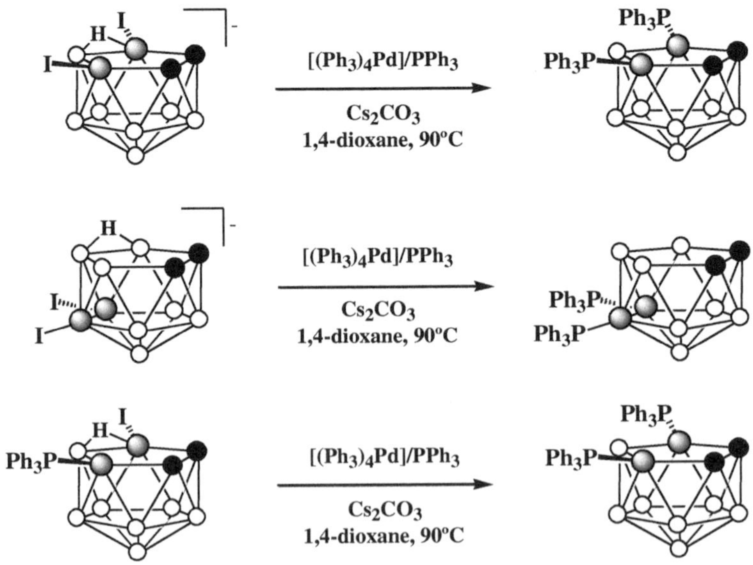

Scheme 42

In the case of triiodo substituted *nido*-carborane [5,6,9-I_3-7,8-$C_2B_9H_9$]$^-$, the substitution proceeds selectively in the open pentagonal belt and in the most distant position in the lower pentagonal belt to form [5-I-6,9-(Ph_3P)$_2$-7,8-$C_2B_9H_8$] (Scheme 43). Alternatively, this compound can be obtained by the reaction of the diiodo derivative of *nido*-carborane [5,6-I_2-7,8-$C_2B_9H_{10}$]$^-$ with PPh_3 in presence of sub-equimolar amount of [(Ph_3P)$_4$Pd] in dioxane at 90 °C (Scheme 43) *(138)*. In this case, both the BX and BH activation takes place.

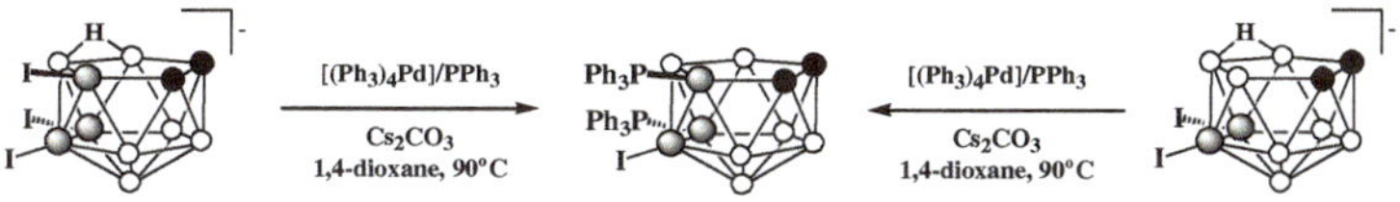

Scheme 43

The reaction of *nido*-carborane [7,8-$C_2B_9H_{12}$] with 0.5 equiv. of [(Me_2PhP)$_2PdCl_2$] in dichloromethane at room temperature followed by heating with $NaBH_4$ in benzene leads to the *BH*-activation of *nido*-carborane with the formation of a mixture of isomeric phosphonium derivatives [9-Me_2PhP-7,8-$C_2B_9H_{11}$] and [10-Me_2PhP-7,8-$C_2B_9H_{11}$] (Scheme 44) *(134)*.

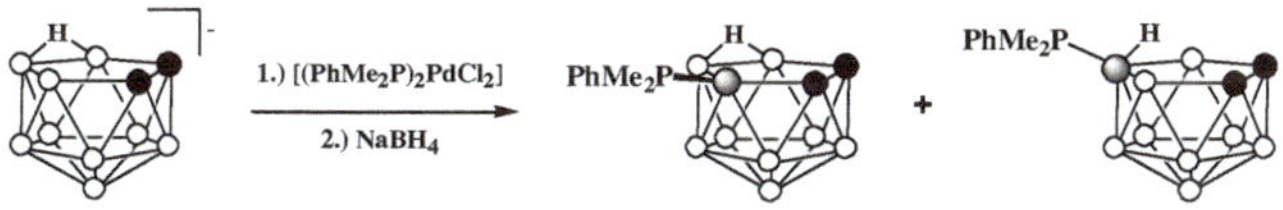

Scheme 44

The reactions of sulfide [7,8-μ-S(CH_2)$_3$-7,8-$C_2B_9H_{10}$]$^-$ with 1 equiv. of palladium(II) complexes [L_2PdCl_2] (L = PPh_3, PPh_2Me) in boiling ethanol lead to selective *BH*-activation at the nearest to sulfur boron atom of the upper pentagonal belt with the formation of [11-L-7,8-μ-(S(CH_2)$_3$)-7,8-$C_2B_9H_9$] (L = PPh_3P, PPh_2Me) (Scheme 45) *(139)*. It is assumed that the alkyl sulfide substituent plays the role of a directing group.

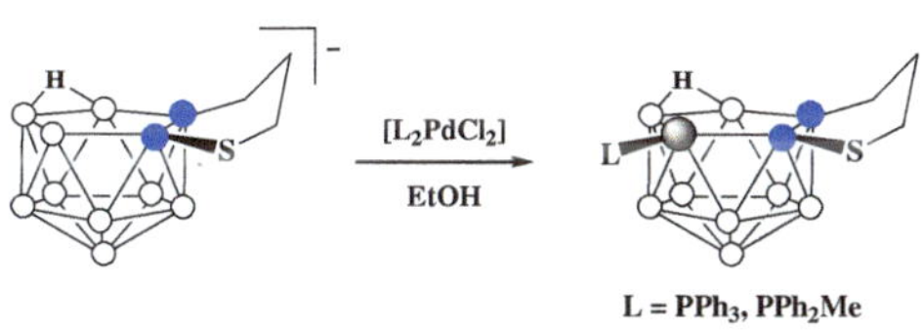

Scheme 45

The 2-pyridyl substituent can also act as a directing group. The reactions 2-pyridyl substituted *nido*-carborane $[7\text{-}(2'\text{-Py})\text{-}7,8\text{-}C_2B_9H_{11}]^-$ with various phosphines in the presence of catalytic amounts of $PdCl_2$ in a mixture of toluene, water and acetonitrile mixture at 120 °C result in the corresponding phosphonium derivatives $[11\text{-L-}7\text{-}(2'\text{-Py})\text{-}7,8\text{-}C_2B_9H_{10}]$ (Scheme 46). Various substituted pyridines (2-, 3-, and 4-picoline, 4-trifluoromethylpyridine), as well as 2-benzoxazolyl and diphenylphosphine groups can also be used as guide groups. The reaction is also tolerant to the presence of alkyl and aryl substituents at the second carbon atom of the carborane cage *(140)*.

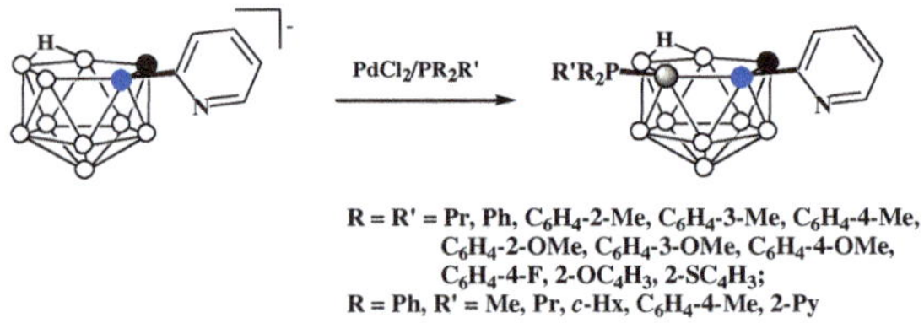

Scheme 46

4. Reactions with formation of the B—O bonds

4.1 Icosahedral carboranes

When studying the reactions of Pd-catalyzed amination of 2-iodo-*para*-carborane, it was found that one of the by-products of amination in the presence of *t*-BuONa is always 2-hydroxy-*para*-carborane. Moreover, the reaction of 2-iodo-*para*-carborane with 5 mol% of $Pd(dba)_2$, 5 mol% of BINAP and 3 equiv. of NaH in 1,4-dioxane at 100 °C results in $2\text{-HO-}1,12\text{-}C_2B_{10}H_{11}$ in 90% yield (Scheme 47) *(78)*.

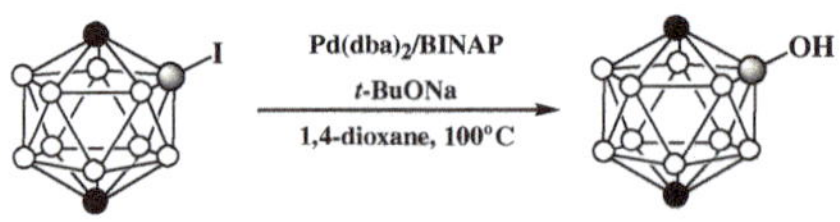

Scheme 47

Later, this approach was used to prepare a series of 2-alkoxy- and 2-aryloxy derivatives of *para*-carborane (Scheme 48) *(141)*.

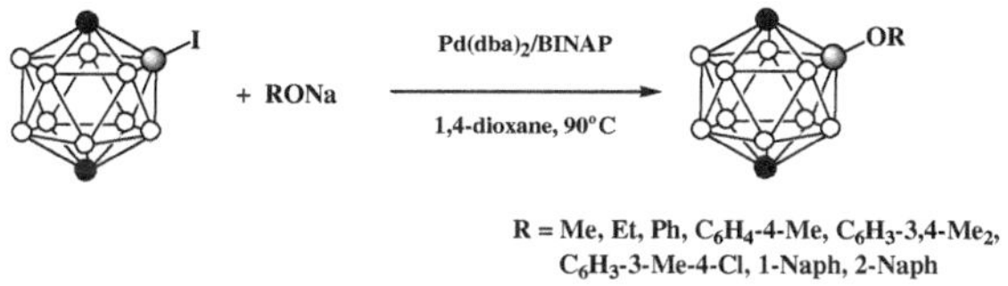

Scheme 48

The reaction of 9-bromo-*meta*-carborane with aqueous K_3PO_4 in the presence of 5 mol% of [(SPhos)(ABP)Pd(OMs)] and 5 mol% of SPhos in 1,4-dioxane at 80 °C results in the formation of the corresponding hydroxy derivative 9-HO-1,7-$C_2B_{10}H_{11}$ in nearly quantitative yield (Scheme 49) *(86)*. Similar reactions with MeONa and with CF_3CH_2ONa lead to the corresponding 9-alkoxy derivatives 9-RO-1,7-$C_2B_{10}H_{11}$ (R=Me, CH_2CF_3) (Scheme 49) *(86)*. The reactions of 9-bromo-*meta*-carborane with 3,5- and 2,6-dimethylphenols in the presence of 5 mol% of [(SPhos)(ABP)Pd(OMs)] or [(DavePhos)(ABP)Pd(OMs)], 5 mol% of SPhos or DavePhos and 3 equiv. of *t*-BuOK in 1,4-dioxane at 80 °C result in the corresponding 9-aryloxy derivatives 9-RO-1,7-$C_2B_{10}H_{11}$ (R=C_6H_3–3,5-Me_2, C_6H_3-2,6-Me_2) (Scheme 49), while the reaction with 3,5-dimethylphenol in the presence of 5 mol% of [(XPhos)(ABP)Pd(OMs)] and 5 mol% of XPhos gives in a mixture of all possible isomers (2-, 4-, 9- and 5-) and the reaction with 2,6-dimethylphenol under the same conditions leads mainly to 2-(2′,6’-$Me_2C_6H_3$)-1,7-$C_2B_{10}H_{11}$ *(86,88)*.

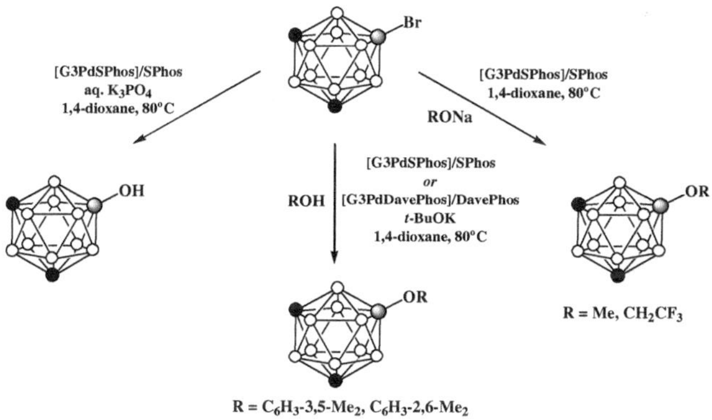

Scheme 49

The reaction of 9,10-dibromo-*meta*-carborane with 3,5-dimethylphenol in the presence of 5 mol% of [(SPhos)(ABP)Pd(OMs)], 5 mol% of SPhos and

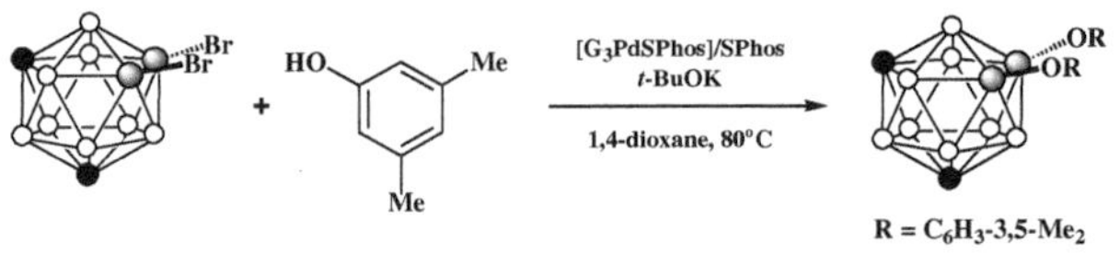

Scheme 50

t-BuOK in 1,4-dioxane at 80 °C results in 9,10-(3′,5′-$Me_2C_6H_3)_2$–1,7-$C_2B_{10}H_{10}$ (Scheme 50) *(86)*.

The reaction of 3-bromo-*ortho*-carborane with 3,5-dimethylphenol in the presence of 5 mol% of [(SPhos)(ABP)Pd(OMs)], 5 mol% of SPhos and 4 equiv. of K_3PO_4 in 1,4-dioxane at 70 °C leads to the corresponding 3-aryloxy derivative 3-(3′,5'-$Me_2C_6H_3$)-1,2-$C_2B_{10}H_{11}$ (Scheme 51) *(86)*.

Much more attention has been given to transition metal catalyzed B—H activation reactions of carboranes with the formation of derivatives with a B—O bond. Reactions of *C,C*′-disubstituted *ortho*-carboranes 1,2-R_2-1,2-$C_2B_{10}H_{10}$ (R = Me, $CH_2C_6H_4$–4-Me, $CH_2C_6H_4$-3-Br, Ph, C_6H_4-4-MeO, C_6H_4-4-Cl; RR = $(CH_2)_4$, $(CH_2)_2C_6H_4$) with acetic acid in the presence of 10 mol% of $Pd(OAc)_2$, 2 equiv. of $Cu(OAc)_2$ and 1.5 equiv. of iodosobenzene PhIO in dichloromethane at 40 °C lead to mixtures of the corresponding 8- and 9-acetoxy derivatives. The similar reactions in the presence of 20 mol% of $Pd(OAc)_2$ and 5 equiv. of PhIO in acetic anhydride at 80 °C result in the corresponding tetraacetoxy derivatives 8,9,10,12-$(AcO)_4$-1,2-R_2-1,2-$C_2B_{10}H_6$ in moderate yields *(142)*.

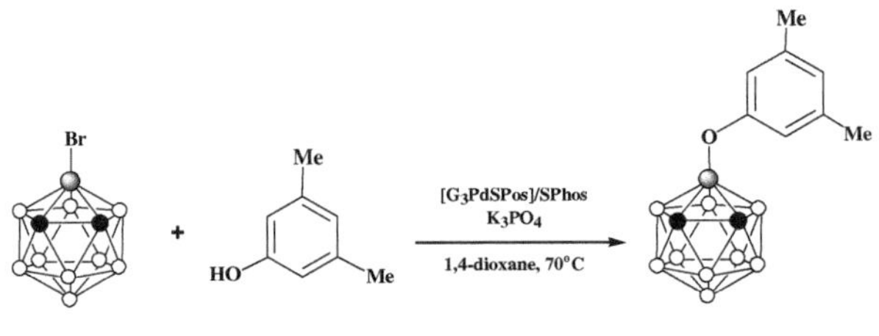

Scheme 51

A series of hydroxy derivatives of *ortho*-carborane was prepared using directing group assisted B—H activation reactions. The reactions of *C*-substituted *ortho*-carboranyl carboxylic acids 1-HOOC-2-R-1,2-$C_2B_{10}H_{10}$ with air O_2 in the presence of 2.5 mol% of $[Cp^*RhCl_2]_2$ and 2 equiv. of KOAc in toluene at 95 °C lead to the corresponding 7-hydroxy derivatives 7-HOOC-1-R-1,2-$C_2B_{10}H_{10}$ (Scheme 52) *(143)*.

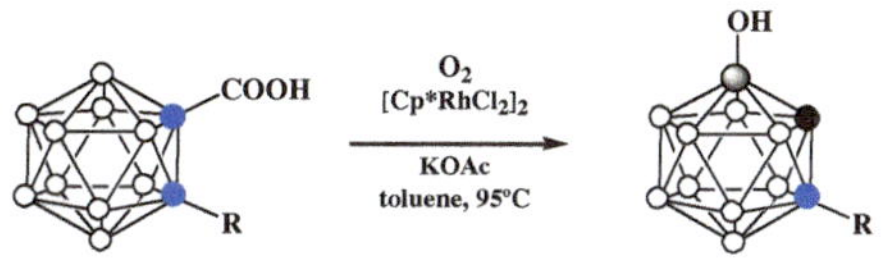

R = Me, Et, *n*-Pr, *i*-Pr, *n*-Bu, *c*-C_5H_9, *c*-C_6H_{11}, Bn, $CH_2C_6H_4$-4-Me, $CH_2C_6H_4$-4-Cl, CH_2CH_2-2-OMe, $SiMe_3$, CH=CHPh, CH=CHC_6H_4-2-Me, CH=CHC_6H_4-3-Me, CH=CHC_6H_4-4-Me, CH=CHC_6H_4-4-OMe, CH=CHC_6H_4-4-F, CH=CHC_6H_4-4-Cl

Scheme 52

The directing group assisted Rh-catalyzed B—H activation of 1-(2′-pyridylsulfenyl)-2-methyl-*ortho*-carborane in the presence of atmospheric oxygen results in the corresponding 4-hydroxy derivative 4-HO-1-(NC_5H_4-2′-S)-2-Me-1,2-$C_2B_{10}H_9$. The reaction proceeds through the formation of cyclometallated complex ***B*** which can be isolated using stoichiometric amount of $[Cp^*RhCl_2]_2$ (Scheme 53) *(144)*.

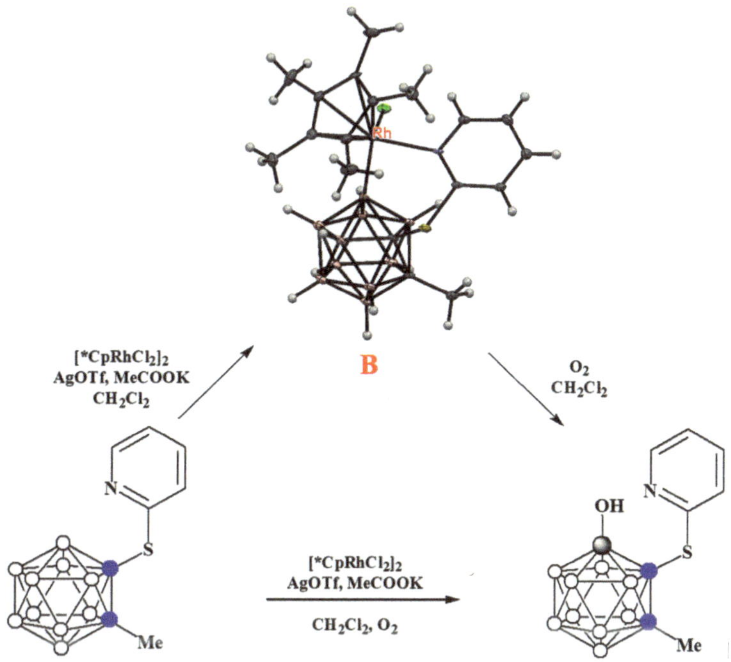

Scheme 53

The electrochemical reactions of amides of *ortho*-carboranyl carboxylic acids and 8-aminoquinoline 1-(8′-$NC_9H_6NH(O)C$)-2-R-1,2-$C_2B_{10}H_{10}$ with *t*-BuOLi in the presence of 20 mol% of $Cu(OTf)_2$ lead to the corresponding 4-*tert*-butoxy derivatives, whereas the reaction of 1-(2′-$NC_5H_4CMe_2NH(O)C$)-2-Me-1,2-$C_2B_{10}H_{10}$ under similar conditions leads to the corresponding 4,5-di(*tert*-butoxy) derivative. The hydrolysis of the *tert*-butoxy derivatives in 12 M HCl at 120 °C leads to the corresponding hydroxy derivatives (Scheme 54) *(145)*, whereas heating with K_2CO_3 in THF leads to the amide hydrolysis and decarboxylation, *i.e.*, complete removal of the directing ligand. The reactions of 1-(8′-$NC_9H_6NH(O)C$)-2-*i*-Pr-1,2-$C_2B_{10}H_{10}$ with lithium phenolates ArOLi under similar conditions lead to the corresponding 4,5-di (aryloxy) derivatives 1-(8′-$NC_9H_6NH(O)C$)-4,5-$(ArO)_2$-2-*i*-Pr-1,2-$C_2B_{10}H_8$ (Scheme 55) *(145)*.

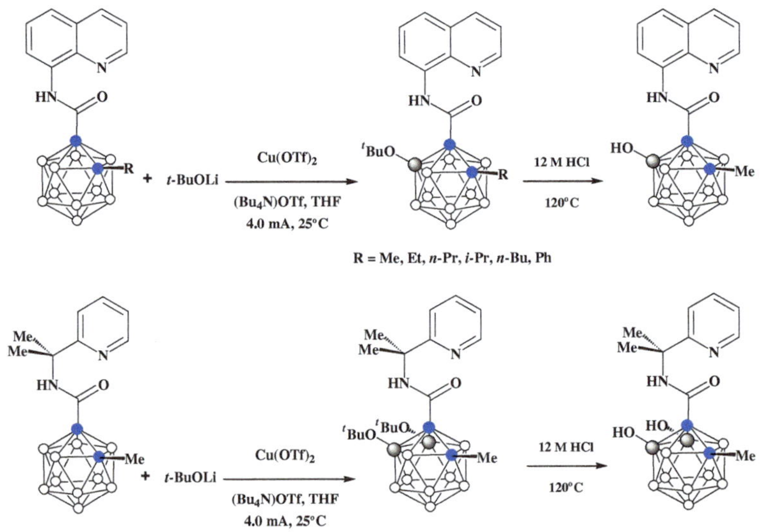

Scheme 54

A series of acyloxy derivatives of *ortho*-carborane was prepared using directing group assisted B—H activation reactions. The reaction of 1-(2′-pyridy)-2-methyl-*ortho*-carborane with $Cu(OAc)_2$ in the presence of 5 mol% of $[Cp^*RhCl_2]_2$, 20 mol% of AgOAc and 2 equiv. of Oxone in 1,2-dichloroethane at 120 °C gives a mixture of the corresponding

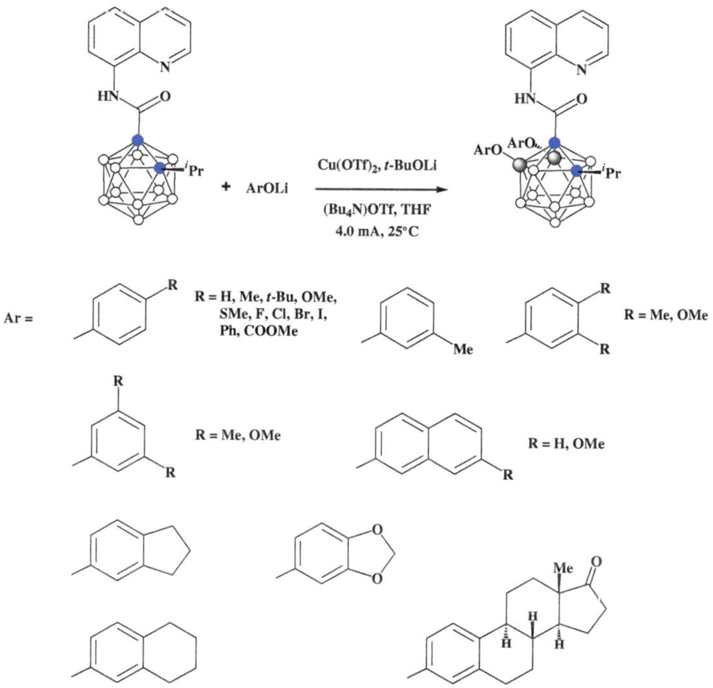

Scheme 55

3-acetoxy- and 3,6-diacetoxy derivatives 3-AcO-1-(2'-NC_5H_4)-2-Me-1,2-$C_2B_{10}H_9$ and 3,6-$(AcO)_2$–1-(2'-NC_5H_4)-2-Me-1,2-$C_2B_{10}H_8$. The reaction is applicable to various alkyl and aryl carboxylic acids in combination with $Cu(OH)_2$ instead of copper carboxylates (Scheme 56). The 2-pyrimidyl group can be used instead of the 2-pyridyl group as the directing group *(146)*.

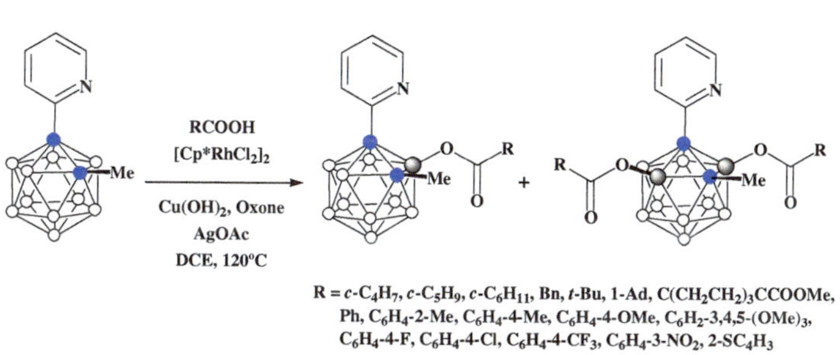

Scheme 56

The reactions of 1-(2′-pyridy)-*ortho*-carborane with carboxylic acids in the presence of 10 mol% of $Pd(OAc)_2$, 20 mol% of *N*-acetyl-*L*-isoleucine and 2 equiv. of cerium ammonium nitrate in 1,2-dichloroethane at room temperature lead to the corresponding 3-acyloxy derivatives 3-RC(O)O-1-(2′-NC_5H_4)-1,2-$C_2B_{10}H_{10}$ in 60–80% yields (Scheme 57) *(147)*. The reaction is also applicable to *ortho*-carboranes bearing substituents at the second carborane carbon atom as well as to *ortho*-carboranes with 4-Me-, 5-Me and 6-Me-substituted pyridyl directing groups *(147)*. Sodium benzoate, benzoic anhydride and iodobenzene diacetate can also be used as sources of the acyloxy group *(147)*.

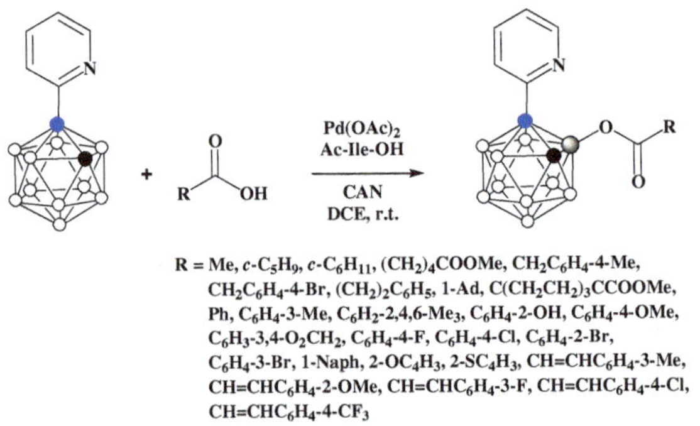

Scheme 57

Due to the use of very mild reaction conditions, this approach can be used for the late-stage functionalization of drug molecules containing alkyl and aryl carboxylic acid groups, such as probenecid, febuxostat, 3-methylflavone-8-carboxylic acid, isoxepac, naproxen, indomethacin, ketoprofen, and gemfibrozil *(146)*.

The reactions of 1-(2′-pyridy)-*ortho*-carborane with carboxylic acids in the presence of 10 mol% of $Pd(OAc)_2$, 20 mol% of $(BnO)_2PO_2H$ and 2 equiv. of cerium ammonium nitrate in 1,2-dichloroethane at room temperature lead to the corresponding 3,4-diacyloxy derivatives 3,4-(RC(O)O)$_2$-1-(2′-NC_5H_4)-1,2-$C_2B_{10}H_9$ (Scheme 58) *(147)*.

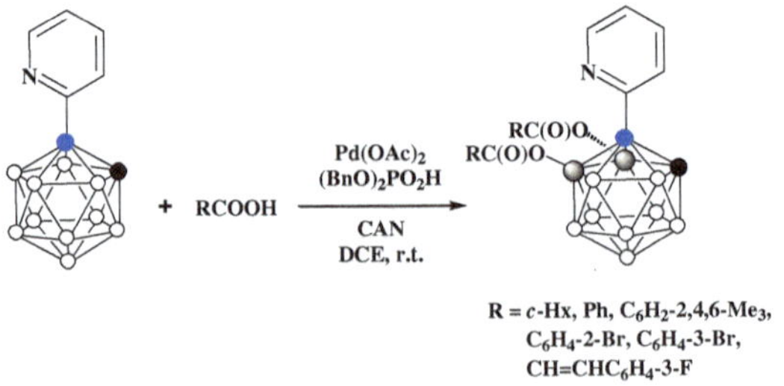

Scheme 58

The reaction of *ortho*-carboranyl carboxylic acid with iodobenzene diacetate in the presence of 10 mol% of $Pd(OAc)_2$, 5 equiv. of acetic acid, 5 equiv. of acetic anhydride and 2 equiv. of K_3PO_4 in 1,4-dioxane at 50 °C leads to 4-acetoxy derivative of *ortho*-carborane 4-AcO-1,2-$C_2B_{10}H_{11}$ *(148)*. In a similar way, the reactions of *C*-substituted *ortho*-carboranyl carboxylic acids result in the corresponding 7-acetoxy derivatives 7-AcO-1-R-1,2-$C_2B_{10}H_{10}$ (Scheme 59) *(148)*.

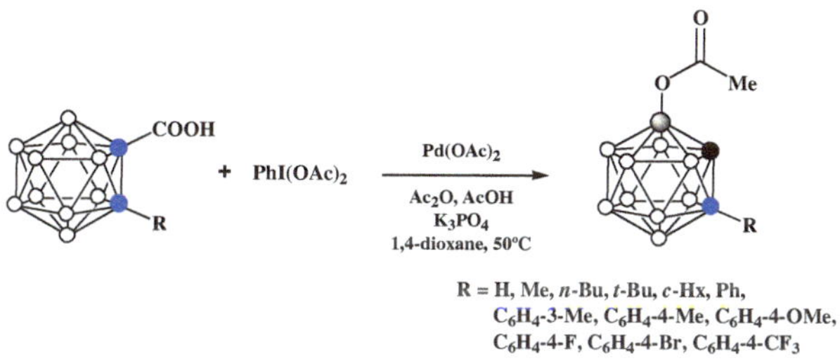

Scheme 59

This approach can be used to synthesize other 7-acyloxy derivatives of *ortho*-carborane as well (Scheme 60) *(148)*.

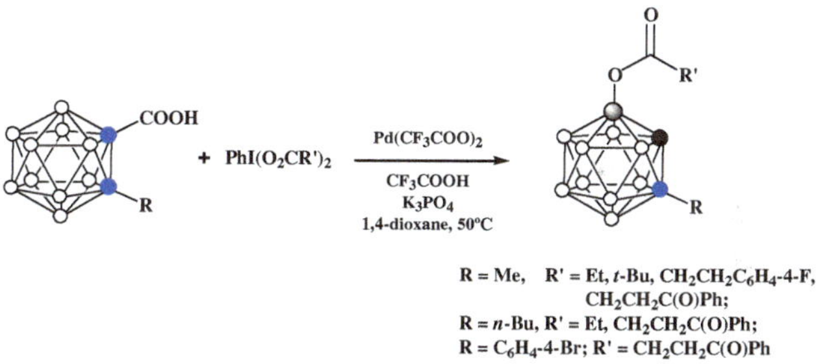

Scheme 60

Similar reactions of *ortho*-carboranyl carboxylic acid and *C*-alkyl *ortho*-carboranyl carboxylic acids with iodobenzene diacetate in toluene using LiOAc instead of K_3PO_4 as a base leads to 4,5-di(acetoxy)-*ortho*-carborane or 7,8-di(acetoxy)-1-alkyl-*ortho*-carboranes, respectively (Scheme 61) *(148)*.

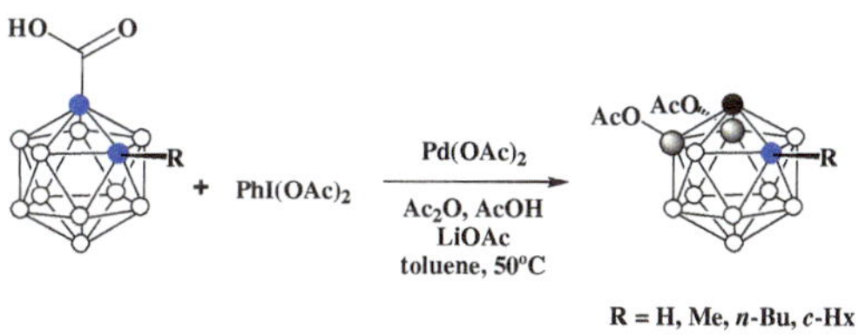

Scheme 61

A special case is the formation of 7,8-substituted carborane analogues of coumarin in the reactions of *ortho*-carboranyl carboxylic acid 1-HOOC-2-Me-1,2-$C_2B_{10}H_{10}$ with benzoic acids in the presence of 5 mol% of $[Cp^*IrCl_2]_2$, 10 mol% of $AgNTf_2$, 2 equiv. of Li_2CO_3 and 5 equiv. of $Cu(OPiv)_2$ in toluene at 160 °C (Scheme 62). In this case, first, the *B*-arylation of the carborane cage occurs, and then cyclization with the formation of the B—O bond *(149)*.

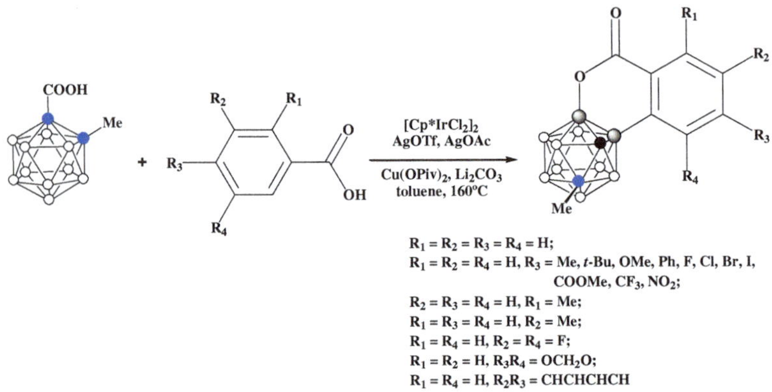

Scheme 62

4.2 1-Carba-*closo*-decaborate anion

The 6-hydroxy derivative of 1-carba-*closo*-decaborate anion is a single product of the Pd-catalyzed reaction of the 6-iodo derivative with *N*-methylacetamide (Scheme 63) *(131)*.

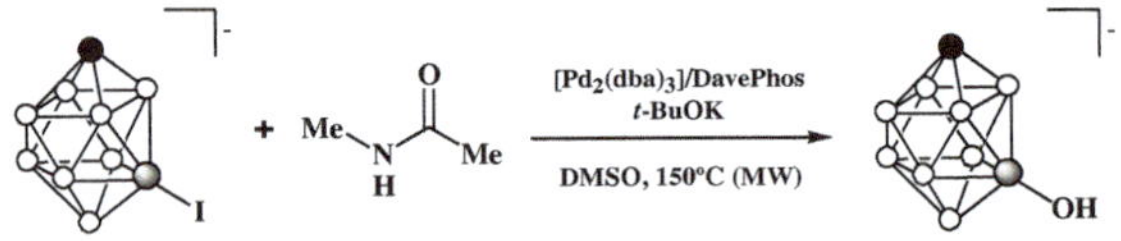

Scheme 63

4.3 *closo*-Dodecaborate anion

There are several examples of the transition metal mediated formation of a B—O bond in anionic polyhedral boron hydrides. Thus, the closure of a five-membered diboroxazole ring with the formation of a boron-oxygen bond occurs upon treatment of *N*-amido derivatives of the *closo*-dodecaborate anion $[B_{12}H_{11}NH_2COR]^-$ (R=Me, Et, NMe_2) with silver acetate in acetonitrile (Scheme 64) *(122,124)*.

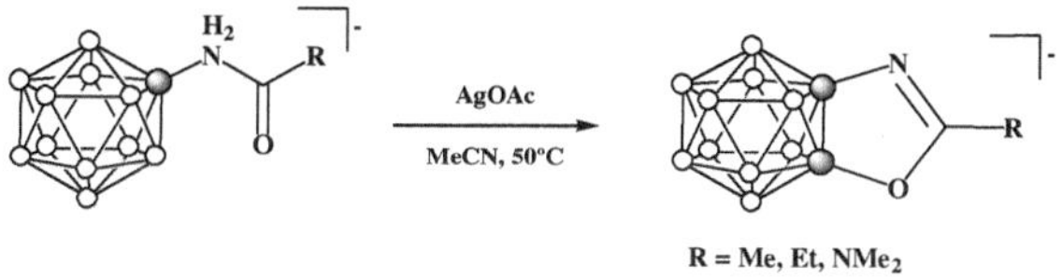

Scheme 64

5. Reactions with formation of the B—S bonds

Pd-catalyzed cross-coupling of iodo derivatives of carboranes with triisopropylsilanethiol was used to synthesize a series of *B*-mercapto carboranes. The reactions of 9-iodo-*meta*-carborane, 2-iodo-*para*-carborane and 2,9-diiodo-*para*-carborane with *i*-Pr_3SiSH in the presence of 20 mol% of $[(Ph_3P)_4Pd]$ and 6 equiv. of K_3PO_4 in 1,4-dioxane at 100 °C lead to the triisopropylsilylmercapto derivatives, which can be easily deprotected by using dilute aqueous HCl to give the corresponding thiols 9-HS-1,7-$C_2B_{10}H_{11}$, 2-HS-1,12-$C_2B_{10}H_{11}$ and 2,9-$(HS)_2$-1,12-$C_2B_{10}H_{10}$ (Scheme 65) *(150)*.

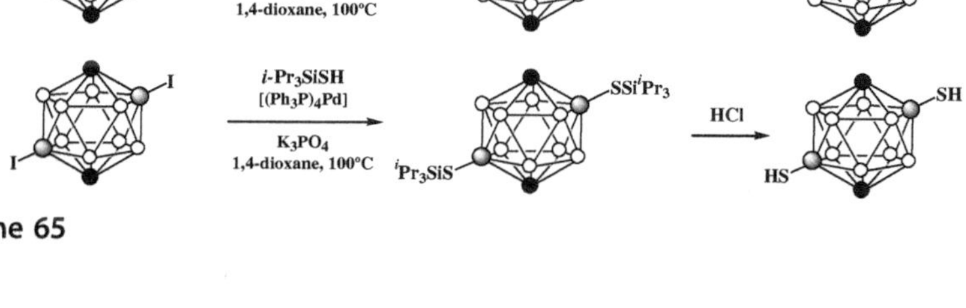

Scheme 65

In a similar way, the reactions of 8-iodo-, 9-iodo- and 9,12-diiodo-*ortho*-carboranes with *i*-Pr_3SiSH under the same conditions result in the corresponding thiols 8-HS-1,2-$C_2B_{10}H_{11}$, 9-HS-1,2-$C_2B_{10}H_{11}$ and 9,12-$(HS)_2$-1,2-$C_2B_{10}H_{10}$ (Scheme 66) *(150)*.

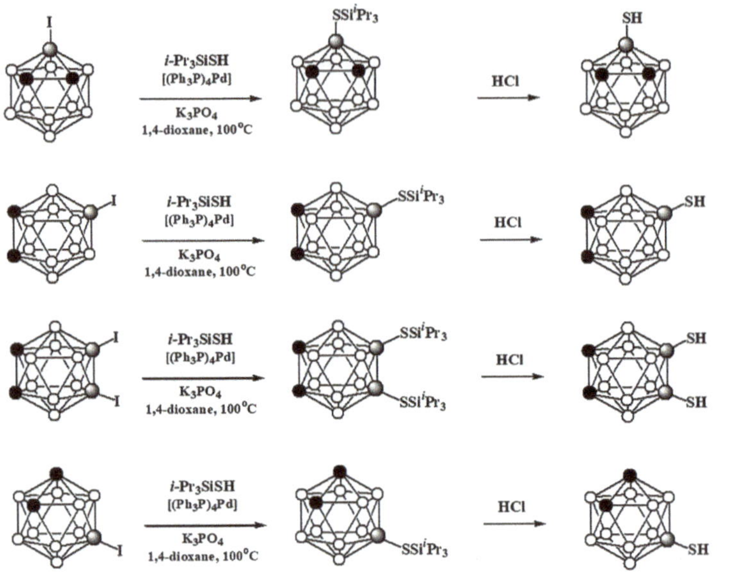

Scheme 66

The reactions of amides of *ortho*-carboranyl carboxylic acids and 8-aminoquinoline 1-(8′-$NC_9H_6NH(O)C$)-2-R-1,2-$C_2B_{10}H_{10}$ with disulfides RSSR in the presence of 1 equiv. of $Cu(OTf)_2$ and 2-phenylpyridine and 4 equiv. of *t*-BuOLi in 1,2-dichloroethane at 130° lead to the corresponding 7,8-di(sulfenyl) derivatives 7,8-$(RS)_2$-1-R-1,2-$C_2B_{10}H_9$ (Scheme 67) *(151)*.

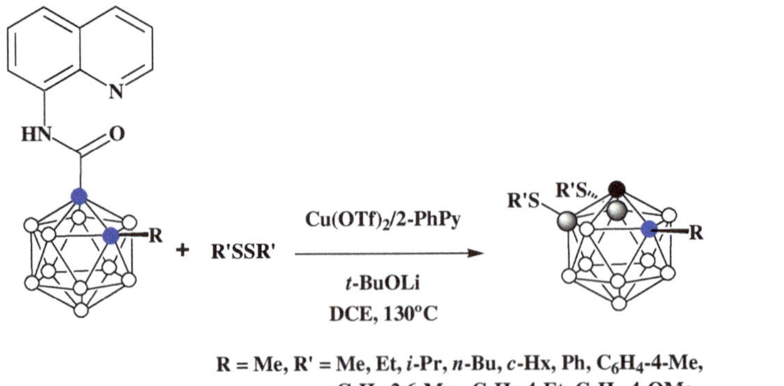

R = Me, R' = Me, Et, *i*-Pr, *n*-Bu, *c*-Hx, Ph, C_6H_4-4-Me, C_6H_3-2,6-Me_2, C_6H_4-4-Et, C_6H_4-4-OMe, C_6H_4-4-SMe, C_6H_4-2-F, C_6H_4-3-F, C_6H_4-4-F, C_6H_4-4-Cl, C_6H_4-2-Br, C_6H_4-4-Br, C_6H_4-4-CF_3, 2-Naph, 2-SC_4H_3;
R = *i*-Pr, R' = Ph; R = *n*-Bu, R' = Ph; R = $CH_2C_6H_4$-4-Me;
R = R' = Ph

Scheme 67

An interesting example of transition metal mediated sulfenylation of *meta*-carborane includes reaction of 9-chloromercuro-*meta*-carborane with $[(Ph_3P)_3Pt]$ in benzene at room temperature that results in the platinated complex **C** which on the treatment with benzyl thiol in the presence of K_3PO_4 gives 9-BnS-1,7-$C_2B_{10}H_{11}$ (Scheme 68) *(99)*.

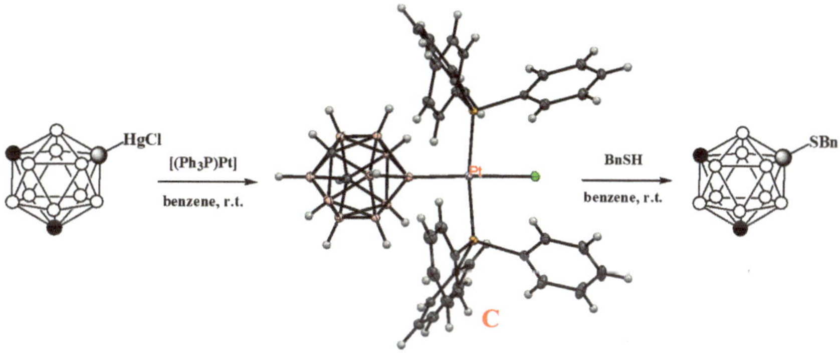

Scheme 68

Due to the mild conditions of sulfenylation, complex **C** has been used to add carborane to cysteine residues of various peptides and proteins *(152)*.

Transition metal-catalyzed reactions of BH-activation of carboranes with the formation of B—N bonds continue to attract increased interest from researchers. More recently, amination and sulfoamidation of *ortho*-carboranes with the formation of B(9)-substituted derivatives in the presence of catalytic amounts of $Pd(OAc)_2$ and $AgPF_6$ or AgOAc in hexafluoropropanol has been described *(153)*.

References

1. Sivaev, I. B.; Bregadze, V. I. *Polyhedral Boron Hydrides in Use: Current Status and Perspectives*; Nova Science Publishers: New York, 2009; p. 85.
2. Boron Science. In *New Technologies and Applications*; Hosmane, N. S., Ed.; CRC Press: Boca Raton, FL, 2012; p. 878.
3. Hosmane, N. S., Eagling, R., Eds., Vol. 1–4; *Handbook of Boron Science with Applications in Organometallics, Catalysis, Materials and Medicine*; World Scientific Publishing Europe: London, 2018; p. 1056.
4. Ochi, J.; Tanaka, K.; Chujo, Y. *Angew. Chem., Int. Ed.* **2020**, *132*, 9925–9939.
5. Anufriev, S. A.; Timofeev, S. V.; Anisimov, A. A.; Suponitsky, K. Y.; Sivaev, I. B. *Molecules* **2020**, *25*, 5745.
6. Zhang, Y.; Wang, L.; Hu, J.; Duttwyler, S.; Cui, X.; Xing, H. *CrystEngComm* **2020**, *22*, 2649–2655.
7. Wang, L.; Sun, W.; Zhang, Y.; Xu, N.; Krishna, R.; Hu, J.; Jiang, Y.; He, Y.; Xing, H. *Angew. Chem., Int. Ed.* **2021**, *60*, 22865–22870.

8. Xia, Q.; Zhang, J.; Chen, X.; Cheng, C.; Chu, D.; Tang, X.; Li, H.; Cui, Y. *Coord. Chem. Rev.* **2021**, *435*, 213783.
9. Hey-Hawkins, E., Viñas, C., Eds. Boron-based compounds: potential and emerging applications. In *Biomedicine*; John Wiley & Sons: Hoboken, NJ, 2018; p. 496.
10. Soloway, A. H.; Tjarks, W.; Barnum, B. A.; Rong, F.-G.; Barth, R. F.; Codogni, I. M.; Wilson, J. G. *Chem. Rev.* **1998**, *98*, 1515–1562.
11. Valliant, J. F.; Guenther, K. J.; King, A. S.; Morel, P.; Schaffer, P.; Sogbein, O. O.; Stephenson, K. A. *Coord. Chem. Rev.* **2002**, *232*, 173–230.
12. Sivaev, I. B.; Bregadze, V. I. *Eur. J. Inorg. Chem.* **2009**, 1433–1450.
13. Issa, F.; Kassiou, M.; Rendina, L. M. *Chem. Rev.* **2011**, *111*, 5701–5722.
14. Scholz, M.; Hey-Hawkins, E. *Chem. Rev.* **2011**, *111*, 7035–7062.
15. Leśnikowski, Z. J. *J. Med. Chem.* **2016**, *59*, 7738–7758.
16. Stockmann, P.; Gozzi, M.; Kuhnert, R.; Sárosi, M. B.; Hey-Hawkins, E. *Chem. Soc. Rev.* **2019**, *48*, 3497–3512.
17. Murphy, N.; McCarthy, E.; Dwyer, R.; Farràs, P. *J. Inorg. Biochem.* **2021**, *218*, 111412.
18. Fink, K.; Uchman, M. *Coord. Chem. Rev.* **2021**, *431*, 213684.
19. Grimes, R. N. *Carboranes*, 3rd ed.; Academic Press: London, 2016.
20. Beller, M., Bolm, C., Eds. *Transition Metals for Organic Synthesis: Building Blocks and Fine Chemicals*; Wiley-VCH Verlag GmbH: Weinheim, 1998.
21. Ribas, X., Ed. *C-H and C-X Bond Functionalization: Transition Metal Mediation*; The Royal Society of Chemistry: Cambridge, 2013.
22. Crawley, M. L., Trost, B. M., Eds. *Applications of Transition Metal Catalysis in Drug Discovery and Development: An Industrial Perspective*; John Wiley & Sons, Inc.: Hoboken, 2012.
23. de Meijere, A., Bräse, S., Oestreich, M., Eds., Vol. 1–3; *Metal-Catalyzed Cross-Coupling Reactions and More*; Wiley-VCH Verlag GmbH & Co. KGaA: Weinheim, 2014.
24. Jutand, A.; Lefèvre, G. In *Arene Chemistry: Reaction Mechanisms and Methods for Aromatic Compounds*; Mortier, J., Ed.; John Wiley & Sons, Inc.: Hoboken, 2016; pp. 511–545.
25. Kannan, M.; Sengoden, M.; Punniyamurthy, T. In *Arene Chemistry: Reaction Mechanisms and Methods for Aromatic Compounds*; Mortier, J., Ed.; John Wiley & Sons, Inc.: Hoboken, 2016; pp. 547–586.
26. King, R. B. *Chem. Rev.* **2001**, *101*, 1119–1152.
27. Chen, Z.; King, R. B. *Chem. Rev.* **2005**, *105*, 3613–3642.
28. Zakharkin, L. I.; Koveredov, A. I.; Ol'shevskaya, V. A.; Shaugumbekova, Z. S. *J. Organomet. Chem.* **1982**, *226*, 217–226.
29. Logan, J.; Logan, C. F.; Jones, M. *Inorg. Chem.* **1991**, *34*. 4866–4848.
30. Zheng, Z.; Jiang, W.; Zinn, A. A.; Knobler, C. B.; Hawthorne, M. F. *Inorg. Chem.* **1995**, *34*, 2095–2100.
31. Jiang, W.; Knobler, C. B.; Curtis, C. E.; Mortimer, M. D.; Hawthorne, M. F. *Inorg. Chem.* **1995**, *34*, 3491–3498.
32. Lee, H.; Knobler, C. B.; Hawthorne, M. F. *Chem. Commun.* **2000**, 2485–2486.
33. Fox, M. A.; Wade, K. *J. Mater. Chem.* **2002**, *12*, 1301–1306.
34. Viñas, C.; Barbera, G.; Oliva, J. M.; Teixidor, F.; Welch, A. J.; Rosair, G. M. *Inorg. Chem.* **2001**, *40*, 6555–6562.
35. Bayer, M. J.; Herzog, A.; Diaz, M.; Harakas, G. A.; Lee, H.; Knobler, C. B.; Hawthorne, M. F. *Chem. Eur. J.* **2003**, *9*, 2732–2744.
36. Beletskaya, I. P.; Bregadze, V. I.; Ivushkin, V. A.; Petrovskii, P. V.; Sivaev, I. B.; Sjöberg, S.; Zhigareva, G. G. *J. Organomet. Chem.* **2004**, *689*, 2920–2929.
37. Beletskaya, I. P.; Bregadze, V. I.; Ivushkin, V. A.; Zhigareva, G. G.; Petrovskii, P. V.; Sivaev, I. B. *Russ. J. Org. Chem.* **2005**, *41*, 1359–1366.
38. Himmelspach, A.; Finze, M. *Eur. J. Inorg. Chem.* **2010**, 2012–2024.

39. Spokoyny, A. M.; Li, T. C.; Farha, O. K.; Machan, C. W.; She, C.; Stern, C. L.; Marks, T. J.; Hupp, J. T.; Mirkin, C. A. *Angew. Chem., Int. Ed.* **2010**, *49*, 5339–5343.
40. Puga, A. V.; Teixidor, F.; Sillanpää, R.; Kivekäs, R.; Viñas, C. *Chem. Commun.* **2011**, *47*, 2252–2254.
41. Safronov, A. V.; Sevryugina, Y. V.; Pichaandi, K. R.; Jalisatgi, S. S.; Hawthorne, M. F. *Dalton Trans.* **2014**, *43*, 4969–4977.
42. Jankowiak, A.; Kaszyński, P. *Inorg. Chem.* **2014**, *53*, 8762–8769.
43. Anufriev, S. A.; Sivaev, I. B.; Bregadze, V. I. *Russ. Chem. Bull.* **2015**, *64*, 712–717.
44. Teixidor, F.; Sillanpää, R.; Pepiol, A.; Lupu, M.; Viñas, C. *Chem. Eur. J.* **2015**, *36*, 12778–12786.
45. Anderson, K. P.; Mills, H. A.; Mao, C.; Kirlikovali, K. O.; Axtell, J. C.; Rheingold, A. L.; Spokoyny, A. M. *Tetrahedron* **2019**, *75*, 187–191.
46. Kelemen, Z.; Pepiol, A.; Lupu, M.; Sillanpää, R.; Hänninen, M. M.; Teixidor, F.; Viñas, C. *Chem. Commun.* **2019**, *55*, 8927–8930.
47. Bennour, I.; Teixidor, F.; Kelemen, Z.; Viñas, C. *Molecules* **2020**, *25*, 2814.
48. Asawa, Y.; Hatsuzawa, S.; Yoshimori, A.; Yamada, K.; Katoh, A.; Kouji, H.; Nakamura, H. *Sci. Rep.* **2021**, *11*, 24101.
49. Zakharkin, L. I.; Ol'shevskaya, V. A. *Synth. React. Inorg. Met. Org. Chem.* **1991**, *21*, 1041–1046.
50. Zakharkin, L. I.; Ol'shevskaya, V. A.; Zhigareva, G. G. *Russ. J. Gen. Chem.* **1998**, *68*, 925–927.
51. Anufriev, S. A.; Shmal'ko, A. V.; Suponitsky, K. Y.; Sivaev, I. B. *Catalysts* **2020**, *10*, 1348.
52. Anufriev, S. A.; Shmal'ko, A. V.; Suponitsky, K. Y.; Sivaev, I. B. *Molecules* **2021**, *26*, 7297.
53. Zakharkin, L. I.; Balagurova, E. V.; Lebedev, V. N. *Russ. J. Gen. Chem.* **1998**, *68*, 922–924.
54. Eriksson, L.; Beletskaya, I. P.; Bregadze, V. I.; Sivaev, I. B.; Sjöberg, S. *J. Organomet. Chem.* **2002**, *657*, 267–272.
55. Aizawa, K.; Ohta, K.; Endo, Y. *Heterocycles* **2010**, *80*, 369–377.
56. Eriksson, L.; Winberg, K. J.; Claro, R. T.; Sjöberg, S. *J. Org. Chem.* **2003**, *68*, 3569–3573.
57. Peymann, T.; Knobler, C. B.; Hawthorne, M. F. *Inorg. Chem.* **1998**, *37*, 1544–1548.
58. Grüner, B.; Bonnetot, B.; Mongeot, H. *Collect. Czech. Chem. Commun.* **1997**, *62*, 1185–1204.
59. Kultyshev, R. G.; Liu, S.; Leung, H. T.; Liu, J.; Shore, S. G. *Inorg. Chem.* **2003**, *42*, 3199–3207.
60. Himmelspach, A.; Finze, M.; Vöge, A.; Gabel, D. *Z. Anorg. Allg. Chem.* **2012**, *638*, 512–519.
61. Rzeszotarska, E.; Novozhilova, I.; Kaszynski, P. *Inorg. Chem.* **2017**, *56*, 14351–14356.
62. Grüner, B.; Janoušek, Z.; King, B. T.; Woodford, J. N.; Wang, C. H.; Všetečka, V.; Michl, J. *J. Am. Chem. Soc.* **1999**, *121*, 3122–3326.
63. Finze, M. *Inorg. Chem.* **2008**, *47*, 11857–11867.
64. Himmelspach, A.; Reiss, G. J.; Finze, M. *Inorg. Chem.* **2012**, *51*, 2679–2688.
65. Hailmann, M.; Herkert, L.; Himmelspach, A.; Finze, M. *Chem. Eur. J.* **2013**, *19*, 15745–15758.
66. Himmelspach, A.; Warneke, J.; Schäfer, M.; Hailmann, M.; Finze, M. *Organometallics* **2015**, *34*, 462–469.
67. Pecyna, J.; Kaszyński, P.; Ringstrand, B.; Pociecha, D.; Pakhomov, S.; Douglass, A. G.; Young, V. G. *Inorg. Chem.* **2016**, *55*, 4016–4025.
68. Franken, A.; Kilner, C. A.; Thornton-Pett, M.; Kennedy, J. D. *Chem. Commun.* **2002**, 2048–2049.

69. Ringstrand, B.; Monobe, H.; Kaszynski, P. *J. Mater. Chem.* **2009**, *19*, 4805–4812.
70. Qiu, Z. *Tetrahedron Lett.* **2015**, *56*, 963–971.
71. Dziedzic, R. M.; Spokoyny, A. M. *Chem. Commun.* **2019**, *55*, 430–442.
72. Zhang, F.; Spring, D. R. *Chem. Soc. Rev.* **2014**, *43*, 6906–6919.
73. Chen, Z.; Wang, B.; Zhang, J.; Yu, W.; Liu, Z.; Zhang, Y. *Org. Chem. Front.* **2015**, *2*, 1107–1295.
74. Gandeepan, P.; Ackermann, L. *Chem* **2018**, *4*, 199–222.
75. Au, Y. K.; Xie, Z. *Bull. Chem. Soc. Jpn.* **2021**, *94*, 879–899.
76. Qiu, Z.; Xie, Z. *Acc. Chem. Res.* **2021**, *54*, 4065–4079.
77. Sivaev, I. B. *Russ. J. Inorg. Chem.* **2021**, *66*, 1289–1342.
78. Beletskaya, I. P.; Bregaze, V. I.; Kabytaev, K. Z.; Zhigareva, G. G.; Petrovskii, P. V.; Glukhov, I. V.; Starikova, Z. A. *Organometallics* **2007**, *26*, 2340–2347.
79. Mukhin, S. N.; Kabytaev, K. Z.; Zhigareva, G. G.; Glukhov, I. V.; Starikova, Z. A.; Bregadze, V. I.; Beletskaya, I. P. *Organometallics* **2008**, *27*, 5937–5942.
80. Sevryugina, Y.; Julius, R. L.; Hawthorne, M. F. *Inorg. Chem.* **2010**, *49*, 10627–10634.
81. Kracke, G. R.; VanGordon, M. R.; Sevryugina, Y. V.; Kueffer, P. J.; Kabytaev, K.; Jalisatgi, S. S.; Hawthorne, M. F. *ChemMedChem* **2015**, *10*, 62–67.
82. Marshall, W. J.; Young, R. J.; Grushin, V. V. *Organometallics* **2001**, *20*, 523–533.
83. Bruno, N. C.; Tudge, M. T.; Buchwald, S. L. *Chem. Sci.* **2013**, *4*, 916–920.
84. Ingoglia, B. T.; Wagen, C. C.; Buchwald, S. L. *Tetrahedron* **2019**, *75*, 4199–4211.
85. Sotnik, S. O.; Mishchenko, A. M.; Rusanov, E. B.; Kozytskiy, A. V.; Gavrilenko, K. S.; Ryabukhin, S. V.; Volochnyuk, D. M.; Kolotilov, S. V. *Molecules* **2021**, *26*, 3507.
86. Dziedzic, R. M.; Saleh, L. M. A.; Axtell, J. C.; Martin, J. L.; Stevens, S. L.; Royappa, A. T.; Rheingold, A. L.; Spokoyny, A. M. *J. Am. Chem. Soc.* **2016**, *138*, 9081–9084.
87. Mu, X.; Hopp, M.; Dziedzic, R. M.; Waddington, M. A.; Rheingold, A. L.; Sletten, E. M.; Axtell, J. C.; Spokoyny, A. M. *Organometallics* **2020**, *39*, 4380–4386.
88. Dziedzic, R. M.; Martin, J. L.; Axtell, J. C.; Saleh, L. M. A.; Ong, T.-C.; Yang, Y.-F.; Messina, M. S.; Rheingold, A. L.; Houk, K. N.; Spokoyny, A. M. *J. Am. Chem. Soc.* **2017**, *139*, 7729–7732.
89. Zakharkin, L. I.; Kalinin, V. N. *Bull. Acad. Sci. USSR Div. Chem. Sci.* **1966**, *15*, 549–551.
90. Andrews, J. S.; Zayas, J.; Jones, M. *Inorg. Chem.* **1985**, *24*, 3715–3716.
91. Safronov, A. F.; Hawthorne, M. F. *Inorg. Chim. Acta* **2011**, *375*, 308–310.
92. Zakharkin, L. I.; Kalinin, V. N.; Gedymin, V. V.; Dzarasova, G. S. *J. Organomet. Chem.* **1970**, *23*, 303–312.
93. Zhao, D.; Xie, Z. *Chem. Sci.* **2016**, 7, 5635–5639.
94. Zakharkin, L. I.; Kalinin, V. N.; Gedymin, V. V. *J. Organomet. Chem.* **1969**, *16*, 371–379.
95. Kasar, R. A.; Knudsen, G. M.; Kahl, S. B. *Inorg. Chem.* **1999**, *38*, 2936–2940.
96. Valliant, J. F.; Schaffer, P. *J. Inorg. Biochem.* **2001**, *85*, 43–51.
97. Xu, T.-T.; Cao, K.; Zhang, C.-Y.; Wu, J.; Jiang, L.; Yang, J. *Chem. Commun.* **2018**, *54*, 13603–13606.
98. Wu, J.; Cao, K.; Zhang, C.-Y.; Xu, T.-T.; Ding, L.-F.; Li, B.; Yang, J. *Org. Lett.* **2019**, *21*, 5986–5989.
99. Saleh, L. M. A.; Dziedzic, R. M.; Khan, S. I.; Spokoyny, A. M. *Chem. Eur. J.* **2016**, *22*, 8466–8470.
100. Au, Y. K.; Zhang, J.; Quan, Y.; Xie, Z. *J. Am. Chem. Soc.* **2021**, *143*, 4148–4153.
101. Lyu, H.; Quan, Y.; Xie, Z. *J. Am. Chem. Soc.* **2016**, *138*, 12727–12730.
102. Li, H.; Bai, F.; Yan, H.; Lu, C.; Bregadze, V. I. *Eur. J. Org. Chem.* **2017**, 1343–1352.
103. Baek, Y.; Kim, S.; Son, J.-Y.; Lee, K.; Kim, D.; Lee, P. H. *ACS Catal.* **2019**, *9*, 10418–10425.

104. Han, G. U.; Baek, Y.; Lee, K.; Shin, S.; Hoh, H. C.; Lee, P. H. *Org. Lett.* **2021**, *23*, 416–420.
105. Han, G. U.; Shin, S.; Baek, Y.; Kim, D.; Lee, K.; Kim, J. G.; Lee, P. H. *Org. Lett.* **2021**, *23*, 8622–8627.
106. Cao, K.; Wu, J.; Zhang, C.-Y.; Ding, L.-F.; Yang, J. *J. Organomet. Chem.* **2021**, *954-955*, 122069.
107. Körbe, S.; Schreiber, P. J.; Michl, J. *Chem. Rev.* **2006**, *106*, 5208–5249.
108. Douvris, C.; Michl, J. *Chem. Rev.* **2013**, *113*, PR179–PR233.
109. Fisher, S. P.; Tomich, A. W.; Lovera, S. O.; Kleinsasser, J. F.; Guo, J.; Asay, M. J.; Nelson, H. M.; Lavallo, V. *Chem. Rev.* **2019**, *119*, 8262–8290.
110. Kanazawa, J.; Kitazawa, Y.; Uchiyama, M. *Chem. Eur. J.* **2019**, *39*, 9123–9132.
111. Pecyna, J. C.; Ringstrand, B.; Kaszynski, P. *Inorg. Chem.* **2012**, *51*, 5353–5359.
112. Konieczka, S. Z.; Himmelspach, A.; Hailmann, M.; Finze, M. *Eur. J. Inorg. Chem.* **2013**, 134–146.
113. Sivaev, I. B.; Bregadze, V. I.; Sjöberg, S. *Collect. Czech. Chem. Commun.* **2002**, *67*, 679–727.
114. Hertler, W. R.; Raasch, M. S. *Inorg. Chem.* **1964**, *86*, 3661–3668.
115. Peymann, T.; Lork, E.; Schmidt, M.; Nöth, H.; Gabel, D. *Chem. Ber.* **1997**, *130*, 795–799.
116. Sivaev, I. B.; Bruskin, A. B.; Nesterov, V. V.; Antipin, M. Y.; Bregadze, V. I.; Sjöberg, S. *Inorg. Chem.* **1999**, *38*, 5887–5893.
117. Justus, E.; Rischka, K.; Wishart, J. F.; Werner, K.; Gabel, D. *Chem. Eur. J.* **2008**, *14*, 1918–1923.
118. Justus, E.; Vöge, A.; Gabel, D. *Eur. J. Inorg. Chem.* **2008**, 5245–5250.
119. Genady, A. R.; Nakamura, H. *Org. Biomol. Chem.* **2010**, *8*, 4427–4435.
120. El-Zaria, M. E.; Genady, A. R.; Nakamura, H. *New J. Chem.* **2010**, *34*, 1612–1622.
121. Zhang, Y.; Sun, Y.; Wang, T.; Liu, J.; Spingler, B.; Duttwyler, S. *Molecules* **2018**, *23*, 3137.
122. Laskova, J.; Ananiev, I.; Kosenko, I.; Serdyukov, A.; Stogniy, M.; Sivaev, I.; Grin, M.; Semioshkin, A.; Bregadze, V. I. *Dalton Trans.* **2022**, *51*, 3051–3059.
123. Hoffmann, S.; Justus, E.; Ratajski, M.; Lork, E.; Gabel, D. *J. Organomet. Chem.* **2005**, *690*, 2757–2760.
124. Zhang, Y.; Sun, Y.; Lin, F.; Liu, J.; Duttwyler, S. *Angew. Chem. Int. Ed.* **2016**, *55*, 15609–15614.
125. Nelyubin, A. V.; Selivanov, N. A.; Bykov, A. Y.; Klyukin, I. N.; Novikov, A. S.; Zhdanov, A. P.; Karpechenko, N. Y.; Grigoriev, M. S.; Zhizhin, K. Y.; Kuznetsov, N. T. *J. Mol. Struct.* **2021**, *22*, 13391.
126. Al-Joumhawy, M. K.; Marei, T.; Shmalko, A.; Cendoya, P.; La Borde, J.; Gabel, D. *Chem. Commun.* **2021**, *57*, 10007–10010.
127. Shmal'ko, A. V.; Sivaev, I. B. *Russ. J. Inorg. Chem.* **2019**, *64*, 1726–1749.
128. Ringstrand, B.; Kaszynski, P.; Young, V. G. *Inorg. Chem.* **2011**, *50*, 2654–2660.
129. Ringstrand, B.; Kaszynski, P.; Young, V. G.; Janoušek, Z. *Inorg. Chem.* **2010**, *49*, 1166–1179.
130. Pecyna, J.; Denicola, R. P.; Ringstrand, B.; Jankowiak, A.; Kaszynski, P. *Polyhedron* **2011**, *30*, 2505–2515.
131. Shmal'ko, A. V.; Cendoya, P.; Anufriev, S. A.; Suponitsky, K. Y.; Gabel, D.; Sivaev, I. B. *Chem. Commun.* **2022**, *58*, 3775–3778.
132. Bernard, R.; Cornu, D.; Luneau, D.; Naoufal, D.; Scharff, J.-P.; Miele, P. *J. Organomet. Chem.* **2005**, *690*, 2745–2749.
133. Jasper, S. A.; Jones, R. B.; Mattern, J.; Huffman, J. C.; Todd, L. J. *Inorg. Chem.* **1994**, *33*, 5620–5624.
134. Jasper, S. A.; Mattern, J.; Huffman, J. C.; Todd, L. J. *Polyhedron* **2007**, *26*, 3793–3798.
135. Yao, H.-J.; Hu, C.-H.; Dou, J.-M.; Sun, J.; Wei, J.-D.; Huang, Z.-E.; Jin, R.-S.; Zheng, P.-J. *Acta Cryst. C* **1998**, *54*, IUC9800008.

136. Yao, H.-J.; Hu, C.-H.; Dou, J.-M.; Jin, R.-S.; Zheng, P.-J. *Acta Cryst. C* **1999**, *55*, IUC9900033.
137. Naoufal, D.; Bonnetot, B.; Mongeot, H.; Grüner, B. *Collect. Czech. Chem. Commun.* **1999**, *64*, 856–864.
138. Kabytaev, K. Z.; Safronov, A. V.; Sevryugina, Y. V.; Barnes, C. L.; Jalisatgi, S. S.; Hawthorne, M. F. *Inorg. Chem.* **2015**, *54*, 4143–4150.
139. Viñas, C.; Cirera, M. R.; Teixidor, F.; Kivekäs, R.; Sillanpää, R.; Llibre, J. *Inorg. Chem.* **1998**, *37* (26), 6746–6750.
140. Yang, Z.; Sun, C.; Wei, X.; Lu, J.; Lu, J.-Y. *ChemCatChem* **2022**, *14*, e202101571.
141. Kabytaev, K. Z.; Mukhin, S. N.; Glukhov, I. V.; Starikova, Z. A.; Bregadze, V. I.; Beletskaya, I. P. *Organometalics* **2009**, *28*, 4758–4763.
142. Xu, T.-T.; Cao, K.; Wu, J.; Zhang, C.-Y.; Yang, J. *Inorg. Chem.* **2018**, *57*, 2925–2932.
143. Lyu, H.; Quan, Y.; Xie, Z. *Angew. Chem. Int. Ed.* **2016**, *55*, 11840–11844.
144. Guo, S.-T.; Cui, P.-F.; Yuan, R.-Z.; Jin, G.-X. *Chem. Commun.* **2021**, *57*, 2412–2415.
145. Au, Y. K.; Lyu, H.; Quan, Y.; Xie, Z. *J. Am. Chem. Soc.* **2020**, *142*, 6940–6945.
146. Li, C.-X.; Zhang, H.-Y.; Wong, T.-Y.; Cao, H.-J.; Yan, H.; Lu, C.-S. *Org. Lett.* **2017**, *19*, 5178–5181.
147. Fu, Y.; Li, Y.; Luo, D.; Lu, Y.; Huang, J.; Yang, Z.; Lu, J.; Jiang, Y.-Y.; Lu, J.-Y. *Inorg. Chem.* **2022**, *61*, 911–922.
148. Ham, H.; Shin, S.; Ko, G. H.; Han, S. H.; Han, G. U.; Maeng, C.; Kim, T. H.; Noh, H. C.; Lee, K.; Yang, H.; Lee, P. H. *J. Org. Chem.* **2021**, *86*, 15153–15163.
149. Au, Y. K.; Lyu, H.; Quan, Y.; Xie, Z. *Chin. J. Chem.* **2020**, *38*, 383–388.
150. Kabytaev, K. Z.; Everett, T. A.; Safronov, A. V.; Sevryugina, Y. V.; Jalisatgi, S. S.; Hawthorne, M. F. *Eur. J. Inorg. Chem.* **2013**, 2488–2491.
151. Chen, Y.; Quan, Y.; Xie, Z. *Chem. Commun.* **2020**, *56*, 12997–13000.
152. Waddington, M. A.; Zheng, X.; Stauber, J. M.; Moully, E. H.; Montgomery, H. R.; Saleh, L. M. A.; Kral, P.; Spokoyny, A. M. *J. Am. Chem. Soc.* **2021**, *143*, 8661–8668.
153. Ma, Y.-N.; Gao, Y.; Ma, Y.; Wang, Y.; Ren, H.; Chen, X. *J. Am. Chem. Soc.* **2021**, *144*, 8371–8378.

About the authors

Igor B. Sivaev was born in 1964 in Kivertsy (Volyn region, Ukraine). After graduation from the Mendeleev Moscow Institute of Chemical Technology (1987) as Chemical Engineer, he studied chemistry of polyhedral boron hydrides in the group of Prof. Nikolai Kuznetsov at the Kurnakov Institute of General and Inorganic Chemistry, Russian Academy of Sciences. Later, Igor pursued his PhD degree at the Department of Organic Chemistry, Uppsala University, Sweden under the guidance of Prof. Stefan Sjöberg (2000) and joined the group of Prof. Vladimir Bregadze at the Nesmeyanov Institute of Organoelement Compounds, Russian Academy of Sciences. In 2014, Igor received a Doctor of Chemical Sciences degree (habilitation) at the Nesmeyanov Institute of

Organoelement Compounds, Russian Academy of Sciences, where currently he is a Leading Researcher (Research Professor) at Laboratory of Organoboron and Organoaluminium Compounds. His scientific interests are the chemistry and application of polyhedral boron hydrides, including boranes, carboranes and metallacarboranes.

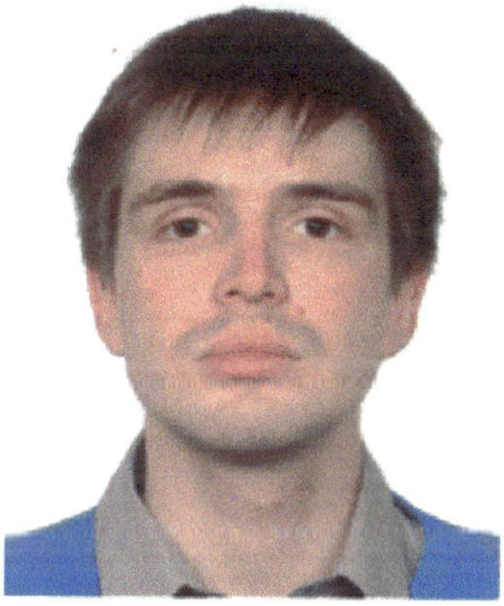

Sergey A. Anufriev was born in 1991 in Ul'yanovsk city, USSR. He graduated from Higher Chemical College of the Russian Academy of Sciences at the Mendeleev Russian University of Chemical Technology in 2013 and joined the group of Dr. Igor Sivaev at the Nesmeyanov Institute of Organoelement Compounds of Russian Academy of Sciences as PhD student. He presented his PhD thesis at the Nesmeyanov Institute of Organoelement Compounds in 2018, and currently he is a senior researcher in the group of Prof. Igor Sivaev. Dr. Sergey Anufriev is author and coauthor of more than 30 scientific papers in the field of chemistry of carboranes and metallacarboranes.

Akim V. Shmal'ko graduated from the Mendeleev Russian University of Chemical Technology in Moscow in 2010 and entered postgraduate school in organic chemistry at the Mendeleev Russian University of Chemical Technology. He performed his PhD project in the group of Dr. Igor Sivaev at the Nesmeyanov Institute of Organoelement Compounds of Russian Academy of Sciences. In 2016 he presented his PhD dissertation under the supervision of Prof. Igor Sivaev and currently he is researcher in the group of Prof. Igor Sivaev. Dr. Akim Shmal'ko is author and coauthor of more than 20 scientific papers, he develops new methods for the functionalization of polyhedral boron hydrides.

CHAPTER THREE

Advances in transition metal catalyzed selective B—H functionalization of *o*-carboranes☆

Jie Zhang and Zuowei Xie*
Department of Chemistry and State Key Laboratory of Synthetic Chemistry, The Chinese University of Hong Kong, Shatin, New Territories, Hong Kong, China
*Corresponding author: e-mail address: zxie@cuhk.edu.hk

Contents

☆Dedicated to Professor Francesc Teixidor and Professor Clara Viñas for their contributions to carborane chemistry.

Advances in Catalysis, Volume 71
ISSN 0360-0564
https://doi.org/10.1016/bs.acat.2022.04.003

Abstract

Carboranes are a class of carbon-boron molecular clusters, possessing extraordinary characteristics including three-dimensional aromaticity conjugated by σ-bonds, icosahedral geometry and inherent robustness. They are finding a variety of applications in medicine, materials and coordination/organometallic chemistry, and more. Thus, selective functionalization of *o*-carboranes has attracted enormous research interests, particularly in regio- and enantio-selective cage BH derivatization among ten chemically similar BH vertices in *o*-carboranes. In recent years, significant progress has been made in catalytic vertex-specific BH functionalization. This review summarizes recent advances in this research realm.

1. Introduction

Icosahedral carboranes, a class of polyhedron molecular clusters consisting of 10 BH and two CH vertices, were generally viewed as 3-D analogs to benzene. Carboranes features unique properties including three-dimensional σ-aromaticity, hydrophobic molecular surface and inherent robustness *(1)*, leading to wide applications as versatile building blocks in boron neutron capture therapy *(2)*, functional materials *(3)* and organometallic/coordination chemistry *(4)*. To this end, functionalization of *o*-carboranes has drawn increasing research interests *(1,5,6)*. In the past decades, a large variety of carborane derivatives decorating with cage C-substituents have been synthesized, via easily removal of the weakly acidic cage C—H bonds (pK_a ~23) *(1,5)*. However, cage-B-substituted *o*-carborane derivatives are relatively less known *(6)*.

Due to the hydridic characteristic of B—H bond(s) and three-dimensional σ-aromaticity of the cage, *o*-carboranes are easily susceptible to attack by electrophiles in the following order B(9,12)-H > B(8,10)-H > B (4,5,7,11)-H >> B(3,6)-H, consisting with the order of vertex charge on *o*-carborane cage *(1,7)* (see Fig. 1 for numbering system and codes). In this

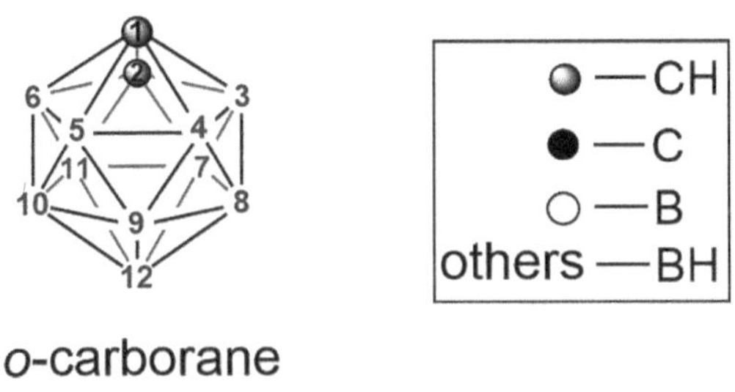

Fig. 1 Numbering system of *o*-carborane.

context, functionalization at cage B-vertices in early days was generally achieved through electrophilic substitution of B—H bonds *(1)*. For instance, in the presence of Lewis acid catalysts, B(9,12)-halogenation and -methylation were achieved through electrophilic substitution reactions at more electron-rich cage B(9,12)-vertices (Scheme 1A) *(1,8)*. Under forced reaction conditions, other six BH vertices could undergo further electrophilic substitution to give 4,5,7,8,9,10,11,12-X_8-*o*-carboranes (X=I, Me), except for the most electron-deficient 3,6-BH ones *(9)*. Complementally, reaction of the dicarbollide (nido-7,8-$C_2B_9H_{11}{}^{2-}$) with boron halides could afford B(3)-substituted *o*-carboranes (Scheme 1B) *(10)*. 4,7-I_2-*o*-carborane could be prepared from electrophilic diiodination of dicarbollide nido-7,8-$C_2B_9H_{12}{}^{-}$, followed by capitation (Scheme 1C) *(11)*. On the other hand, reactions of B-substituted decaboranes with

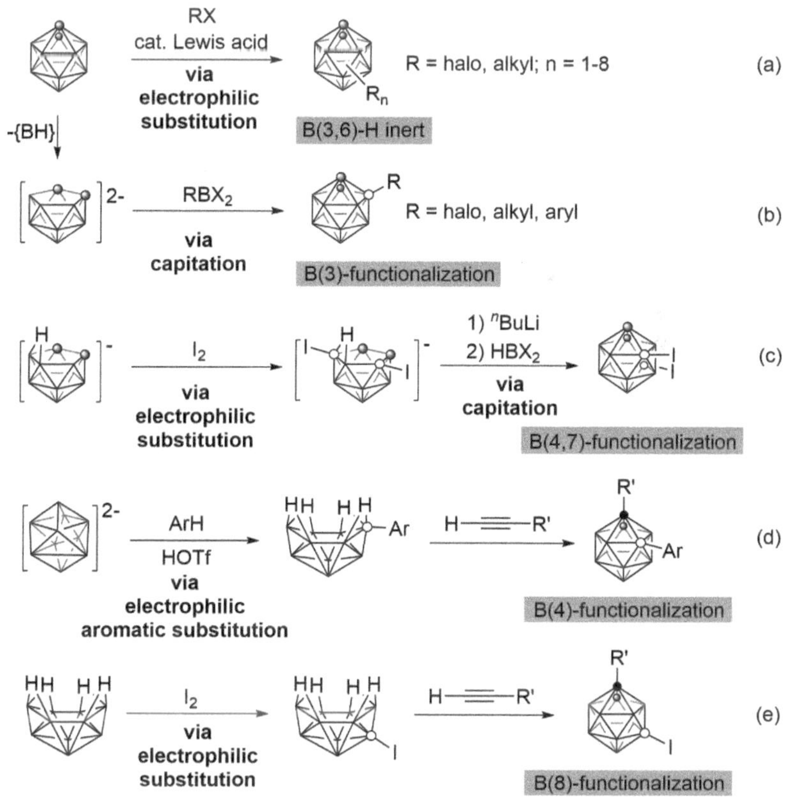

Scheme 1 Synthetic approaches to cage B-substituted *o*-carboranes.

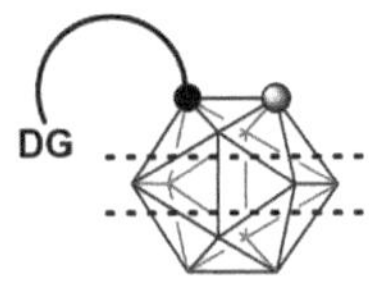

Fig. 2 General strategy for transition metal-catalyzed selective B–H activation of *o*-carboranes.

alkynes produced *o*-carboranes with substituents at the corresponding B-vertex *(1)*. A route from closo-$B_{10}H_{10}^{2-}$ involving acid-induced cage opening, electrophilic aromatic substitution and dehydrogenative alkyne-insertion reaction of decaborane furnished 4-aryl-*o*-carboranes (Scheme 1D) *(12)*. Additionally, treatment of iododecaborane with alkynes gave access to 8-I-*o*-carboranes (Scheme 1E) *(13)*. The above procedures illustrated in Scheme 1B–E are neither direct nor atom-economical methods.

For B(3,6)-H functionalization, metalation of the B(3)—H bond, followed by treatment with electrophiles served as an alternative approach *(1,14)*. Several stoichiometric reactions of *o*-carboranes with transition metal compounds were explored to accomplish cage B(3,6)—H functionalization *(14,15)*. It was noteworthy that cage B(3,6)-deuteration and B(3)-propenylation of *o*-carboranes could be realized in a catalytic manner with transition metal catalysts *(16,17)*.

Stimulated by the aforementioned results, the Xie group proposed a general strategy for *o*-carborane cage B—H functionalization by transition metal catalysis (Fig. 2) *(18)*. This strategy contains three rules: (1) electron-rich transition metal catalysts, which prefer oxidative addition reactions, should work well to activate the most electron-deficient B(3,6)—H vertices (bonded to both cage CH units), (2) electron-rich B(8,9,10,12)—H vertices (without connection to any cage CH units) prefer to be activated by electrophilic transition metal catalysts, and (3) directing group assisted electrophilic transition metal catalysis would realize selective functionalization of B(4,5,7,11)—H vertices (bonded to only one cage CH unit). Successful applications of this strategy lead to the development of a set of methods in the laboratories of the Xie and other groups for regioselective catalytic cage B—H functionalization *(18)*, resulting in the synthesis of a large class of *o*-carborane derivatives at specific cage B sites. This chapter summarizes remarkable advancements in this realm.

2. B(3,6)—H functionalization

2.1 B(3,6)-Diborylation

According to the strategy illustrated in Fig. 2, The Xie group realized an efficient and regioselective diborylation of cage B(3,6)—H of *o*-carboranes with bis(pinacolato)diboron (B_2pin_2) in the presence of $[(cod)IrCl]_2$ (cod = 1,5-cyclooctadiene) as catalyst and 2-methylpyridine as ligand (Scheme 2) *(19)*. Alkyl-, alkenyl-, aryl-, benzyl- and iodine-substituted *o*-carboranes successfully underwent the Ir-catalyzed borylation with high efficiency. It was noted that the steric hindrance could significantly affect the regioselectivity of borylation. When the sterically hindered substituents were introduced to cage B(4) position, the adjacent B(3) position was blocked to generate B(6)-monoborylated *o*-carboranes in high isolated yields. In addition, when a bulkier silyl group was introduced to cage C(1) vertex, the adjacent B(3) and B(6) positions were blocked to generate B(4)-monoborylated *o*-carboranes in high isolated yields. With the borylated *o*-carboranes in hand, subsequent multiple transformations of Bpin unit were performed to afford B(3,6)-disubstituted carborane derivatives.

A plausible mechanism of the Ir-catalyzed regioselective borylation was proposed on the basis of control experiments. As shown in Scheme 2, Ir(III) intermediate **5** is generated by oxidative addition of B_2pin_2 to the Ir(I)-Bpin **4**, which is in situ produced from $[(cod)IrCl]_2$. Subsequent further oxidative addition of the intermediate **5** at the electron-deficient cage B(3)-H bond forms an Ir(V) intermediate **6**, which undergoes reductive elimination to yield the intermediate **7** and HBpin (path A). Alternatively, the Ir(III) intermediate **5** undergoes regioselective electrophilic substitution at cage B(3)-H to afford the intermediate **7** and HBpin (path B). Reductive elimination of the intermediate **7** yields the mono-borylated product **2**, which undergoes another catalytic borylation cycle to offer cage B(3,6)-diborylated *o*-carboranes **3** *(20)*.

2.2 B(3,6)-oxygenation

The Yan group developed a Rh(III)-catalyzed selective B(3)-acyloxylation of *o*-carboranes with the assistance of a pyridyl directing group (Scheme 3) *(21)*.

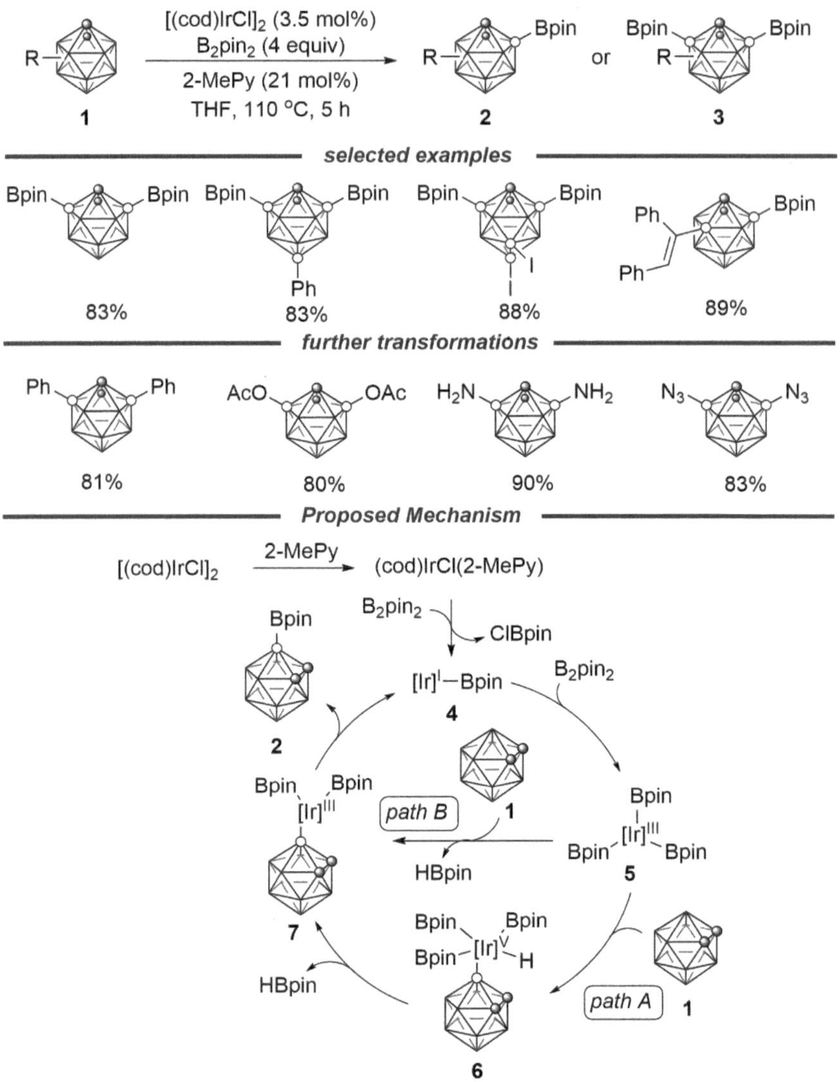

Scheme 2 Ir-catalyzed B(3,6)-borylation of *o*-carboranes.

Mixtures of B(3)-acyloxylated *o*-carboranes **9** and B(3,6)-diacyloxylated *o*-carboranes **10** were prepared from the reaction of 1-pyridyl-*o*-carboranes **8** with carboxylic acids using $[Cp^*RhCl_2]_2$ as catalyst. The reaction conditions and substrates could significantly affect the ratio of mono/di-selectivity. A wide variety of carboxylic acids containing different functional groups, such as alkyl, phenyl, thienyl and indolyl substituents were compatible in such acyloxylation with no observable electronic effects.

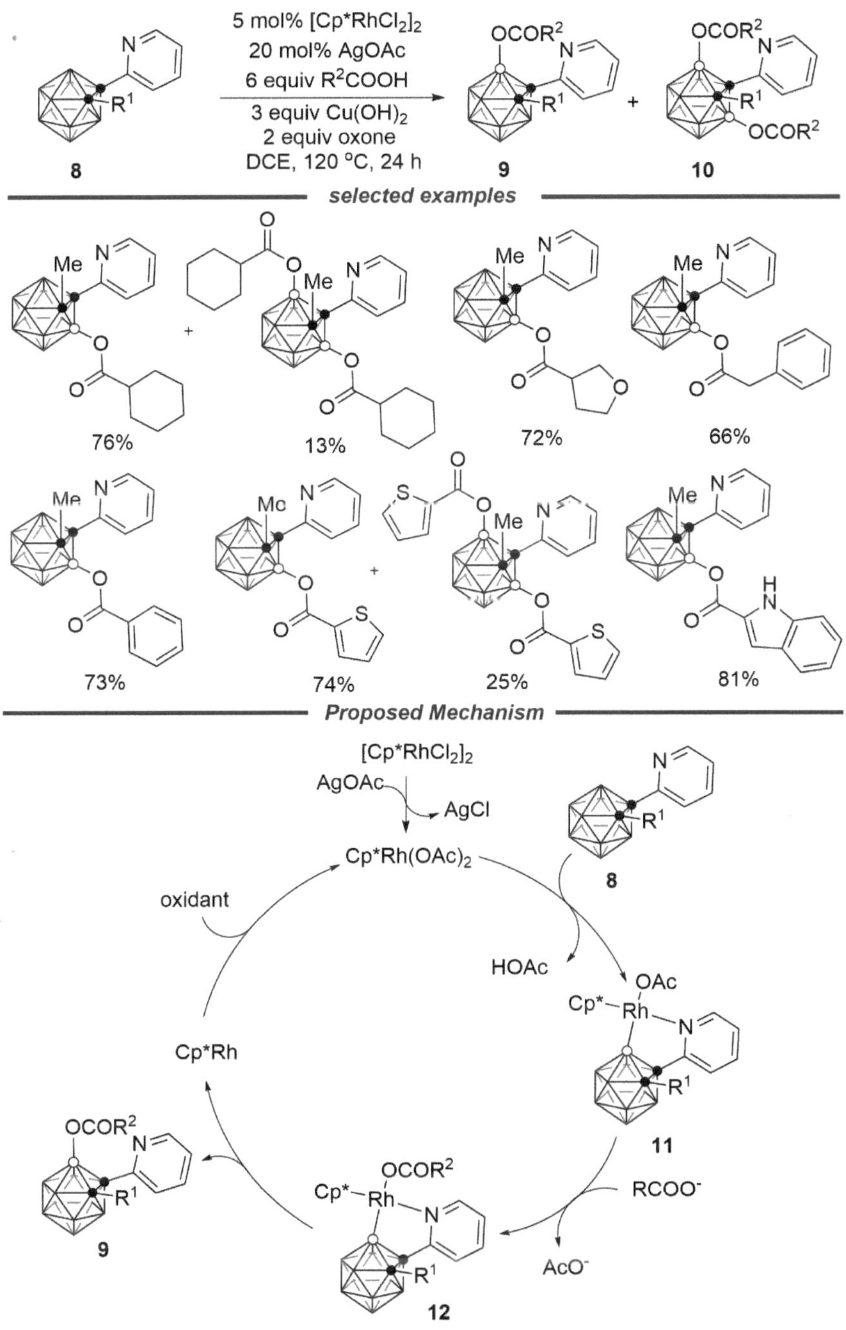

Scheme 3 Rh-catalyzed B(3)-acyloxylation of *o*-carboranes.

On the basis of both experimental and theoretical studies, a plausible mechanism of Rh-catalyzed B(3)-acyloxylation was proposed in Scheme 3. The reaction is initiated by the coordination of pyridinyl group to the active catalyst $Cp^*Rh(OAc)_2$, which is in situ generated from $[Cp^*RhCl_2]_2$. Subsequent B(3)—H activation forms a five-membered rhodacycle complex **11**, which undergoes ligand exchange at Rh(III) center and reductive elimination to release B(3)-acyloxylated *o*-carboranes **9** and Rh(I) species. Oxidation of Rh(I) species by oxone or/and copper salt completes the catalytic cycle. Another acyloxylation cycle from monoacyloxylated *o*-carboranes **9** affords diacyloxylated *o*-carboranes **10**.

An efficient Pd-catalyzed intramolecular dehydrogenative coupling of cage BH and phenol OH for the construction of cage B—O bond was developed in the Xie group (Scheme 4), yielding a series of 1,3- and 1,4-*o*-carboranobenzoxaborole derivatives (**14** and **15**) *(22)*. The products were conveniently separated by column chromatography with 1,3-*o*-carboranobenzoxaboroles **14** as major products and 1,4-*o*-carboranobenzoxaboroles **15** as minor ones. The ratio of two products was determined by the steric effects of substituents on the phenyl ring. Employment of various substrates bearing either electron-donating or -withdrawing substituents on phenyl ring resulted in high coupling efficiency with no observable electronic effects. It was found that fluoro-substituted substrates could generally promote the reactions by enhancing the acidity of phenols. A possible Pd(II)–Pd(0)–Pd(II) catalytic cycle was illustrated in Scheme 4. Deprotonation of phenol substrate **13** with $Pd(OAc)_2$ generates a complex **16**, which undergoes intramolecular B—H bond activation to form a six-membered palladacycle **17**. Subsequently, the final B—O coupling product is generated by reductive elimination with the release of Pd(0). Oxidation of Pd(0) by silver salt regenerates the active Pd(II) catalyst to complete the cycle.

2.3 B(3,6)-arylation

The Cao group presented an efficient B(3,6)-diarylation of *o*-carboranes, catalyzed by in situ formed $[Pd(NHC)_2]^{2+}$ complex (NHC=*N*-heterocyclic carbene) *(23)*. As displayed in Scheme 5, using $Pd(OAc)_2$ as precatalyst and IPrHCl (1,3-bis-(2,6-diisopropylphenyl)imidazolinium chloride) as ligand, the selective coupling of *o*-carboranes with aryl iodides afforded B(3,6)-diarylated *o*-carboranes in good yields. Furthermore, when 3-phenyl-*o*-carborane **19'** was utilized as a substrate, unsymmetrical

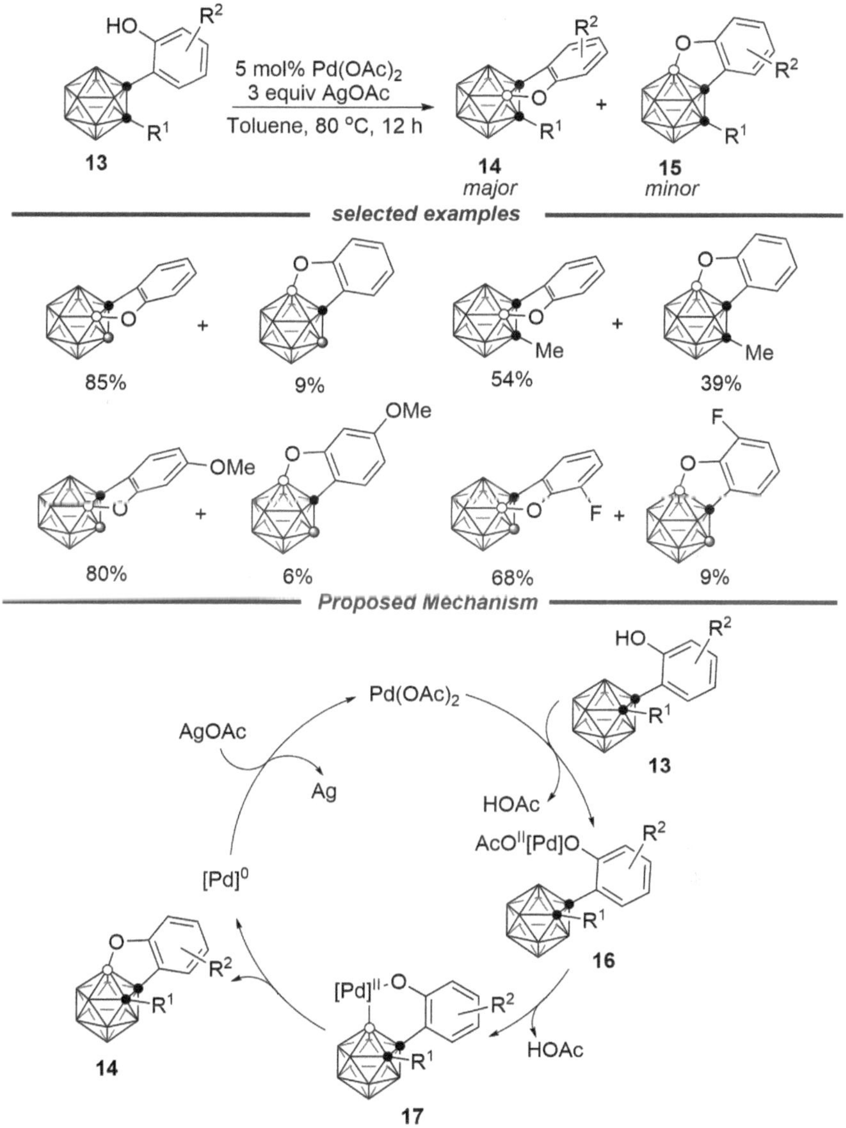

Scheme 4 Pd-catalyzed intramolecular B-O dehydrogenative coupling.

B(3,6)-diarylated *o*-carboranes **18'** could be generated via selective B(6)—H activation. A variety of functional groups such as halo, trifluoromethyl, cyano, aldehyde, ester and alkyl were compatible in such arylation under mild conditions. As shown in Scheme 5, a plausible Pd(II)–Pd(IV)–Pd(II) catalytic cycle starting from in situ generated [Pd(NHC)$_2$]

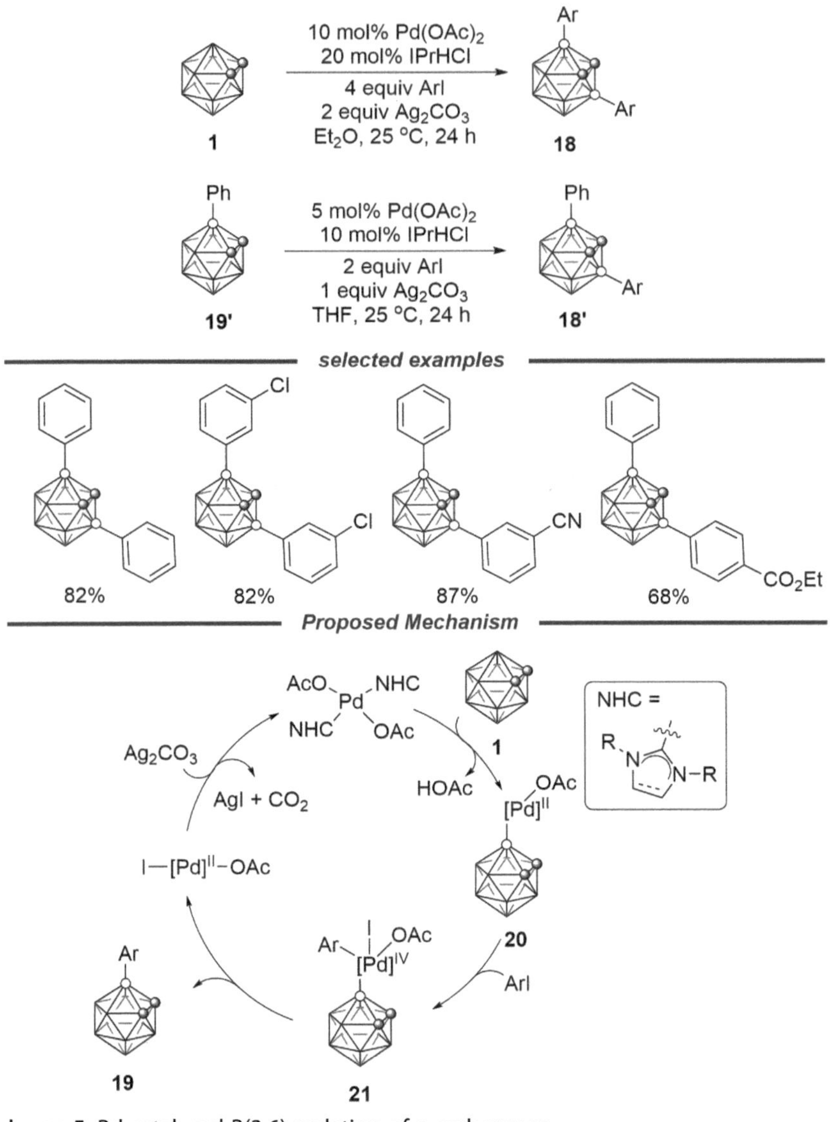

Scheme 5 Pd-catalyzed B(3,6)-arylation of *o*-carboranes.

$(OAc)_2$ was proposed. Electrophilic substitution at B(3) position, followed by oxidative addition with iodobenzenes generates a Pd(IV) intermediate **21**. Subsequently, the product 3-aryl-*o*-carborane **19** is yielded via reductive elimination along with the release of Pd(II) which reacts with Ag_2CO_3 in the presence of IPrHCl to regenerate the active palladium species for the next catalytic cycle. Another arylation cycle from 3-aryl-*o*-carborane **19** gives cage B(3,6)-diarylated carboranes **18**.

The Yu and Che group disclosed a facile approach to the preparation of monoaryl- and diaryl-substituted monophosphino-*o*-carboranes via Rh(I)-catalyzed phosphine-assisted B(3,6)-H functionalization of *o*-carboranes (Scheme **6**) *(24)*. The reaction demonstrated good tolerance toward alkyl,

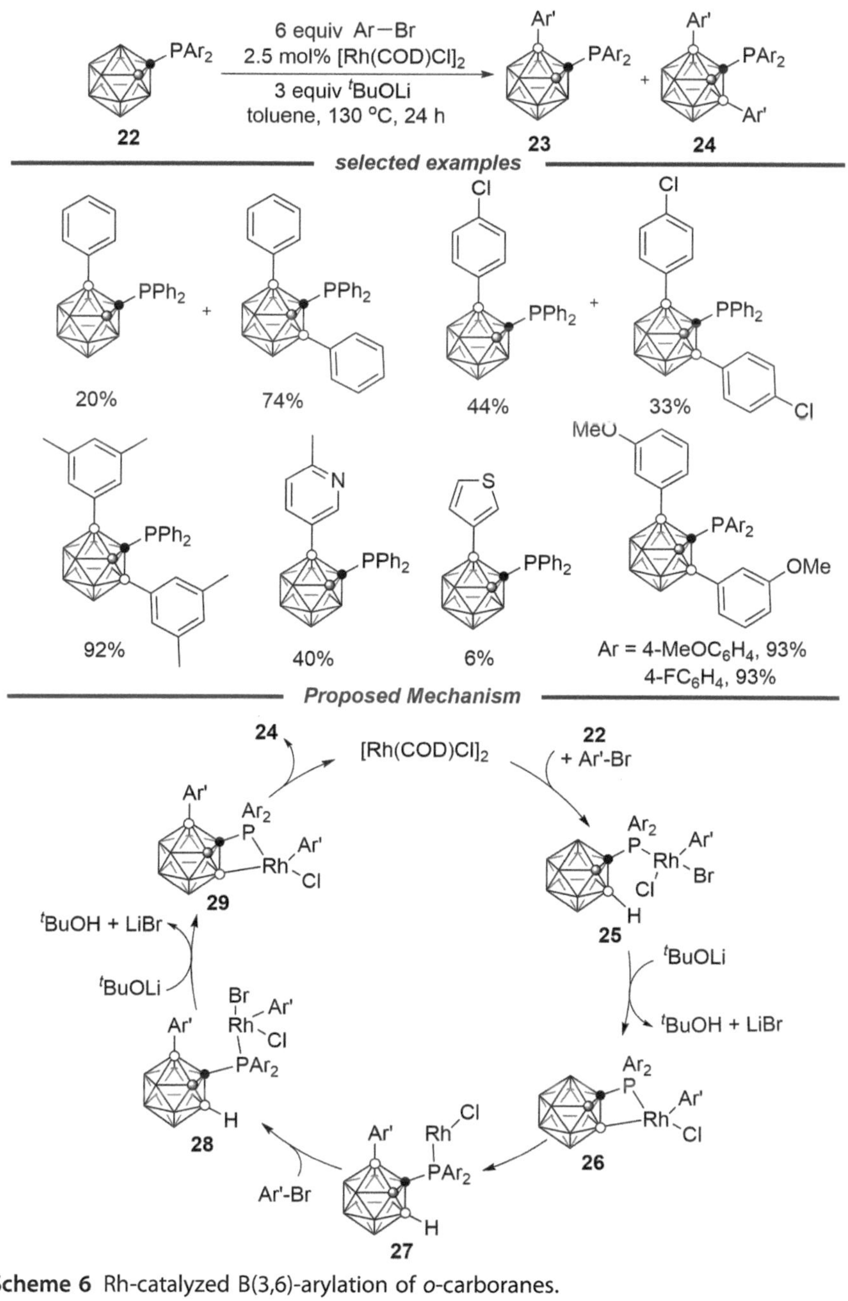

Scheme 6 Rh-catalyzed B(3,6)-arylation of *o*-carboranes.

methoxy, halo and pyridyl groups, giving the B(3)-arylation and B(3,6)-diarylation products (**23** and **24**) in moderate to good yields. A tentative mechanism for such arylation of B(3,6)—H bonds was described in Scheme **6**. The coordination of $[Rh(cod)Cl]_2$ with monophosphino-*o*-carboranes, followed by oxidative addition of aryl bromide to the Rh center yields an intermediate **25**. Electrophilic substitution and reductive elimination give mono-arylated species **27**. **27** undergoes another catalytic cycle to produce the diarylation products.

2.4 B(3,6)-alkenylation

The Xie group achieved an efficient and regioselective Ir-catalyzed alkenylation of *o*-carboranes by direct B—H activation (Scheme 7) *(25)*. The $[Cp^*IrCl_2]_2$-catalyzed alkenylation of *o*-carboranes with alkylacetylenes

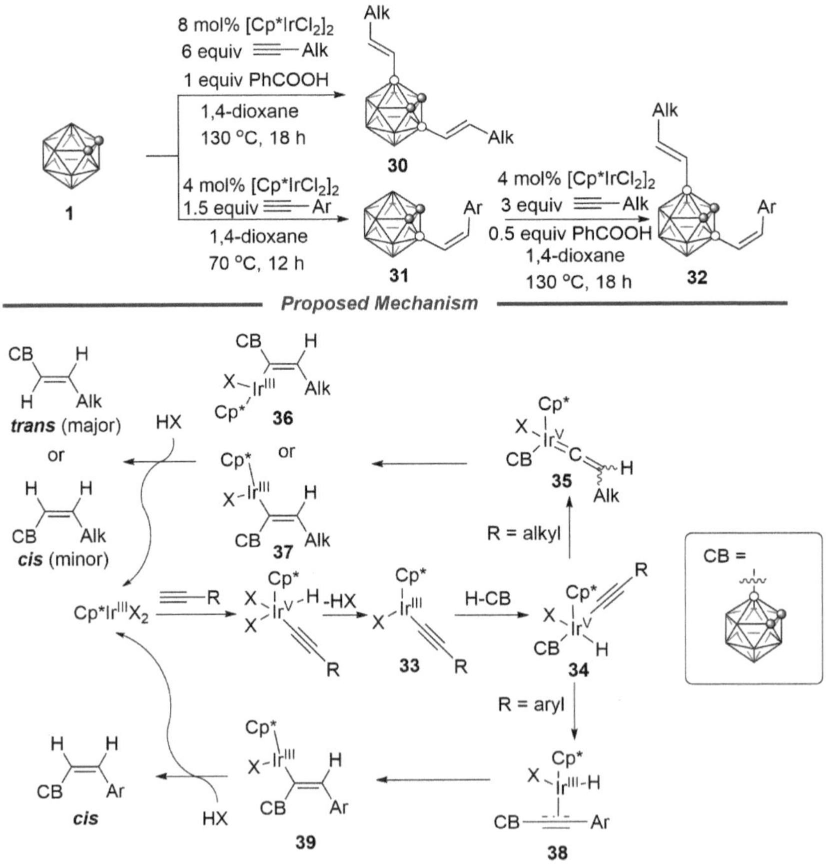

Scheme 7 Ir-catalyzed B(3)-alkenylation or B(3,6)-dialkenylation of *o*-carboranes.

in 1,4-dioxane yielded B(3,6)-[*trans*-(AlkCH=CH)]$_2$-*o*-carboranes **30**, while alkenylation with bulky arylacetylenes yielded B(3)-*cis*-(ArCH=CH)-*o*-carboranes **31** that underwent B(6)-alkenylation with alkylacetylenes to yield B(3)-*cis*-(ArCH=CH)-B(6)-*trans*-(AlkCH=CH)-*o*-carboranes **32**. These results demonstrated that the configuration of the resultant olefins can be controlled by the alkyne substituents. A plausible mechanism for alkenylation was proposed in Scheme 7. Oxidative addition of the terminal alkyne to the Ir catalyst, followed by reductive elimination leads to the formation of the iridium acetylide **33**, which undergoes another oxidative addition of cage B(3)—H bond to generate Ir(V) hydride **34**. For alkylacetylenes, intermediate **34** undergoes 1,3-hydrogen migration to afford a vinylidene intermediate **35**. Subsequent 1,2-migration and protonation afford *trans* olefin as a major product and *cis* olefin as a minor product. It is noted that the ratio of *trans*/*cis* products is dependent on the steric hinderance of the alkyl groups. For bulky arylacetylenes, intermediate **34** prefers to undergo reductive elimination to generate **38**. Subsequent *cis*-addition and protonation give *cis*-olefin

Thereafter, the same group achieved a palladium catalyzed oxidative annulation of 1-hydroxy-*o*-carborane **40** with a series of internal alkynes for the synthesis of C,B-substituted *o*-carborane-fused oxaboroles **41** (Scheme 8) *(26)*. It was found that both electron-withdrawing and -donating groups in diphenylacetylenes were compatible with such annulation. Substrates bearing naphthyl, thienyl and furanyl also successfully underwent the annulation to afford *o*-carborane-fused oxaboroles in acceptable yields. Moreover, oxidative cyclization of the *o*-carborane-fused oxaboroles **41** was achieved to generate *o*-carborane-fused large π systems, which may have potential applications in organic materials *(27)*. Based on experimental results and theoretical analysis, a plausible mechanism for the oxidative annulation was proposed in Scheme 8. Deprotonation of 1-hydroxy-*o*-carborane **40** with Pd(OAc)$_2$, followed by alkyne insertion affords **43**. Regioselective electrophilic attack at cage B(3)-H generates a six membered palladacycle **44**. The final products **41** are furnished by reductive elimination of palladacycle **44** along with the release of Pd(0) complex that is oxidized by PhI(OAc)$_2$ to regenerate the active catalyst.

2.5 B(3)-amination

An efficient and regioselective iridium-catalyzed cage B(3)—H amination of *o*-carboranes with ammonia gas was achieved in the Xie group (Scheme 9) *(28)*.

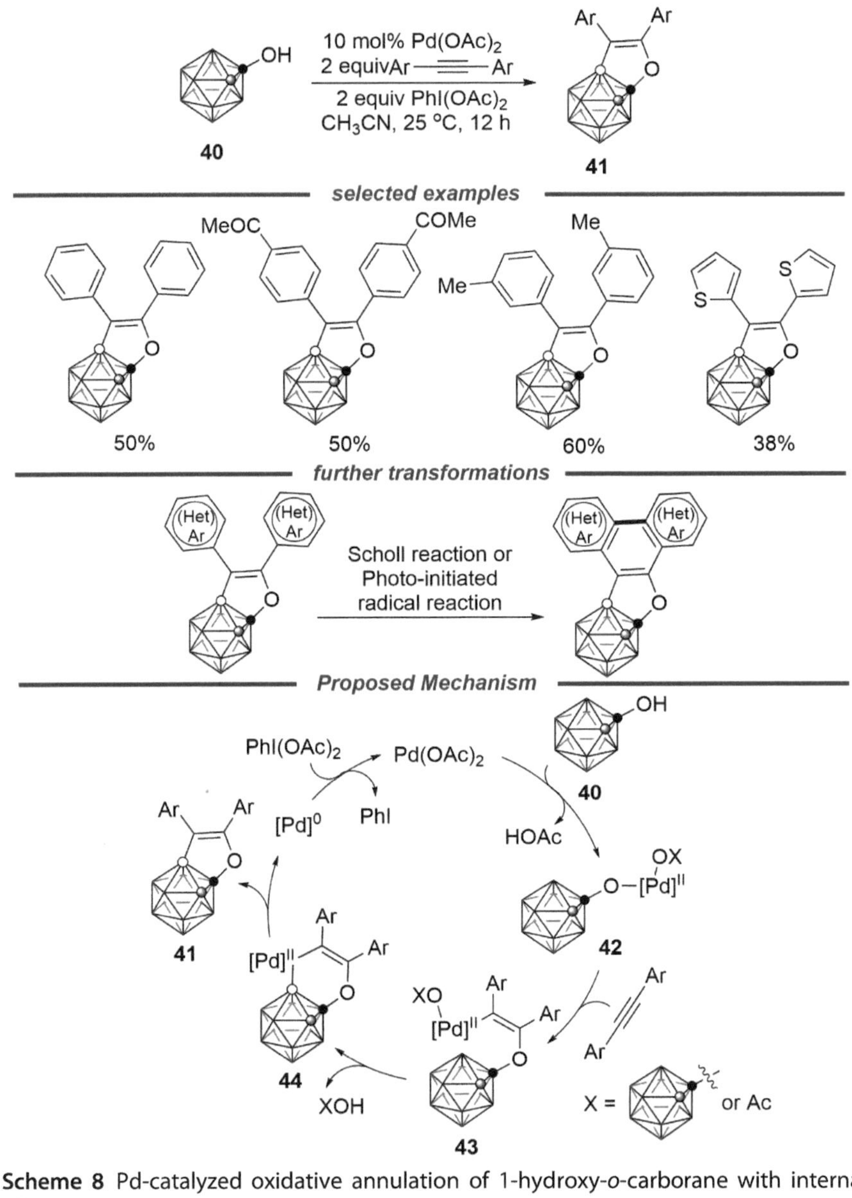

Scheme 8 Pd-catalyzed oxidative annulation of 1-hydroxy-*o*-carborane with internal alkynes.

In the presence of $[(cod)Ir(OMe)]_2$ catalyst, the B-H/N-H dehydrogenative coupling of *o*-carboranes with NH_3 afforded a series of 3-NH_2-*o*-carborane derivatives **45** with a broad substrate scope. Both Cy_3P (Cy = cyclohexyl) ligand and K_2CO_3 additive played crucial roles for the smooth occurrence of this

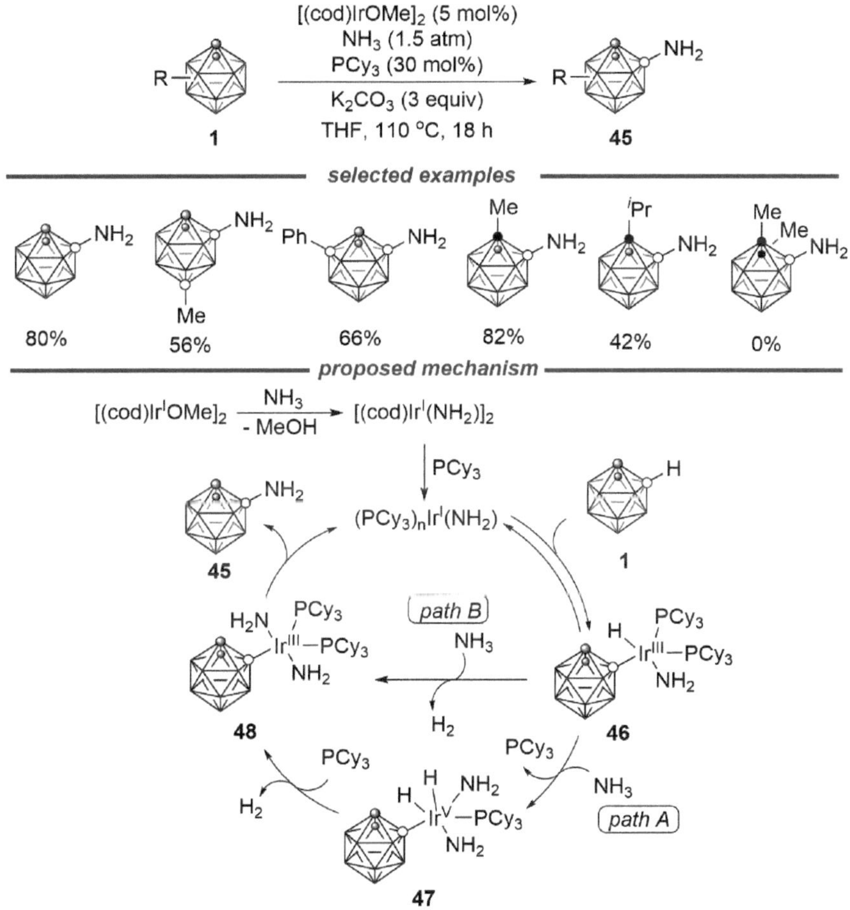

Scheme 9 Ir-catalyzed B(3)-amination of *o*-carboranes.

reaction. A plausible reaction mechanism of such amination was proposed on the basis of a series of control experiments. The initial catalyst [(cod)Ir(OMe)]$_2$ reacts with NH_3, followed by ligand exchange with Cy_3P, leading to a monomeric active species $(Cy_3P)_nIr(NH_2)$. The oxidative addition of $(Cy_3P)_nIr(NH_2)$ at the electron-deficient cage B(3)-H bond of *o*-carborane affords an Ir(III) complex **46**. Another oxidative addition of complex **46** with NH_3 generates an Ir(V) complex **47**. Reductive elimination affords the Ir(III) complex **48** and H_2 (path A). Alternatively, the Ir(III) complex **46** undergoes directly acid – base reaction with NH_3 to afford **48** (path B). The final product **45** is generated by reductive elimination of **48**.

3. B(8,9,10,12)-H functionalization

3.1 B(8,9,10,12)-halogenation

In 2013, the Xie group achieved a Pd(II)-catalyzed regioselective and safe tetrafluorination of *o*-carboranes through B—H activation (Scheme 10) *(29)*. The tetrafluorination of *o*-carboranes with F^+ reagent in the presence of $Pd(MeCN)_4(BF_4)_2$ catalyst delivered B(8,9,10,12)-tetrafluorinated *o*-carboranes **49** with a broad range of functional groups such as alkyl, alkenyl, aryl and benzyl. Although parent *o*-carborane (o-$C_2B_{10}H_{12}$) resulted in a low isolated yield (31%), *o*-carboranes with electron-donating groups on cage C were found to promote the reaction, leading to the products in good to high

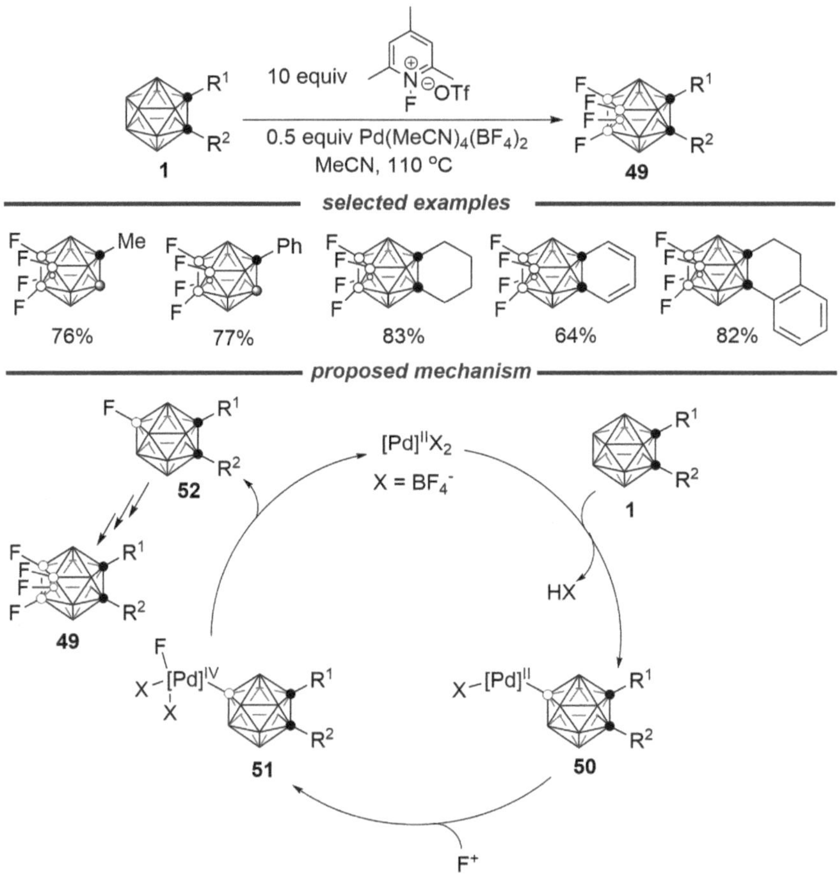

Scheme 10 Pd-catalyzed B(8,9,10,12)-tetrafluorination of *o*-carboranes.

isolated yields probably due to the increased electron density of the cage B—H vertices. A plausible mechanism involving Pd(II)—Pd(IV)—Pd(II) catalytic cycle was proposed in Scheme 10. The catalytic cycle is initiated with electrophilic palladation of B—H to afford an intermediate **50**. Oxidation by F^+ agent generates a Pd(IV) complex **51**. Reductive elimination of **51** gives the product **52** and regenerates the Pd(II) catalyst to complete the catalytic cycle. Such a catalytic cycle is repeated three times to afford the final product B(8,9,10,12)-tetrafluorinated *o*-carboranes **49**.

Later on, the Cao group developed a Pd-catalyzed monochlorination of *o*-carboranes, affording B(8/9)-chlorinated *o*-carboranes with various substituents (Scheme 11) *(30)*. Treatment of *o*-carboranes with $FeCl_3$ in the presence of $Pd(OAc)_2$ catalyst triggered the chlorination to afford B(8/9)-chlorinated *o*-carboranes **53** in moderate to good yields. A possible catalytic cycle was proposed in Scheme 11, involving electrophilic palladation of B—H bond, oxidation with $FeCl_3$ and reductive elimination.

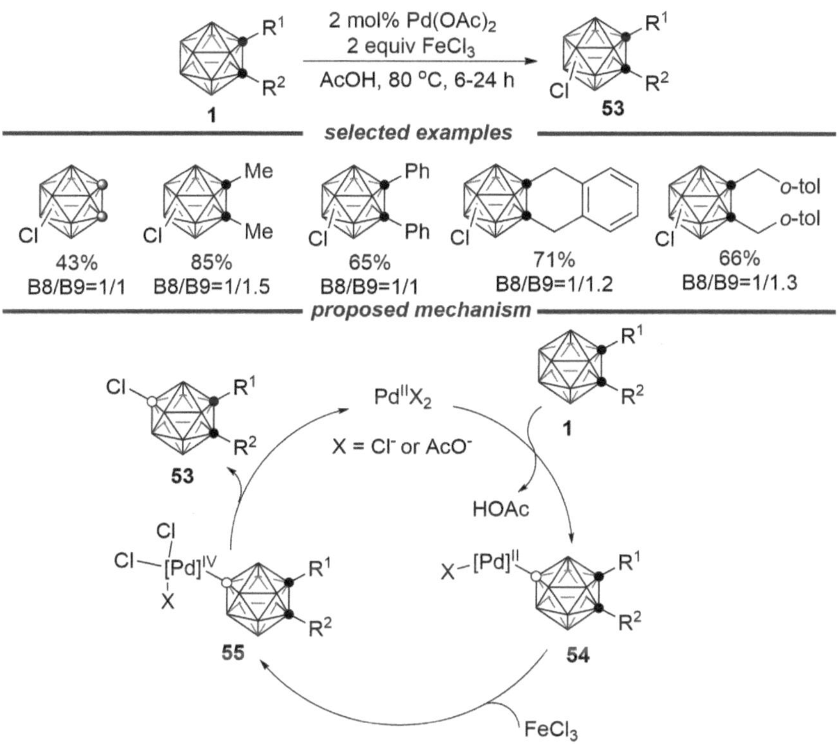

Scheme 11 Pd-catalyzed B(8/9)-chlorination of *o*-carboranes.

3.2 B(8/9)-alkenylation

A B(8/9)-alkenylation of *o*-carboranes via Pd-catalyzed Heck type coupling was achieved in the Cao group (Scheme 12) *(31)*. Treatment of *o*-carboranes with 10 mol% of $Pd(OAc)_2$ catalyst, 2 equiv. of AgOAc oxidant and olefin, the desired coupling product B(8/9)-alkenylated *o*-carboranes **56** were obtained in moderate yields with a ratio of B(8)/B(9) isomers being around 1:1. The scope of this transformation was explored by using diverse *o*-carboranes and terminal alkenes. A possible catalytic cycle was shown in Scheme 12. Electrophilic palladation of B—H bond, followed by alkene insertion into the B—Pd bond affords a complex **58**. β—H Elimination gives the alkenylation product **56** with the release of Pd(0). Oxidation of Pd(0) by Ag(I) regenerates the active catalyst to complete the catalytic cycle.

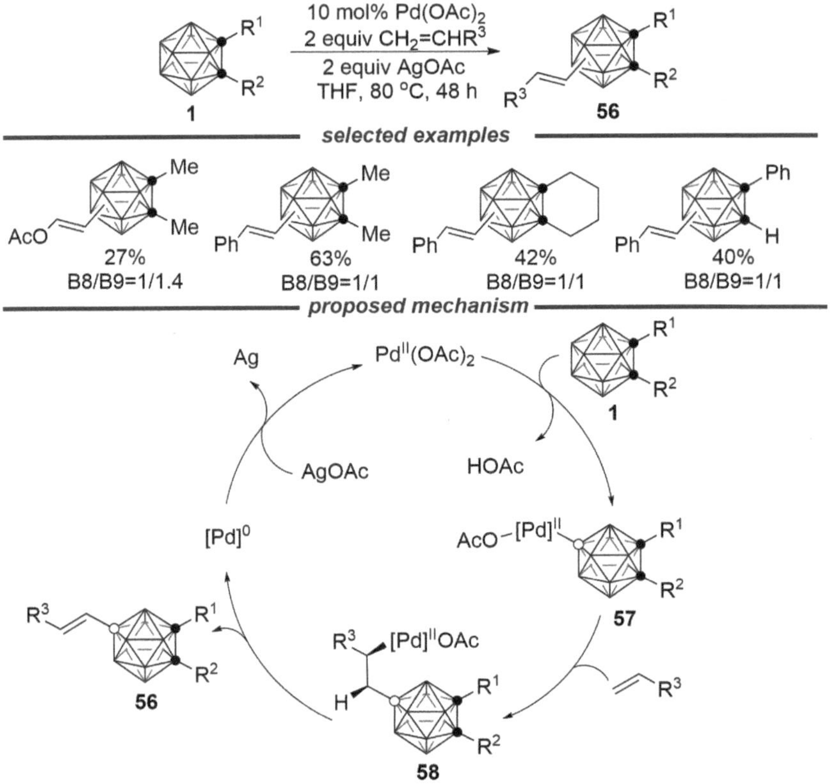

Scheme 12 Pd-catalyzed B(8/9)-alkenylation of *o*-carboranes.

3.3 B(8/9)-arylation

The Cao group disclosed a Pd-catalyzed B(8/9)-arylation of *o*-carboranes with aryl iodides (Scheme 13) *(32)*. Similar to the B(8/9)-alkenylation reported by the same group, this reaction also suffered from an issue of regioselectivity, resulting in a batch of mixture of B(8)- and B(9)-arylated *o*-carboranes **59** in moderate yields with a ratio of ca.1:2. Both electron-withdrawing and -donating substituents in aryl iodides were well tolerated in the reaction. A plausible reaction mechanism involving Pd(II)-Pd(IV)-Pd(II) cycle was proposed in Scheme 13.

Due to the extremely small difference between the electron density of B(8)- and B(9)-vertex, transition metal catalyzed regioselective B(8)-H functionalization of *o*-carboranes remains very challenging. In 2019, the Xie group reported a proof-of-concept study on regioselective B(8)-arylation of *o*-carboranes via palladium catalysis with the assistance of a B(3)-acylamino directing group (Scheme 14) *(33)*. Using $Pd(OAc)_2$ as

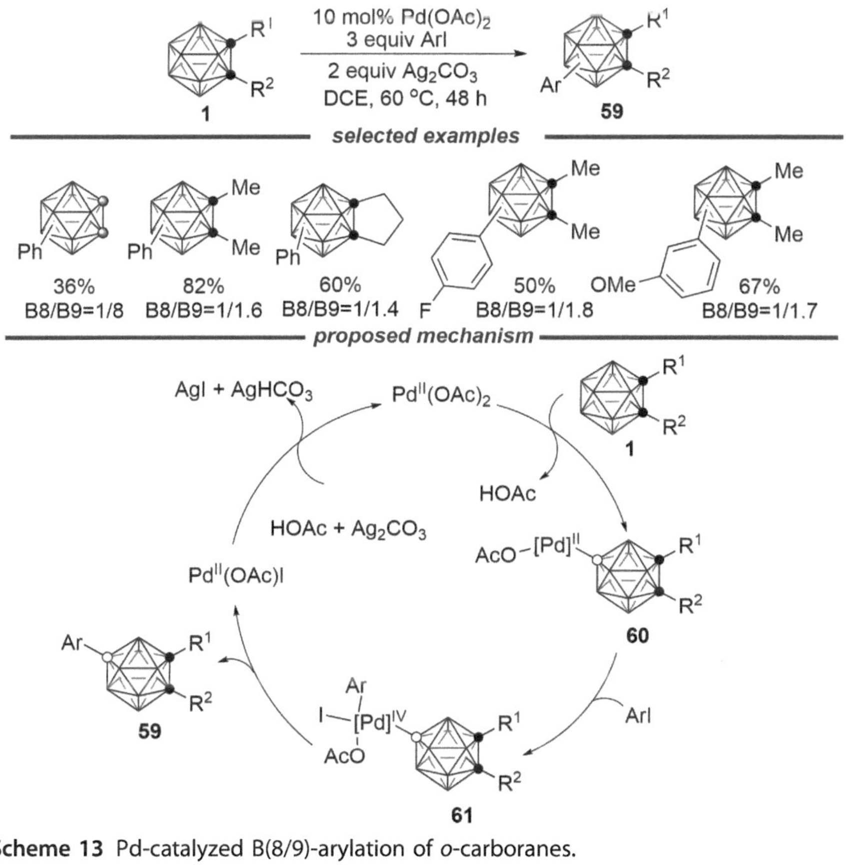

Scheme 13 Pd-catalyzed B(8/9)-arylation of *o*-carboranes.

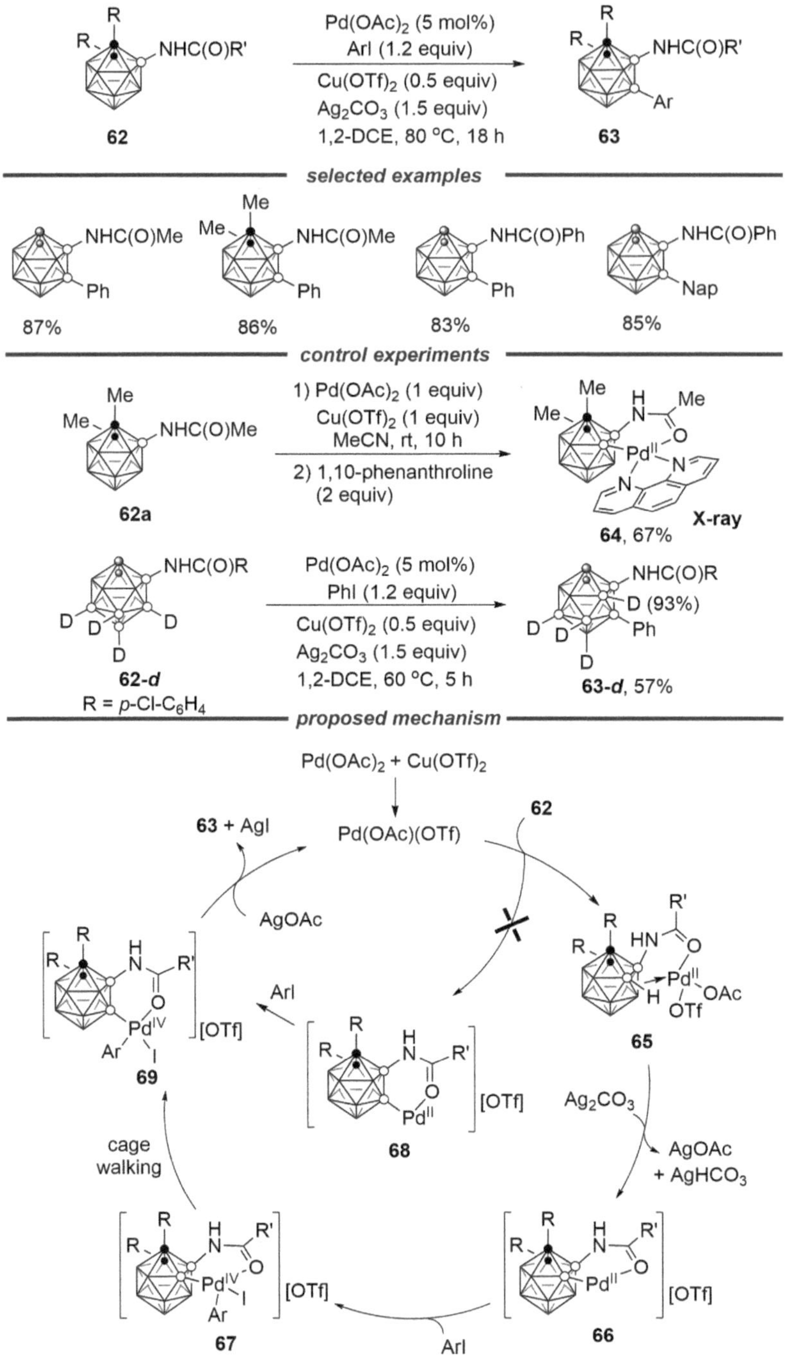

Scheme 14 Pd-catalyzed B(8)-arylation of *o*-carboranes.

catalyst, 3-acylamino-*o*-carboranes **62** reacted with various aryl iodides smoothly to exclusively generate a series of B(8)-arylated *o*-carboranes **63** in excellent yields.

Several control experiments were performed to gain some insight into the reaction mechanism. The reaction of 1,2-dimethyl-3-acetylamino-*o*-carborane **62a** with 1 equiv. of $Pd(OAc)_2$ and $Cu(OTf)_2$ afforded a B—H palladation intermediate that was not isolable. Under the help of 1,10-phenanthroline ligand, a B(4)-palladacyclic intermediate **64** was isolated and confirmed by single-crystal X-ray analysis, proving that the initial palladation of B—H bond occurred preferentially at B(4)-vertex rather than the anticipated B(8)-vertex (Scheme 14). On the other hand, B(8)-phenylated product **63-*d*** with 93% B(4)-deuteration was obtained in 57% yield in the reaction of B(8,9,10,12)-tetradeuterated *o*-carborane **62-*d*** with iodobenzene. These results indicated that a "cage-walking" process *(34)* from Pd-B(4) to Pd-B(8) species might proceed in this reaction. DFT calculations suggested that "cage walking" process was driven by the difference in stabilization energy between Pd(IV)-B(4) intermediate **67** and Pd(IV)-B(8) intermediate **69**. On the basis of these results, a plausible reaction mechanism was proposed in Scheme 14. A Pd-catalytic B(4)—H activation directed by acyl amino group occurs to form a Pd(II)—B(4) intermediate **66**. Oxidative addition of **66** with aryl iodides generates a Pd(IV)-B (4) intermediate **67**. Migration of Pd moiety from B(4) to B(8) vertex gives a Pd(IV)-B(8) intermediate **69**, which is driven by thermodynamic stability. Reductive elimination of **69** yields the B(8)-arylated products **63**.

3.4 B(8,9,10,12)-tetraacetoxylation

Pd-catalyzed B(8,9,10,12)-tetraacetoxylation of *o*-carboranes with oxidants $PhI(OAc)_2$ or PhIO in a mixed solvent of $AcOH/Ac_2O$ (1,1 in v/v) was reported by the Cao group (Scheme 15) *(35)*. A broad substrate

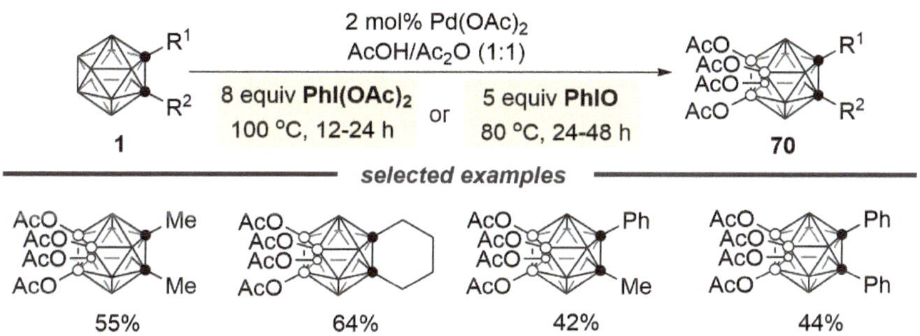

Scheme 15 Pd-catalyzed B(8,9,10,12)-H tetraacetoxylation of *o*-carboranes.

scope of cage C-substituted *o*-carboranes was examined, offering a variety of B(8,9,10,12)-tetraacetoxylated *o*-carboranes **70** in fair to good yields. However, neither *o*-carborane nor mono-substituted *o*-carboranes underwent this reaction to give any desired products. It was proposed that this reaction proceeded in a similar pathway to that of B(8,9,10,12)-tetrafluorination as shown in Scheme 10.

4. B(4,5,7,11)-H functionalization

Since the electron density of B(4,5,7,11)—H vertices lays in between B(3,6)—H and B(8,9,10,12)—H ones, selective functionalization of the middle group B(4,5,7,11)—H bonds is highly challenging. In 2014, the Xie group reported the first example of regioselective cage B(4)—H alkenylation of *o*–carboranes via a carboxy-assisted transition metal catalysis *(36)*. Since then, significant progress in this area has been made.

4.1 Noble metal catalyzed B(4,5,7,11)—H functionalization

4.1.1 B(4,5)-alkenylation

Ir-catalyzed selective B(4)-alkenylation was achieved by treatment of carboranyl carboxylic acids **71** with diaryl alkynes in the presence of an Ir catalyst (Scheme 16) *(36)*. The carboxy group on cage C served as a traceless directing group to control the regioselectivity. Both electronic and steric effects were observed in this reaction. Diaryl alkynes bearing electron-withdrawing substituents gave desired products **72** in higher yields, comparing to electron-donating ones. On the other hand, the yields of **72** were decreased successively when aryl group was changed from *p*- to *m*- to *o*-tolyl moiety. An array of functional groups ranging from halo, thiomethyl, alkoxy, ester to thienyl was tolerated in this reaction. A key iridacycle intermediate **73'** was isolated from the reaction of 1-Me-2-COOH-*o*-carborane **71a** with 1 equiv. of Cp*Ir$(OAc)_2$(DMSO). Single-crystal X-ray analyses unambiguously confirmed the selective formation of a B(4)—Ir bond. A plausible catalytic cycle involving electrophilic B(4)-iridation, insertion of alkyne into the Ir—B bond, protonation and decarboxylation was outlined in Scheme 16.

The Xie group subsequently realized a Pd-catalyzed B(4,5)-dialkenylation of *o*-carboranes (Scheme 17) *(37)*. With the employment of 10 mol% $Pd(OAc)_2$, carboranyl carboxylic acids **71** reacted with various styrenes to generate a batch of B(4,5)-dialkenylated *o*-carborane derivatives **76** featuring *trans*-configuration in high yields. Such reaction was tolerant

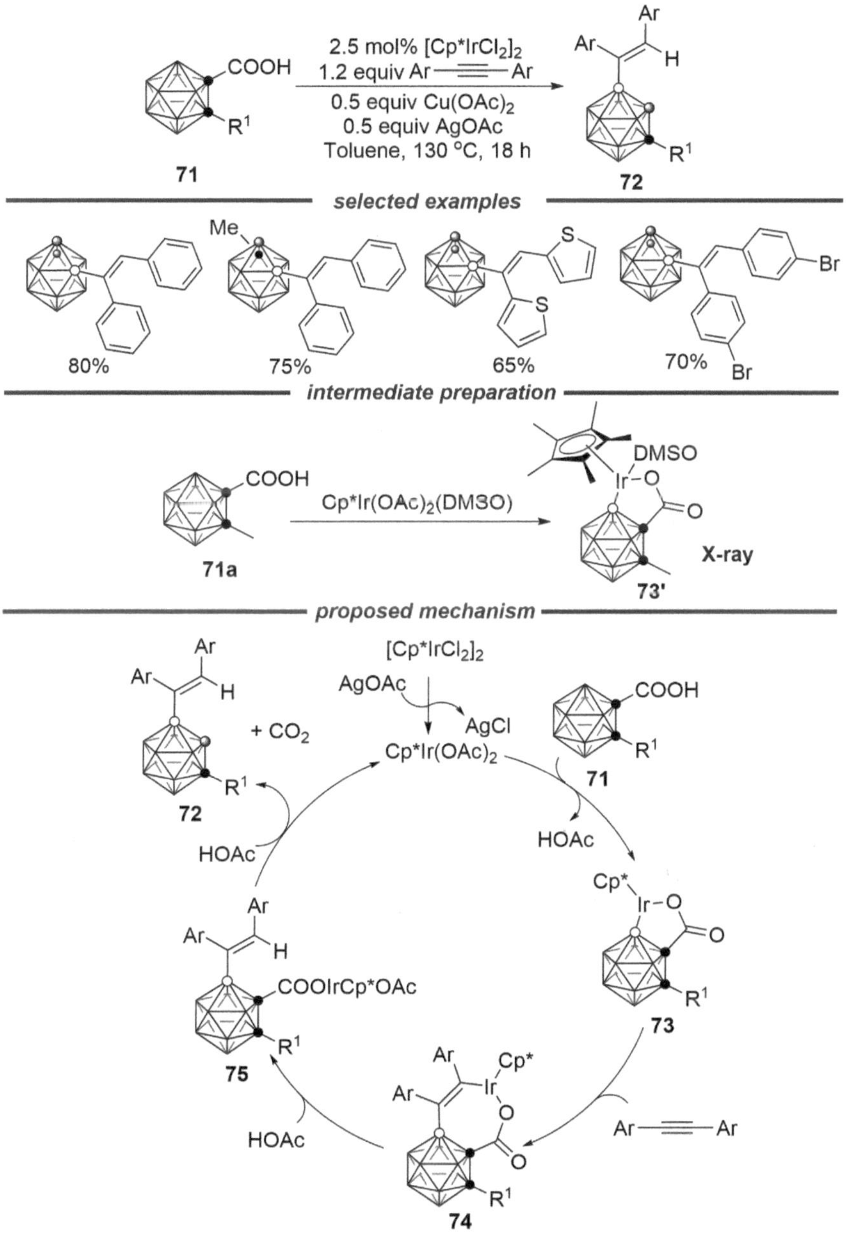

Scheme 16 Ir-catalyzed B(4)-alkenylation of *o*-carboranes with internal alkynes.

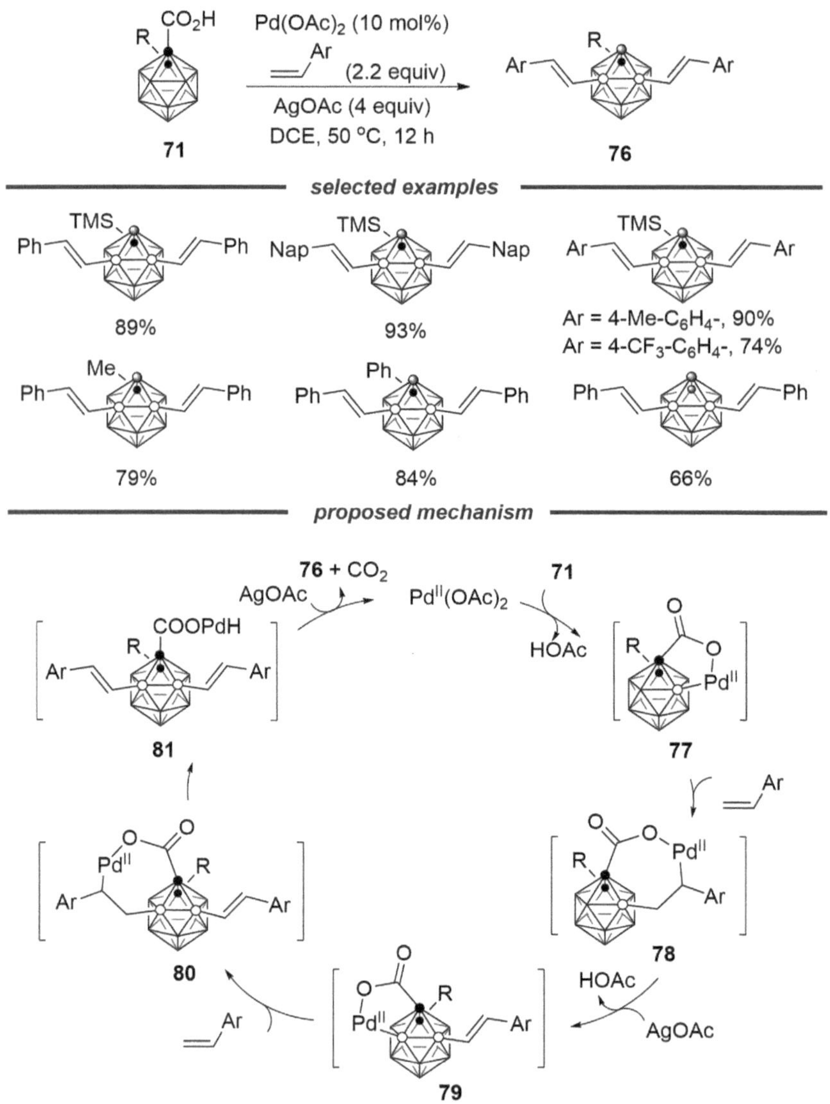

Scheme 17 Pd-catalyzed B(4,5)-dialkenylation of *o*-carboranes with styrenes.

of a range of substituents including methyl, trifluoromethyl, halo, naphthyl and methoxy groups. Scheme 17 displayed a Pd(II)-initiated reaction mechanism, where the first catalytic cycle involving electrophilic substitution, alkene insertion and β–H elimination takes place to install a B(4)-alkenyl group with the retention of the carboxy directing group. Repeating of this catalytic cycle with the help of carboxy group, followed by in-situ decarboxylation, leads to the formation of B(4,5)-dialkenylated *o*-carboranes **76**. It was undoubted that the carboxy group was indispensable for regio- and di-selectivity.

Similarly, the Lu group reported a carboxy group assisted B(4,5)-H alkenylation of *o*-carboranes with α,β-unsaturated carbonyl compounds in the presence of $Pd(TFA)_2$ catalyst, giving access to a variety of B(4)- and B(4,5)-substituted *o*-carboranes (**82** and **83**) in moderate to good yields (Scheme 18) *(38)*. α,β-Unsaturated carbonyls bearing amino, aryl or *tert*-butyl groups mainly offered the monosubstituted *o*-carboranes **82**, while α,β-unsaturated esters gave disubstituted *o*-carboranes **83** as major products. Such a reaction was proposed to proceed in a similar pathway as demonstrated in Scheme 17.

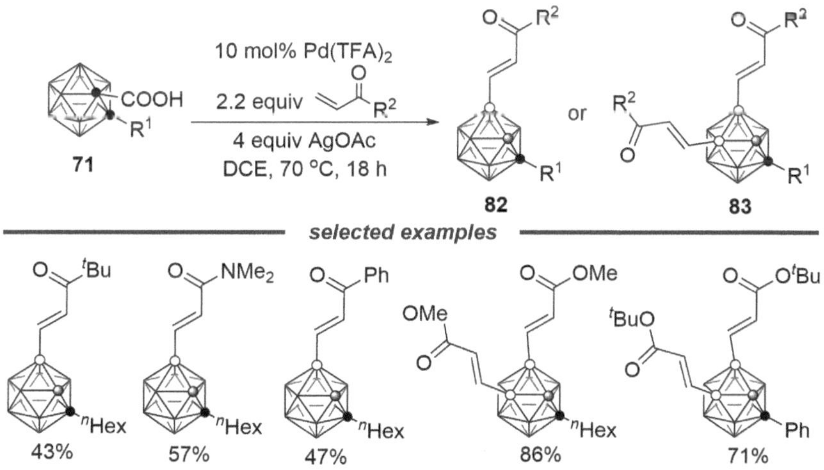

Scheme 18 Pd-catalyzed B(4,5)-alkenylation of *o*-carboranes with α,β-unsaturated carbonyls.

An efficient three-component [3+2+1] annulation of 1-hydroxyl-*o*-carboranes **32b**, CO and internal alkynes via Pd-catalysis was achieved in the Xie group, approaching to facile synthesis of C,B-substituted carborano-coumarin derivatives **84** in a single step in good yields, which are potential candidates in pharmaceuticals (Scheme 19) *(39)*. An array of functional groups such as methoxy, halo, trifluoromethyl and trimethylsilyl on phenyl rings of alkynes were compatible with this reaction. Replacement of aryl ring of alkynes with thiophenyl, naphthyl or alkyl groups also gave access to the desired products **84**. A plausible Pd(II)-Pd(0)-Pd(II) catalytic

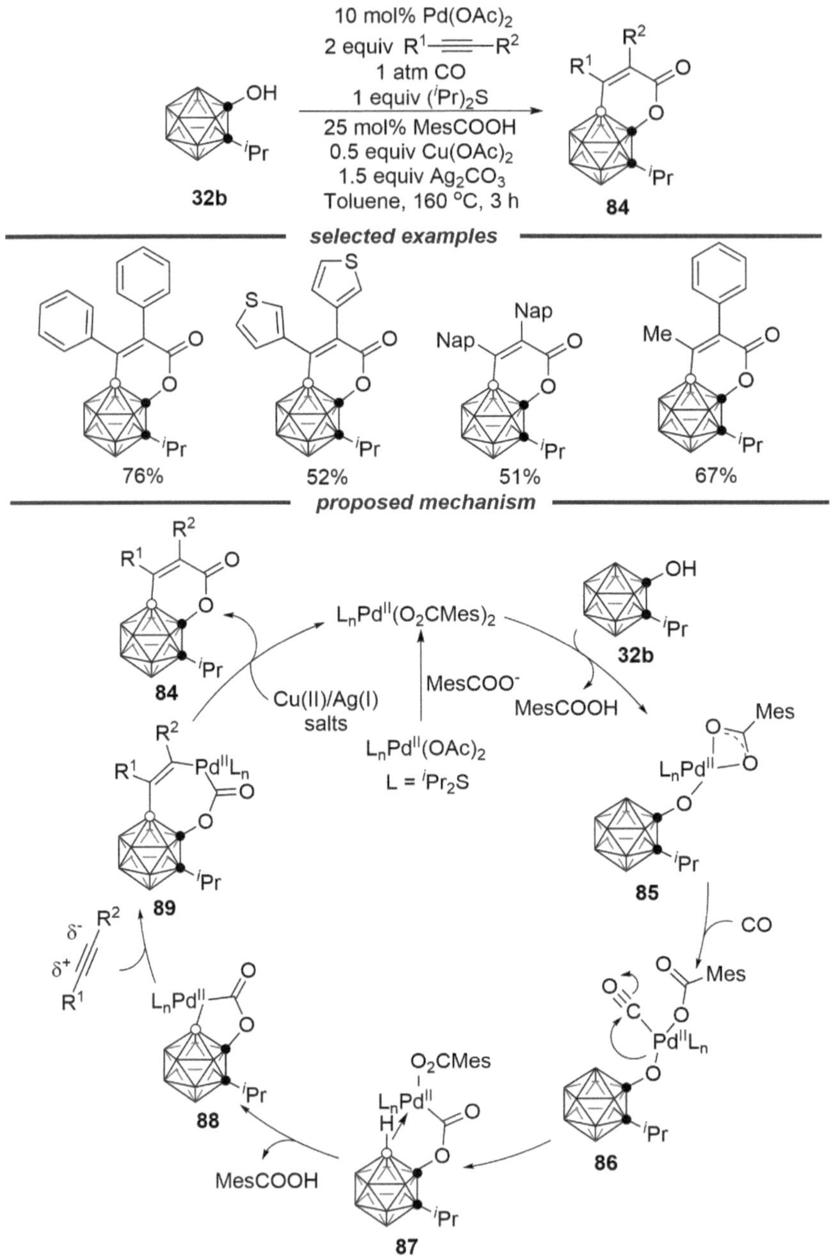

Scheme 19 Pd-catalyzed carbonylative annulation of 1-hydroxy-*o*-carboranes with internal alkynes and CO.

cycle was exhibited in Scheme 19. The acid-base reaction of 1-hydroxyl-*o*-carboranes **32b** with Pd(II) catalyst gives an intermediate **85**. The insertion of the coordinated CO into the Pd—O bond in **86**, followed by electrophilic B(4)-palladation offers a five-membered palladacycle species **88**. Subsequent alkyne insertion, reductive elimination and oxidation of Pd(0) complete the catalytic cycle, leading to the production of carborano-coumarins **84** and regeneration of the reactive Pd(II) catalyst. It was noteworthy that thioether ligand played an indispensable role in two ways: (1) preventing the poisoning of Pd center by CO, and (2) protecting Pd(0) from aggregation.

Very recently, the Xie group developed an efficient iridium-catalyzed formal [4+2] annulation of carboranyl carboxylic acids with alkynes in the presence of 5 mol% of $[Cp^*IrCl_2]_2$ catalyst, 20 mol% of $AgNTf_2$, and 3 equiv. of $Cu(OPiv)_2$, resulting in the facile synthesis of a new class of carboranoisocoumarin derivatives **90** (Scheme 20) *(40)*. Aryl alkynes with electron-donating or electron-withdrawing groups on the benzene ring, such as alkyl, methoxyl and halo groups, were well compatible with this reaction to deliver the desired annulated products in high yields. The carboxyl group not only served as a directing group to control the regioselectivity but also ingeniously became a part of the final products. The reaction mechanism was proposed in Scheme 20, involving sequential carboxyl-directed B(4)-H metalation, alkyne insertion, and reductive elimination.

4.1.2 B(4)-alkynylation

Two catalytic systems were developed to realize cage B(4) alkynylation of *o*-carboranes (Scheme 21) *(41)*. In the presence of 5 mol% $Pd(OAc)_2$, either 1 equiv. of alkyne bromides or 2 equiv. of terminal alkynes were able to react with carboranyl carboxylic acids **71** to give the B(4)-alkynylated *o*-carboranes **93** in moderate to good yields. Various terminal alkynes bearing silyl, aryl or carboranyl groups were well compatible with the dehydrogenative cross-coupling reaction, which involved a Pd(II)-Pd(0)-Pd(II) catalytic cycle. On the other hand, only alkyne bromides with bulky silyl groups offered access to B(4)-alkynylated *o*-carboranes **93** via a Pd(II)-Pd(IV)-Pd(II) cycle. It was noted that the five-membered palladacycles with a B(4)-Pd bond (**94**/**97**) was proposed as an intermediate in both reaction pathways.

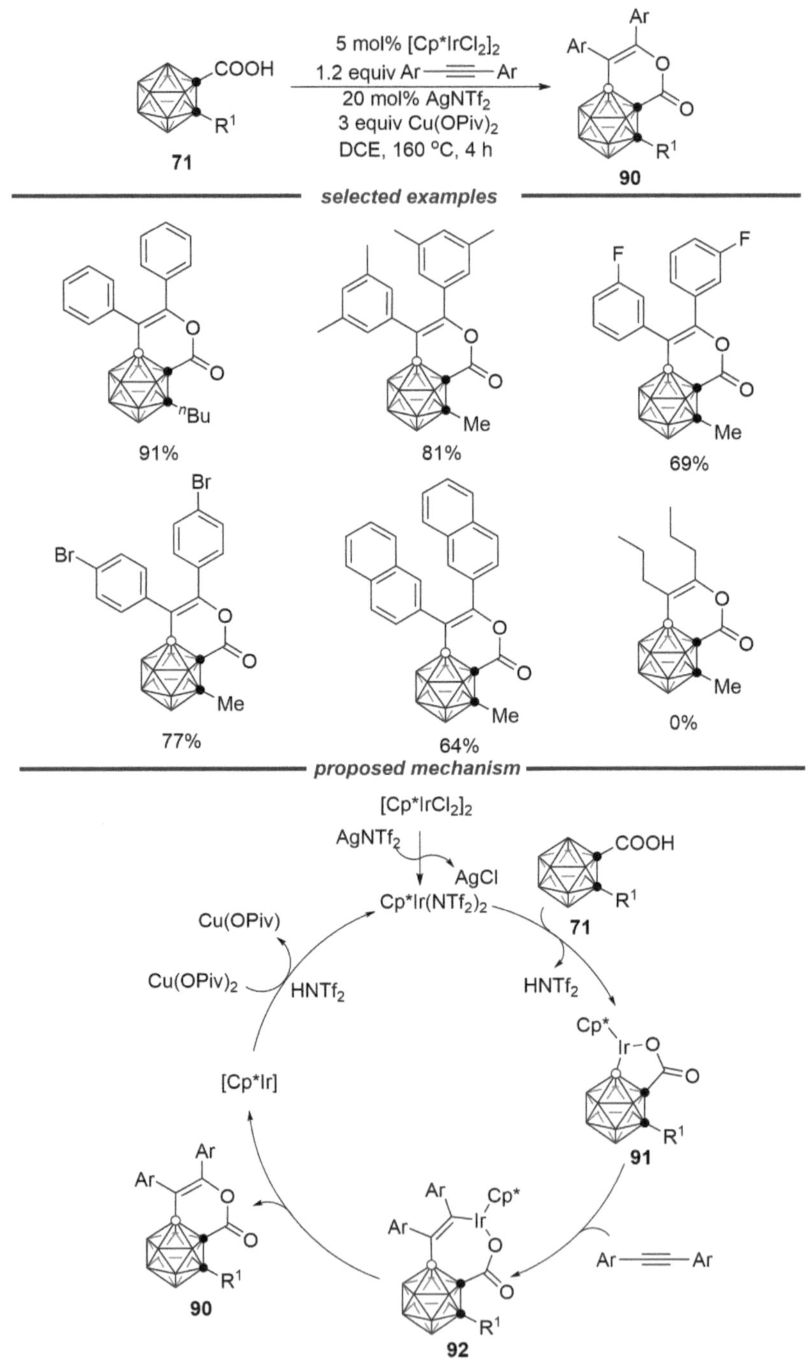

Scheme 20 Ir-catalyzed annulation of *o*-carboranyl carboxylic acids with alkynes.

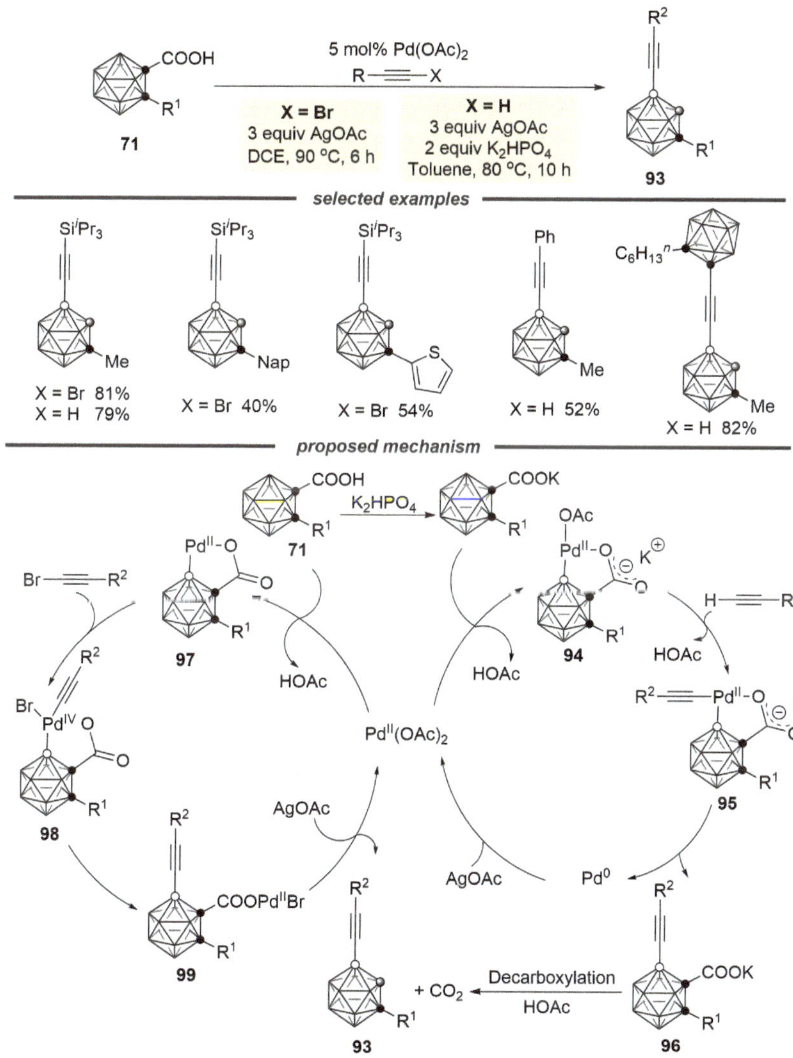

Scheme 21 Pd-catalyzed B(4)-alkynylation of *o*-carboranes.

4.1.3 B(4,5)-arylation

The Xie group disclosed a B(4,5)-diarylation of *o*-carboranes via Pd-catalyzed intermolecular coupling reactions of B—H in carboranyl carboxylic acids **71** with aryl iodides (Scheme 22) *(42a)*. The results showed that aryl iodides with electron-donating substituents worked better than those with the electron-withdrawing ones. A class of aryl iodides decorating with various functional groups including halo, alkyl, aryl, methoxy, acyl

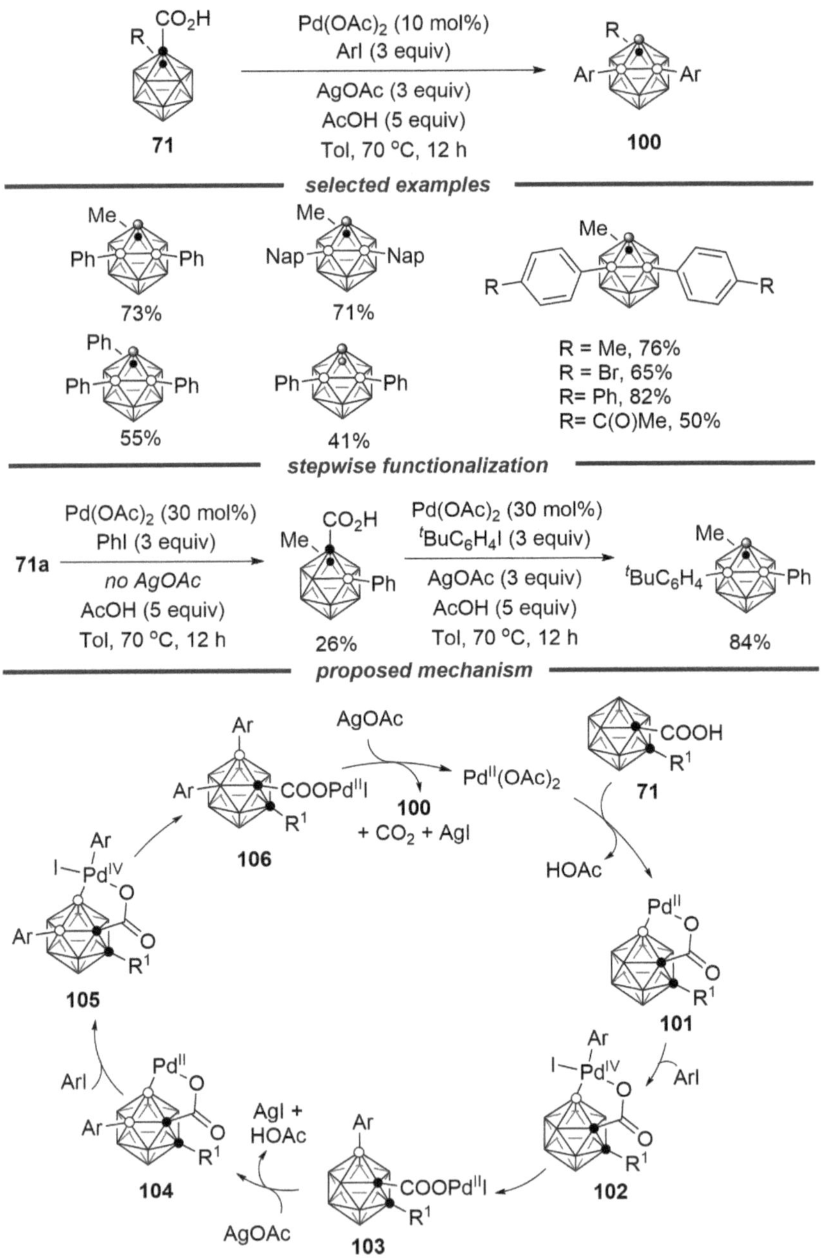

Scheme 22 Pd-catalyzed B(4,5)-diarylation of *o*-carboranes.

and ester groups were compatible with this diarylation reaction. Treatment of 1-Me-2-COOH-*o*-carborane **71a** with two different aryl iodides step by step gave unsymmetrical B(4,5)-diarylated *o*-carboranes. Control experiments and DFT calculations *(42b)* implied a Pd(II)-Pd(IV)-Pd(II) catalytic cycle, where a five-membered palladacycle containing a B-Pd(IV) bond was proposed as an intermediate (Scheme 22).

Later on, the Yan group developed a transient directing group guided palladium-catalyzed protocol for B(4,5)-diarylation of carboranyl aldehydes **107** (Scheme 23) *(43)*. A range of aryl iodides bearing alkyl, aryl, trifluoromethyl, halo, nitro, acyl and ester functional groups afforded B(4,5)-diarylated *o*-carboranyl aldehydes **109** in good yields. On the other hand, lowering the amount of the catalysts, Ag salts and aryl iodides produced the B(4)-monoarylated *o*-carboranyl aldehydes **110** in high yields. The stoichiometric reaction of in situ formed imine **111** with $Pd(OAc)_2$ offered a bicyclic Pd species **112'** featuring a B(4)-Pd bond as confirmed by single-crystal analyses. A Pd(II)-Pd(IV)-Pd(II) catalytic cycle involving in situ formation of imine directing group, electrophilic substitution, oxidative addition and reductive elimination was proposed as plausible reaction mechanism (Scheme 23).

With the help of another transient directing group $—CH_2N=CHCOOH$, B(4,5)-diarylation of *o*-carboranyl methyleneamine **117** was realized by the same group under similar reaction conditions, resulting in the isolation of B(4,5)-diarylated *o*-carboranyl methylenenamine **119** in high yields (Scheme 24) *(44)*. A key bicyclic Pd(II) intermediate **121** containing two carboranyl moieties was obtained via the reaction of $Pd(OAc)_2$ with 2 equiv. of *o*-carboranyl methyleneamine **117a**.

A B(9)-acylamino group guided B(4)-arylation of *o*-carboranes with aryl boronic acids in the presence of $Pd(MeCN)_4(BF_4)_2$ catalyst was disclosed by the Cao group (Scheme 25) *(45)*. Various B(4)-arylated products **123** decorated with a broad substituents ranging from halo, hydroxy, methoxy, acetyl, to trifluoromethyl groups were prepared in moderate to good yields. The calculated Mulliken charge of B—H vertices highlighted that the pre-installed B(9)-acylamino group made the B(4/5)—H bonds more electron-rich, which might be responsible for the unexpected B(4)-selectivity in such a reaction. Accordingly, a plausible reaction mechanism was proposed in Scheme 25. Selective electrophilic substitution of B(4)—H, followed by transmetalation and reductive elimination affords the final products **123** and Pd(0). Oxidation of Pd(0) to Pd(II) by Cu salts completes the catalytic cycle.

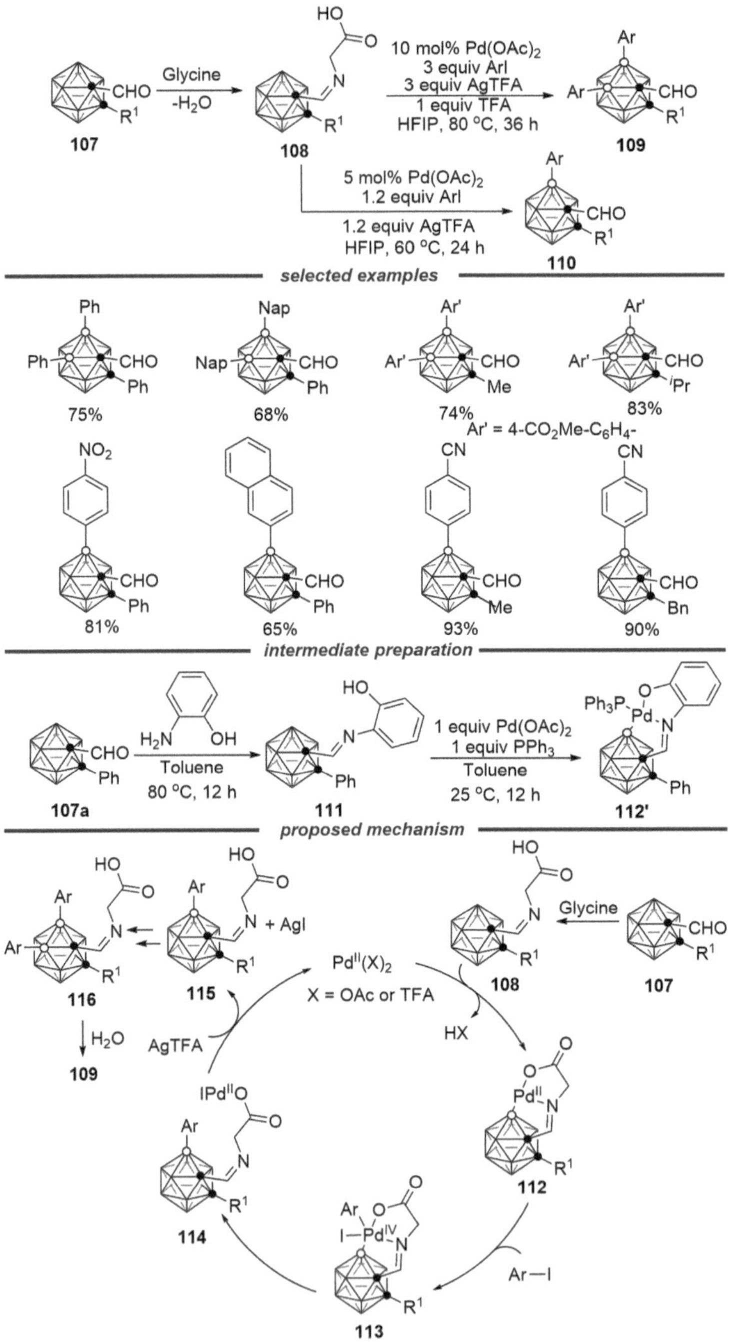

Scheme 23 Transient directing group guided B(4)-arylation and B(4,5)-diarylation of *o*-carboranyl aldehydes.

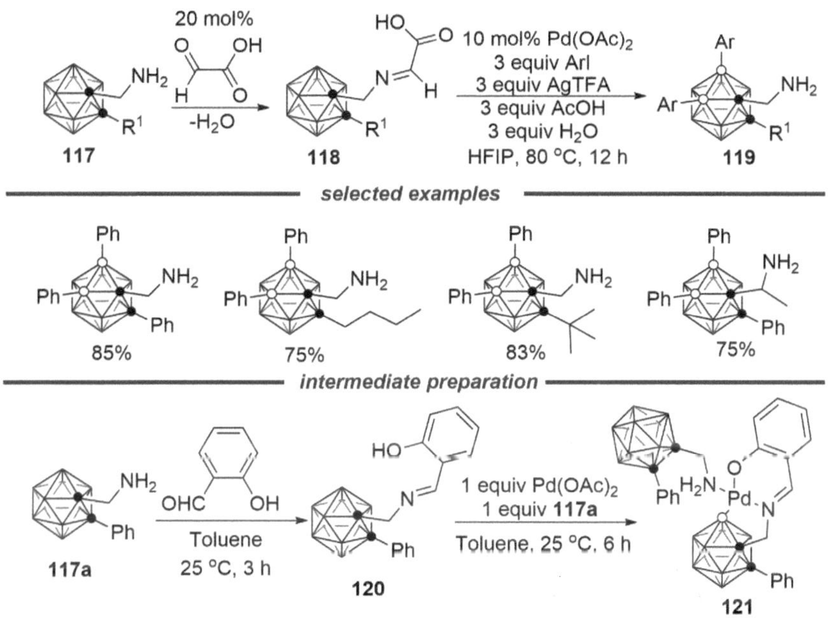

Scheme 24 Transient directing group guided B(4,5)-diarylation of *o*-carboranyl methyleneamines.

To improve atom- and step-economy in the aforementioned coupling reactions, the Xie group developed a dehydrogenative cross-coupling reaction of *o*-carboranes with non-prefunctionalized arene coupling partners (Scheme 26) *(46)*. Using 5 mol% Ir(III) catalyst, carboranyl carboxylic acids **71** reacted with a range of (hetero)arenes bearing various nitrogen-based directing groups, such as quinoline, pyrimidine, thiazole, pyrazole and indole, to produce a series of B(4)-(hetero)arylated *o*-carboranes **126** in very high yields. This reaction showed high tolerance towards various functional groups on aryl rings, including —F, —Cl, —Br, —CF_3, —OMe, —CHO and —CN groups. Moreover, treatment of 1-Me-2-COOH-*o*-carborane **71a** with an equimolar amount of Cp*$Ir(OAc)_2$ and 2-[4-(trifluoromethyl)phenyl]pyridine generated a key iridacycle intermediate **127'** featuring a C—Ir(III) bond as confirmed by single crystal X-ray analyses. This result demonstrated that aryl C—H iridation occurred preferentially over the cage B(4)—H bond. Accordingly, a plausible reaction mechanism involving sequential aryl C—H iridation, oxidation of Ir(III) to Ir(IV), B(4)—H iridation and reductive elimination was proposed in Scheme 26.

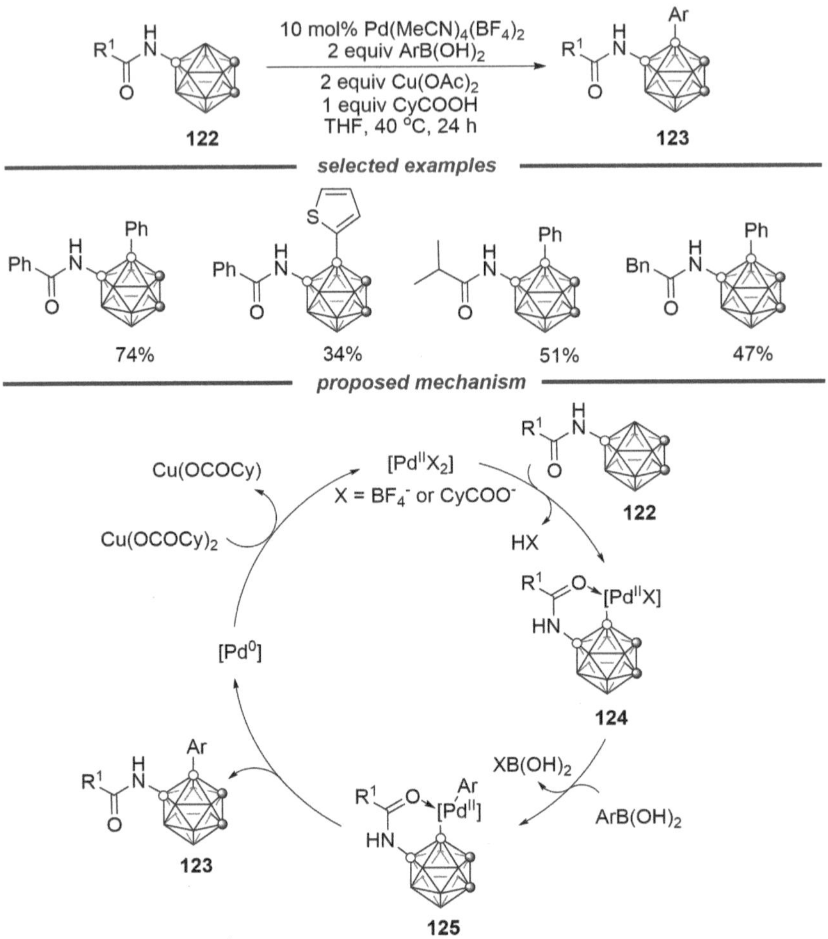

Scheme 25 Acyl amino group guided Pd-catalyzed B(4)-arylation of *o*-carboranes.

4.1.4 B(4)-thienylation

The Xie group reported an efficient regioselective B(4)-H/C(sp^2)-H dehydrogenative cross-coupling reaction of carboranyl carboxylic acids **71** with thiophenes via iridium catalysis (Scheme 27) *(47)*. Various thiophenes bearing halo, alkyl, alkenyl and ester groups were compatible with this thienylation. A class of benzothiophenes, thienothiophenes and oligothiophenes served as competent coupling partners, giving B(4)-thienylated *o*-carboranes **131** in good yields. On the basis of control experiments, a plausible reaction pathway was proposed in Scheme 27. This catalytic cycle is initiated with the electrophilic iridation of B(4)—H to give a five-membered iridacycle intermediate **132**. Subsequent activation of

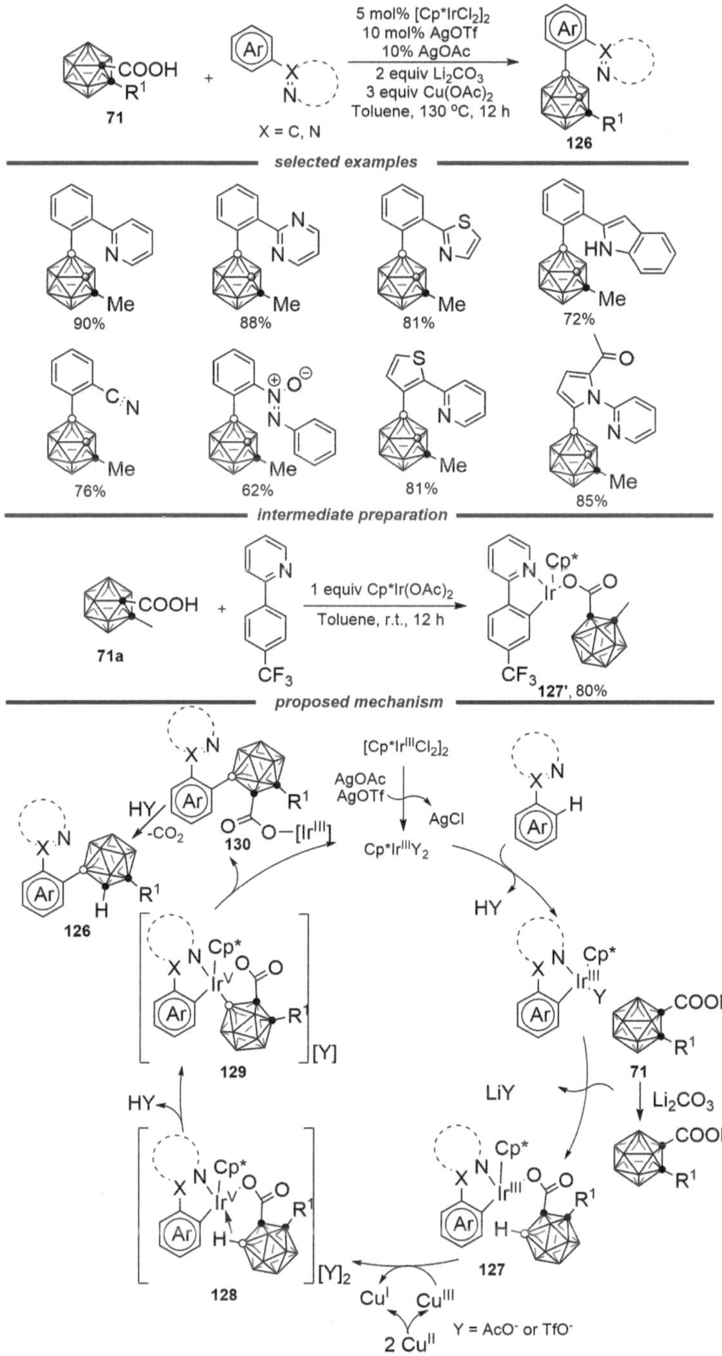

Scheme 26 Ir-catalyzed dehydrogenative B-H/C-H cross-coupling.

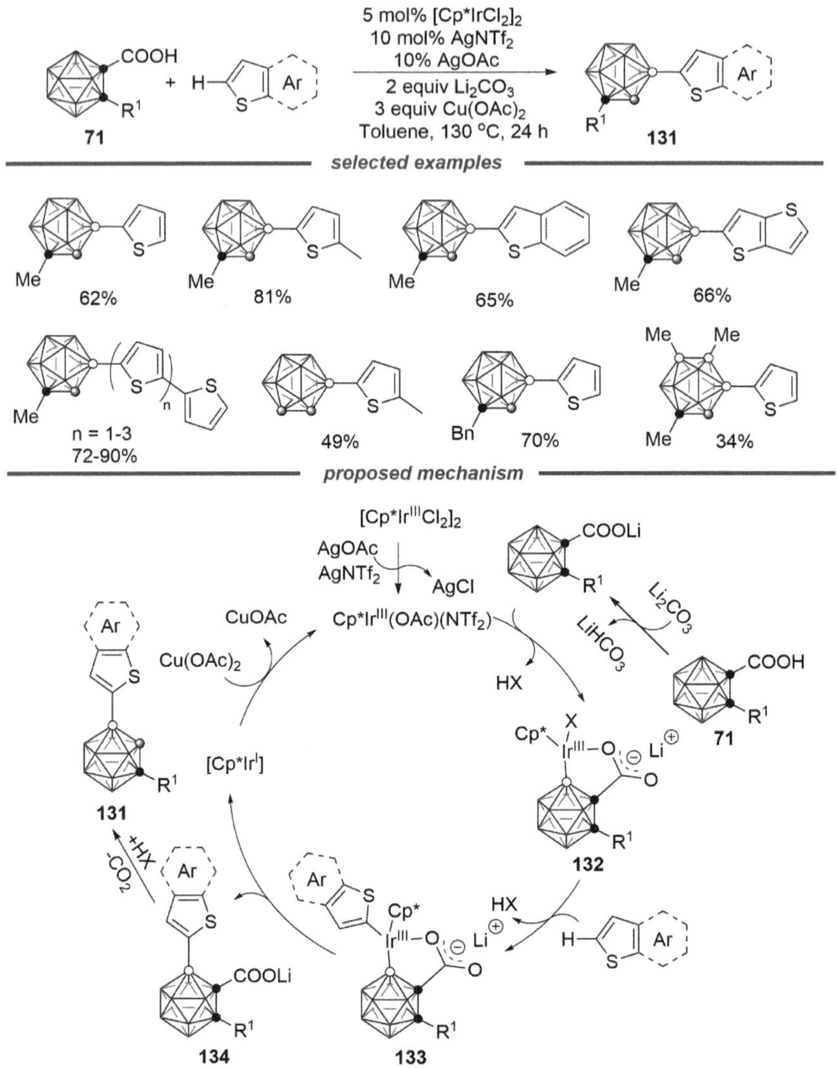

Scheme 27 Ir-Catalyzed dehydrogenative cross-coupling of *o*-carboranes with thiophenes.

thienyl α-C—H generates a thienyl iridacycle intermediate **133**, which undergoes a sequential reductive elimination, protonation and decarboxylation to afford the final products **131**.

4.1.5 B(4,5)-alkylation

The Lu group realized a Rh-catalyzed regioselective B(4,5)-dialkylation of carboranyl carboxylic acids **71** with allylic alcohols, giving access to

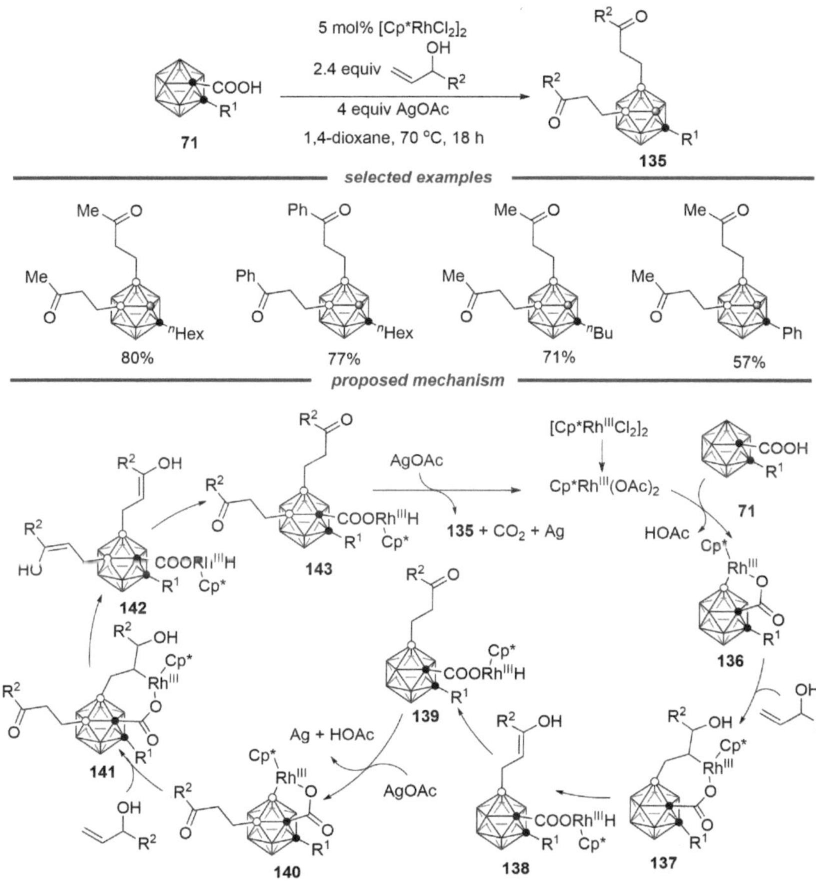

Scheme 28 Rh-catalyzed B(4,5)-dialkylation of *o*-carboranes.

B(4,5)-dialkylation products **135** in good yields (Scheme 28) *(48)*. An array of allyl alcohols tethering with different aryl or alkyl groups were well compatible in this dialkylation reaction. A possible reaction pathway was proposed in Scheme 28. A sequential of cage B(4)—H metalation, alkene insertion, β—H elimination and enol isomerization, followed by decarboxylation gives B(4)-alkylated products **139** that are subjected to another catalytic cycle, affording B(4,5)-dialkylated products **135**.

With the help of B(9)-amide directing group, a Pd-catalyzed B(4)-methylation of *o*-carboranes **122** with methyl iodide was reported by the Cao group (Scheme 29) *(49)*. This B(4)-methylation reaction was applicable to various 9-amide-*o*-carboranes with electron-withdrawing and -donating substituents, affording B(4)-methylation products **144** in fair yields.

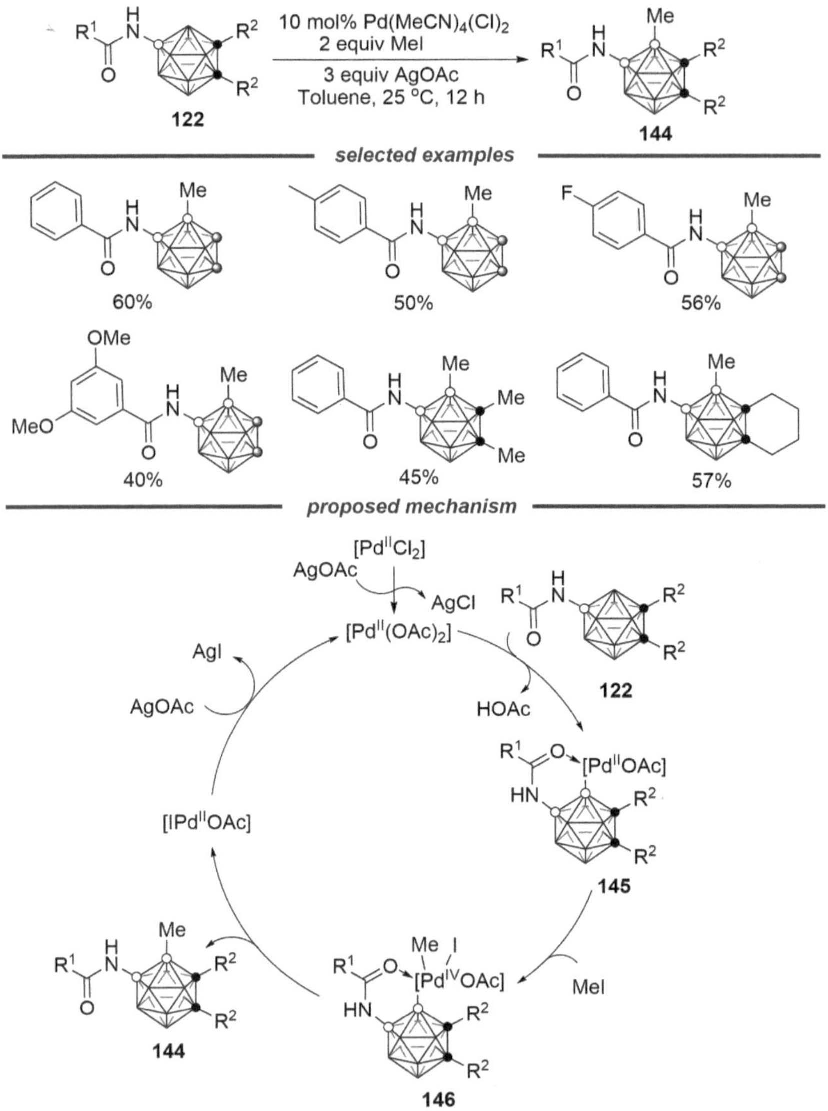

Scheme 29 Acyl amino group guided Pd-catalyzed B(4)-methylation of *o*-carboranes.

A Pd(II)-Pd(IV)-Pd(II) catalytic cycle was proposed, in which a sequential of electrophilic substitution, oxidative addition and reductive elimination was accounted for the construction of **144** (Scheme 29).

The Yan group developed a Rh-catalyzed B(4)-acylmethylation of 1-pyridyl-2-methyl-*o*-carborane **8a** (Scheme 30) *(50)*. In the presence of 6 mol% of $[RhCp^*Cl_2]_2$, the pyridyl-directed reaction of 1-pyridyl-2-methyl-*o*-carboranes **8a** with 2 equiv. of α-carbonyl sulfoxonium ylides

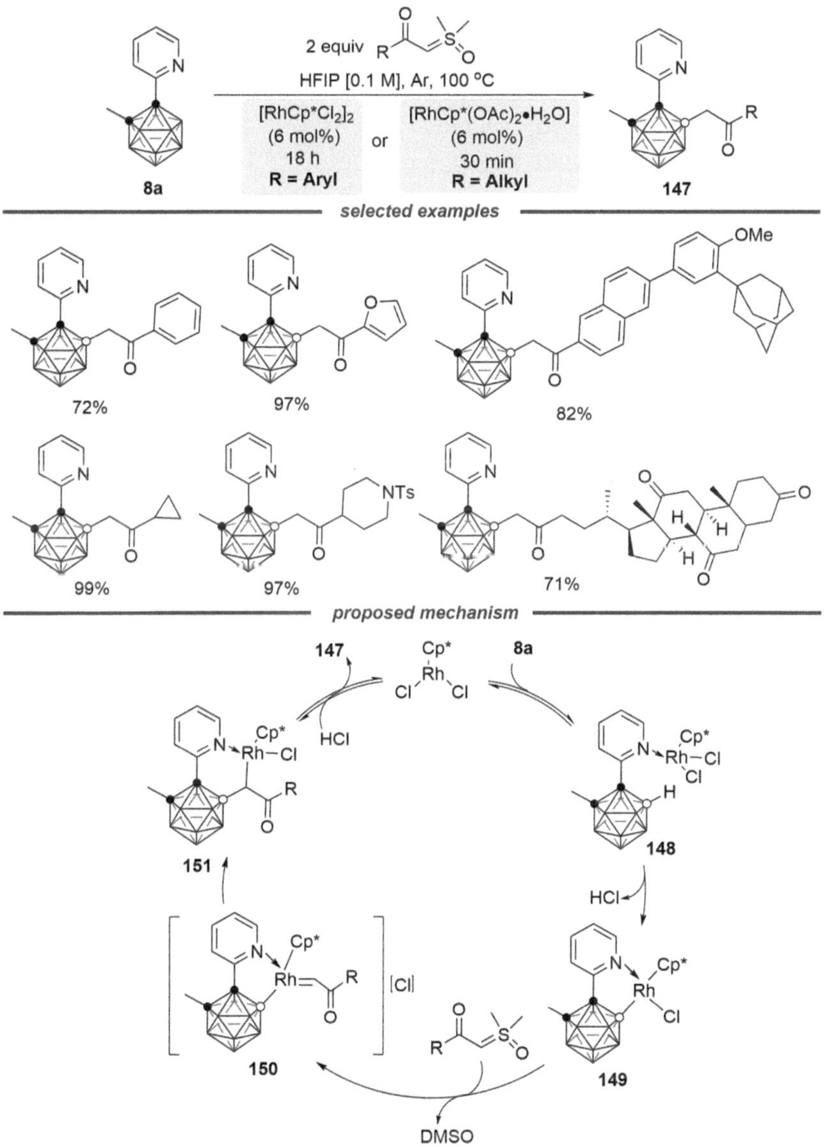

Scheme 30 Pyridyl group guided Pd-catalyzed B(4)-acylmethylation of *o*-carboranes.

in the absence of any additives afforded a class of mono-acylmethylated *o*-carboranes **147** in high yields with excellent B(4)-vertex selectivity. This reaction was compatible with both alkyl and aryl sulfoxonium ylides, bearing sterically or electronically diverse substituents, resulting in efficient synthesis of mono-acylmethylated *o*-carboranes **147** in moderate to excellent yields. A plausible reaction mechanism involving coordination, electrophilic substitution, migratory insertion of α-carbonyl sulfoxonium ylides, and protodemetallation was depicted in Scheme 30.

During that time period, the Xie *(51)* and Lee *(52)* groups independently reported an Ir-catalyzed B(4)-acylmethylation of 1-COOH-*o*-carboranes **71** using sulfoxonium ylides as alkylating agents, affording a series of B(4)-acylmethylated *o*-carboranes in high yields. The associated reaction mechanism is proposed to be similar to that shown in Scheme 30.

4.1.6 B(4)-oxygenation

The Xie group disclosed a carboxy-assisted B(4) hydroxylation of carboranyl carboxylic acids **71** with O_2 via rhodium catalysis, in which O_2 served as hydroxylating reagent and oxidant. A batch of 4-hydroxy-*o*-carboranes **152** was efficiently prepared in a sustainable, economical and environmentally benign way (Scheme 31) *(53)*. Diverse substrates decorating with chloro, fluoro, silyl and methoxy functional groups were compatible in the construction of B—O bond. The B(4)—OH group could undergo various transformations for further synthetic exploitations. The isolation of ^{18}O-labeled product **152'** from the control experiment with $^{18}O_2$ under the same reaction conditions unambiguously confirmed the role of molecular oxygen as oxygen source. The presence of the key rhodacycle intermediate **154** with a O—O bond was validated by MALDI-TOF MS and a parallel experiment with $^{18}O_2$. Accordingly, a reaction mechanism was proposed in Scheme 31. Directing by the carboxy group, electrophilic rhodation of B(4)—H bond generates a five-membered Rh(III) species **153**. The insertion of O_2 into the B(4)—Rh bond forms a key intermediate **154**. Protonation, decarboxylation and deoxygenation give 4-hydroxy-*o*-carboranes **152**.

4.1.7 B(4)-amination

The Xie group developed two catalytic systems for the construction of B(4)—N bond in *o*-carboranes enabled by carboxy directing group and transition metal catalysts (Scheme 32) *(54)*. A series of tertiary B(4)-aminated *o*-carboranes **156** were prepared in high yields through the reaction of carboranyl carboxylic acids **71** with O-benzoyl hydroxylamines promoted by $Pd(OAc)_2$ catalyst. In the presence of Pd/C, hydrogenolysis of 4-N$(CH_2Ph)_2$-*o*-carborane **156a** took place to form 4-NH_2-*o*-carborane **158**, which could be further functionalized for diverse utilizations. On the other hand, a range of B(4)-aminated *o*-carboranes **157** were synthesized in

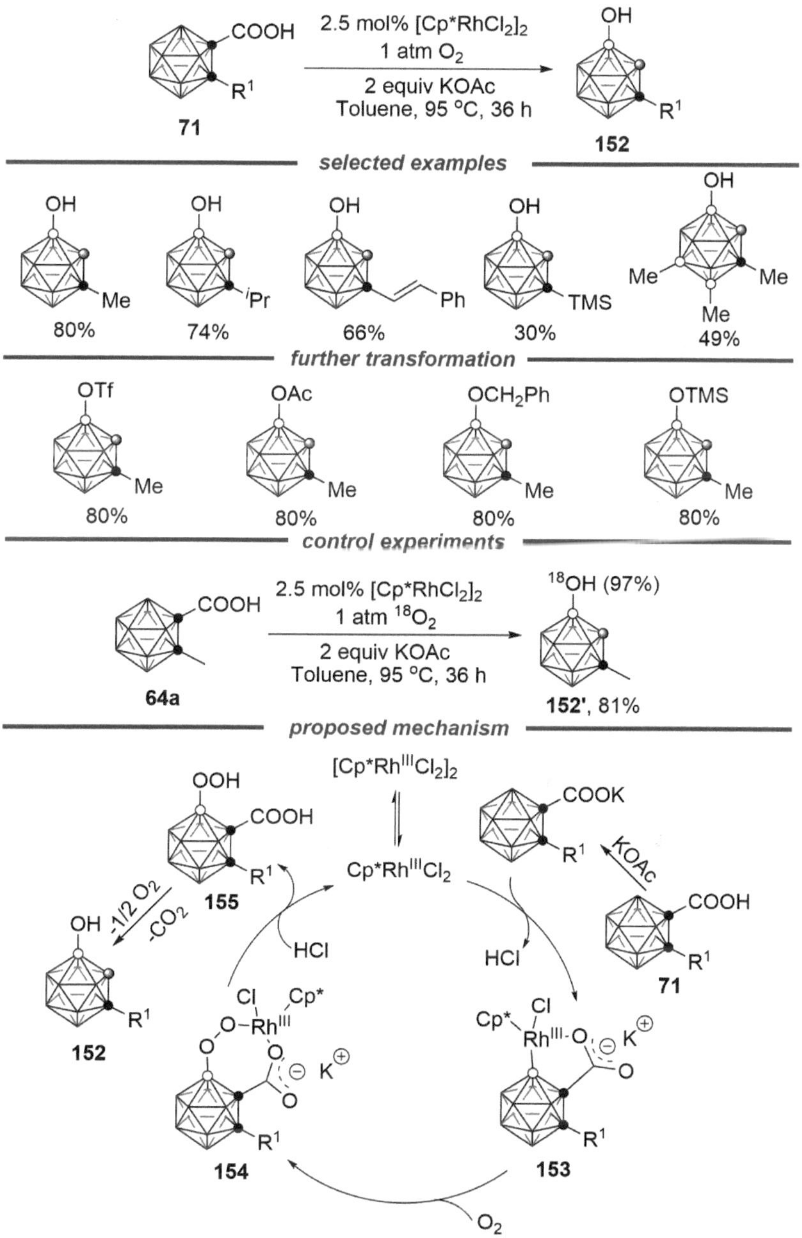

Scheme 31 Rh-catalyzed B(4)-hydroxylation of *o*-carboranes.

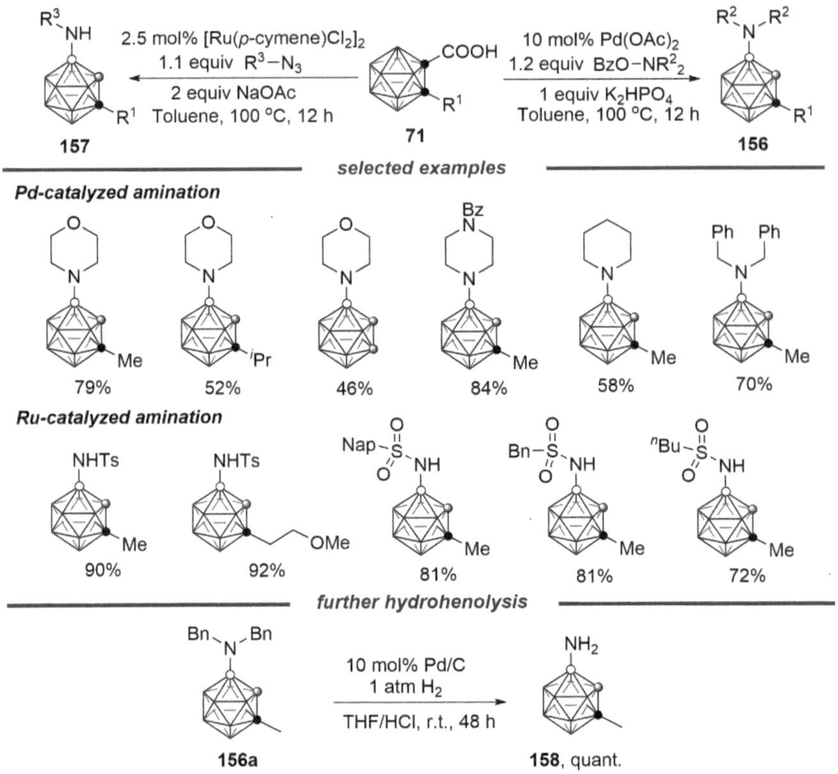

Scheme 32 Pd- and Ru-catalyzed B(4)-amination of *o*-carboranes.

excellent yields from the reaction of carboranyl carboxylic acids **71** with sulfonyl azides as amination reagents and [Ru(*p*-cymene)Cl_2]$_2$ as a catalyst.

The aforementioned two amination reactions proceeded in different reaction pathways, in which five-membered metallacycles might be involved. For Pd-catalyzed amination, the palladacycle **159** undergoes oxidative addition with *O*-benzoyl hydroxylamines to generate a Pd(IV) species **160**, followed by a sequential of reductive elimination, protonation and decarboxylation to give B(4)-aminated *o*-carboranes **156** (Scheme 33). For Ru-catalyzed system, the reaction of metallacycle **163** with organic azides forms a Ru-nitrene intermediate **165**, whose existence is verified by the norbornene-capture-experiment. Subsequent migratory insertion, protonation and decarboxylation afford the target product **157** (Scheme 34).

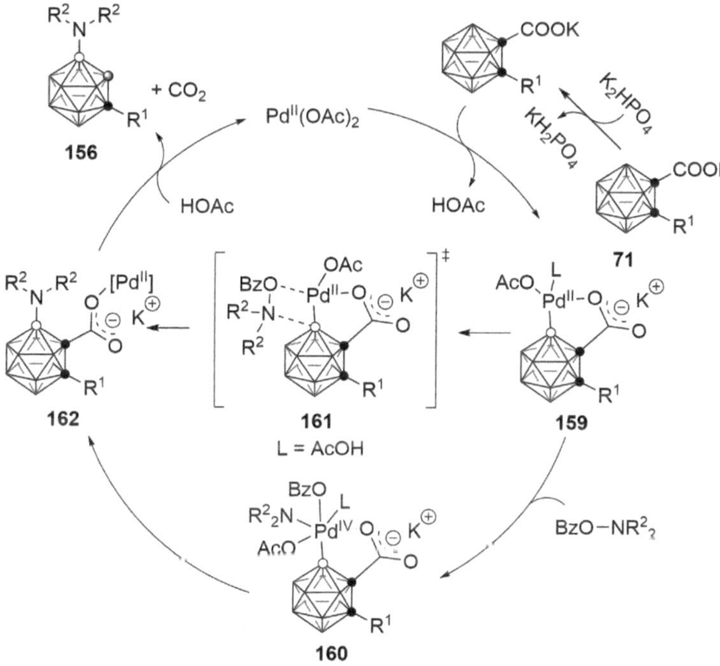

Scheme 33 Plausible reaction mechanism of Pd-catalyzed B(4)-amination.

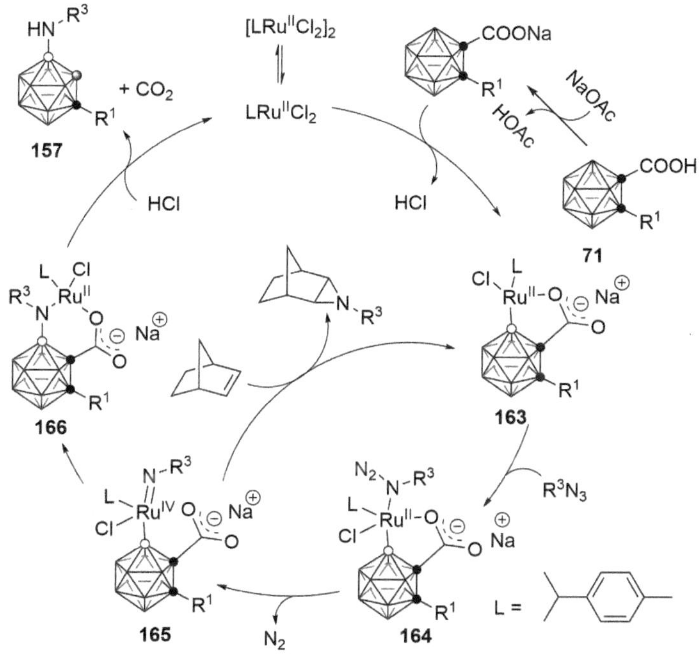

Scheme 34 Plausible reaction mechanism of Ru-catalyzed B(4)-amination.

Later on, the Yan group reported a similar Ir-catalyzed B(4)-H amination reaction of carboranyl carboxylic acids **71** with sulfonyl azides (Scheme 35) *(55)*. Various sulfonyl azides bearing halo, methoxy and trifluoromethyl groups worked well, offering the corresponding B(4)-sulfonamidated *o*-carboranes **167** in high yields. It was proposed that an Ir-nitrenoid intermediate might be involved in a similar pathway as described in Scheme 34.

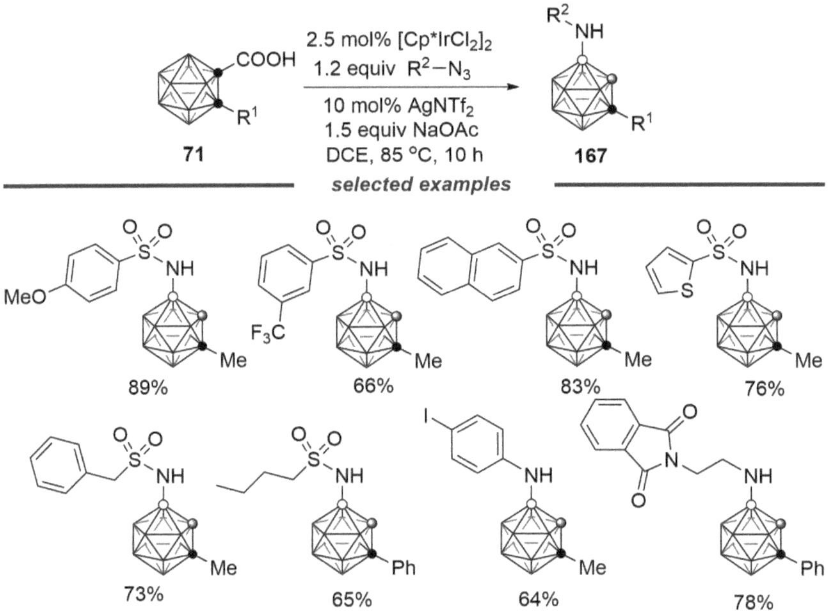

Scheme 35 Ir-catalyzed B(4)-amination of *o*-carboranes.

The Lee group demonstrated a COOH-directed Rh-catalyzed amidation of *o*-carboranes at B(4)-vertex with dioxazolones as agents (Scheme 36) *(56)*. A broad range of dioxazolones decorating with alkyl, halo, cyano, nitro and thienyl groups were competent substrates in this reaction. A plausible catalytic cycle was proposed in Scheme 37. A successive sequence of electrophilic B—H rhodation, coordination of dioxazolones to Rh center, electrophilic amidation, protonation and decarboxylation offers the final products **168**. It was noted that such a B(4)-amidation of carboranyl carboxylic acids **71** with dioxazolones was also achieved under mechanochemical conditions without organic solvents using $[Cp^*IrCl_2]_2$ as a catalyst *(57)*. The results were very comparable to Rh-catalyzed ones as shown in Scheme 36.

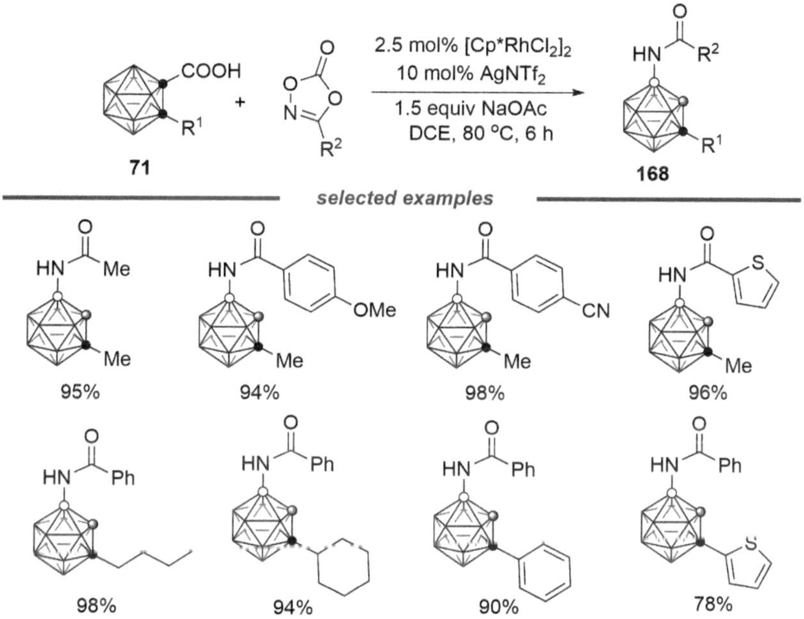

Scheme 36 Rh-catalyzed B(4)-amidation of *o*-carboranes.

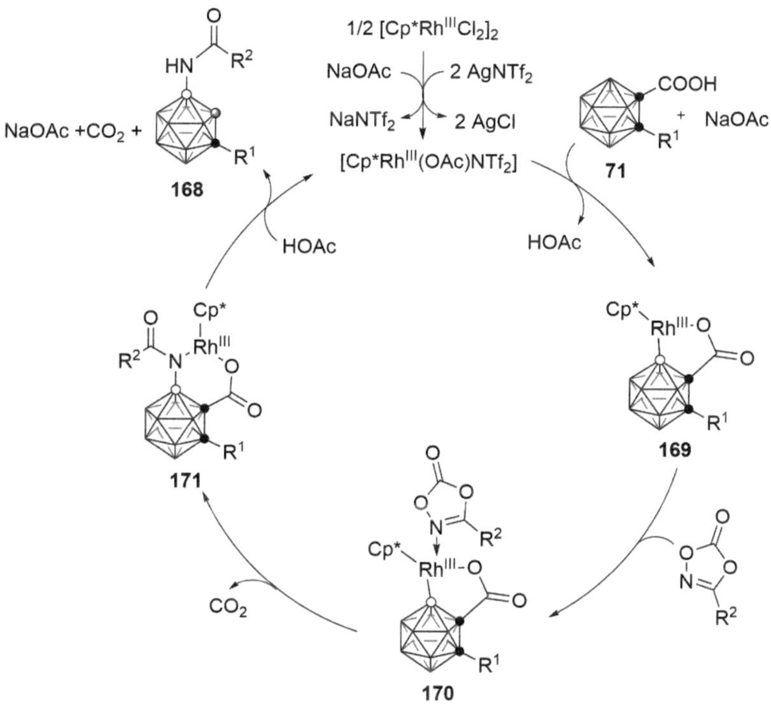

Scheme 37 Plausible reaction mechanism of Rh-catalyzed B(4)-amidation.

On the other hand, an intramolecular amidation of 3-benzamido-*o*-carborane **172** was developed by the Cao group, using $Pd(MeCN)_4(BF_4)_2$ as a catalyst (Scheme 38) *(58)*. A series of *o*-carborane-fused isoquinolinones **173** substituted with various groups, such as alkyl, methoxyl, halo, CF_3, and nitro groups, were synthesized in moderate to good yields. A plausible mechanism involving a sequential of electrophilic B—H activation, oxidative addition of N—H, and reductive elimination was proposed in Scheme 38.

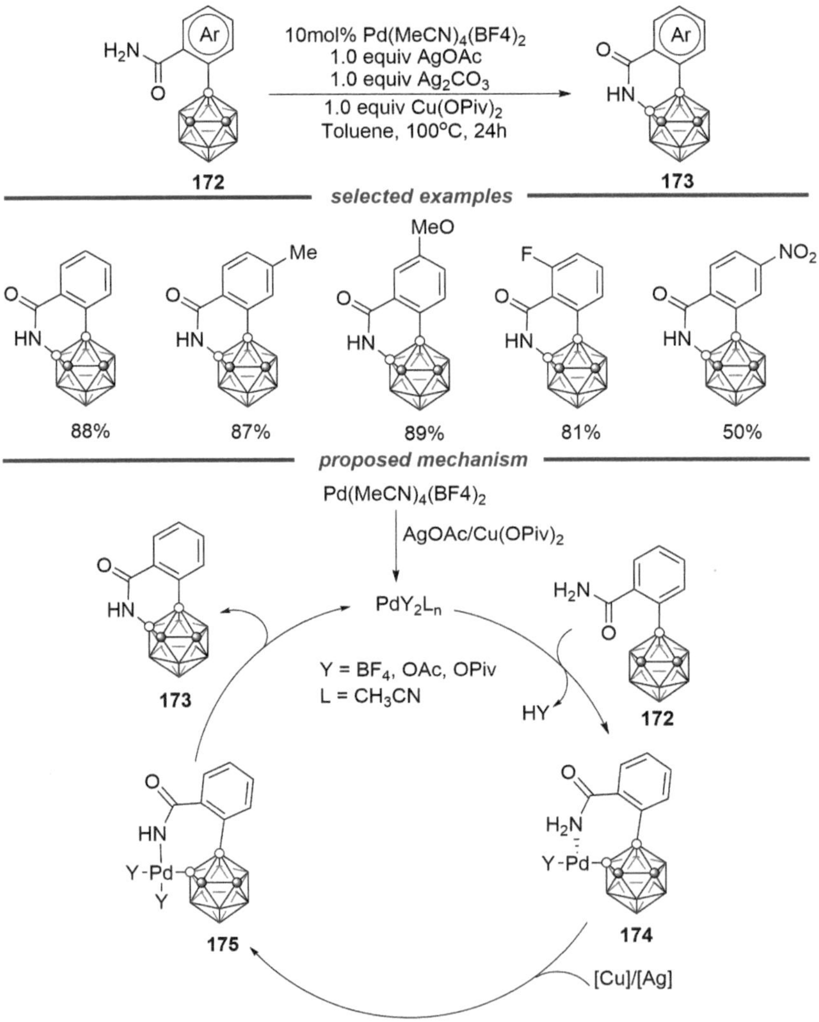

Scheme 38 Pd-catalyzed intramolecular B(4)-amidation of *o*-carboranes.

4.1.8 B(4,5)-halogenation

The Xie group reported transition metal catalyzed B(4)-halogenation and B(4,5)-diiodination of *o*-carboranes using different halogenating agents (Scheme 39) *(59)*. By employing *N*-iodosuccinimide (NIS),

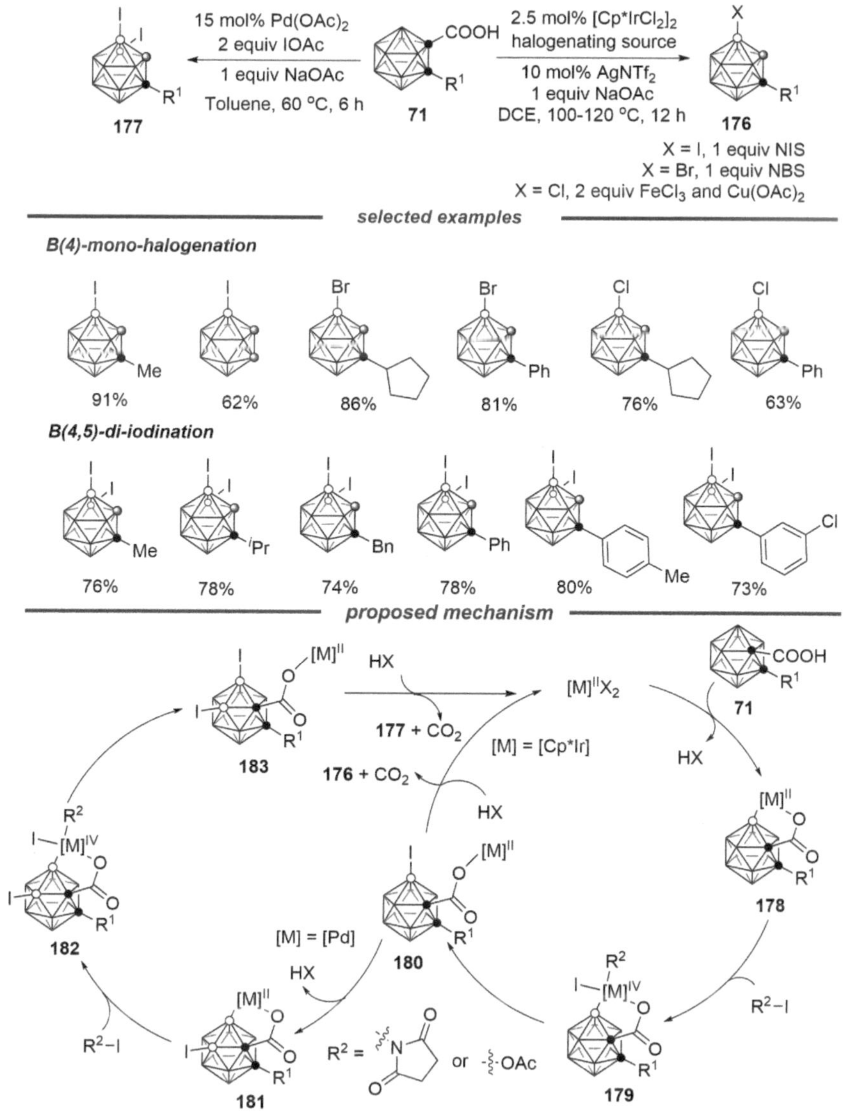

Scheme 39 Ir-catalyzed B(4)-halogenation and Pd-catalyzed B(4,5)-diiodination of *o*-carboranes.

N-bromosuccinimide (NBS) or $FeCl_3$ as halogen sources and $[Cp^*IrCl_2]_2$ as a catalyst, the mono-halogenation of carboranyl carboxylic acids **71** resulted in the corresponding B(4)-mono-halogenated *o*-carboranes **176**. On the other hand, a variety of B(4,5)-diiodinated *o*-carboranes **177** were obtained using IOAc as iodinating agent and $Pd(OAc)_2$ as a catalyst. It was noted that halogenated *o*-carboranes could be converted to a wide variety of functionalized *o*-carboranes derivatives. The plausible catalytic cycles involving five-membered metallacycle intermediates for the Ir-catalyzed B(4)-halogenation and Pd-catalyzed B(4,5)-diiodination of *o*-carboranes were proposed in Scheme 39.

4.1.9 B(4/5)-borylation

The Cao group achieved a Pd-catalyzed oxidative dehydrogenative homo coupling of B—H/B—H bonds of 9-acylamino-*o*-carboranes **122**, leading to the formation of biscarboranes with B(4)—B(4'/5') bond linkage (Scheme 40) *(60)*. A series of biscarboranes **184** containing alkyl, halo and trifluoromethyl groups were synthesized in good yields. Unfortunately, the regioselectivity was poor. A Pd(II)-Pd(IV)-Pd(II) catalytic cycle was proposed, which involves electrophilic substitution of B(4)-H bond, oxidation, electrophilic substitution of B(4/5)—H bond in another *o*-carborane and reductive elimination as described in Scheme 40.

4.2 Base metal catalyzed B(4,5)—H functionalization

4.2.1 Copper-catalyzed B(4)-alkenylation/annulation and -alkynylation

Regioselective cage B—H functionalization of *o*-carboranes discussed in Section 4.1 relies on noble transition metal catalysts such as Ir, Rh and Pd. In view of the advantages of base metal catalysis, the Xie group has uncovered a chelation-assisted strategy for cage B—H activation using Cu catalyst, in which the installation of a bidentate directing group at the cage C(1) position can stabilize a Cu(III) center at B(4) position and allow subsequent B(4)—H functionalization *(61)*.

In the presence of copper catalyst $Cu(OTf)_2$, the alkenylation-annulation of *N*-8-quinolinyl carboranyl amide **188** with internal alkynes yielded carborane-fused pyridones **189** (Scheme 41). Both symmetrical and unsymmetrical internal alkynes were compatible in this reaction to give the corresponding annulated products with high regioselectivity. On the other hand, employing terminal alkynes resulted in B—H/C—H dehydrocoupling to afford B(4)-alkynylated carboranes **191**.

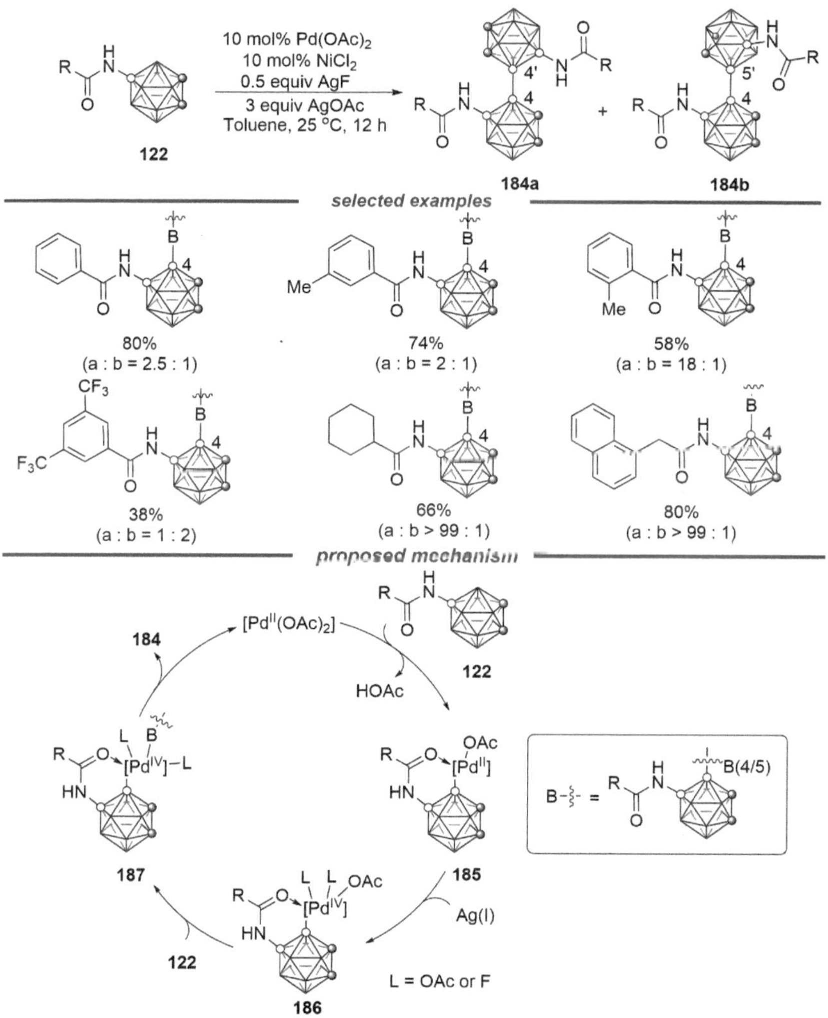

Scheme 40 Acyl amino group guided Pd-catalyzed dehydrogenative cross-coupling of B-H/B-H bonds of *o*-carboranes.

The results of control experiments indicated the necessity of copper catalyst and silver oxidant in such B—H functionalization, suggesting that the reaction pathway might involve a high-valent Cu(III) complex (Scheme 42). The Cu(II) catalyst initially chelates with the bidentate ligand to afford complex **192**, which is oxidized by silver salt to give Cu(III) complex **193**. Electrophilic substitution at the electron-rich B(4)—H vertex generates the intermediate **194**, which undergoes alkyne insertion to

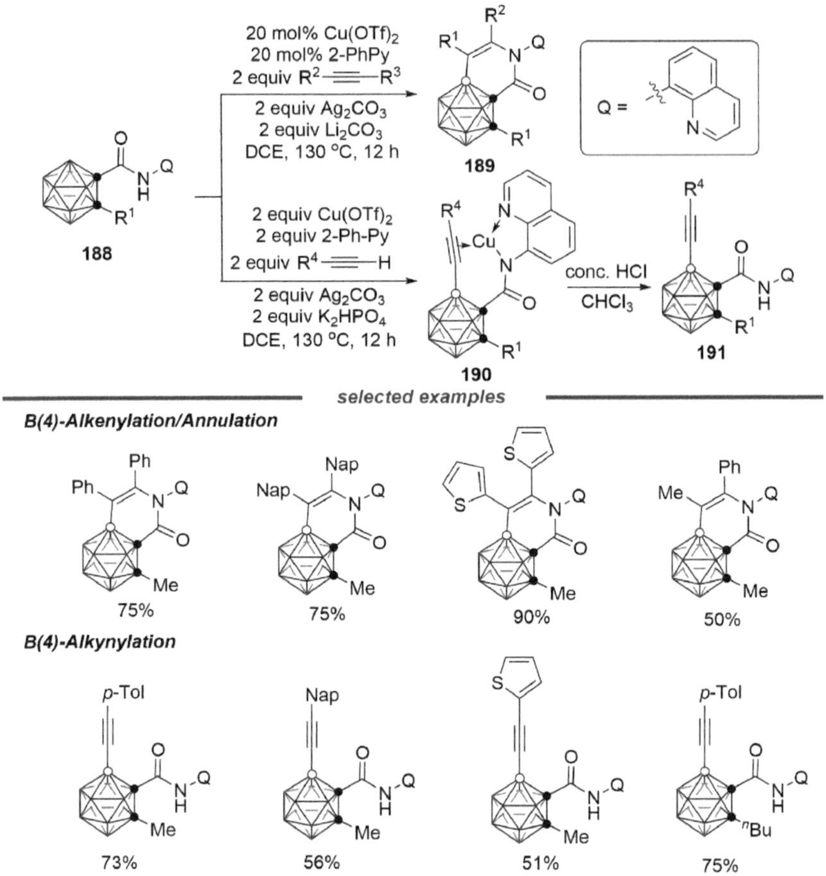

Scheme 41 Cu-catalyzed B(4)-alkenylation/annulation and B(4)-alkynylation.

generate a seven-membered metallacycle **195**. Reductive elimination of **195** yields the product **189** with the release of Cu(I) species. Oxidation of Cu(I) by silver salt regenerates Cu(II) species and complete the catalytic cycle. On the other hand, the Cu(II) catalyst and terminal alkynes initially undergo ligand exchange reaction to give the copper acetylide complex, which coordinates with **188** to afford complex **196**. Oxidation of **196** by silver salt, followed by electrophilic substitution gives an Cu(III) intermediate **198**, which undergoes subsequent reductive elimination and protonation to yield the product **191**.

4.2.2 Copper-catalyzed B(4,5)-disulfenylation

The Xie group described a copper-catalyzed regioselective B(4,5)—H disulfenylation of *o*-carboranyl amides **188** with various disulfides to give

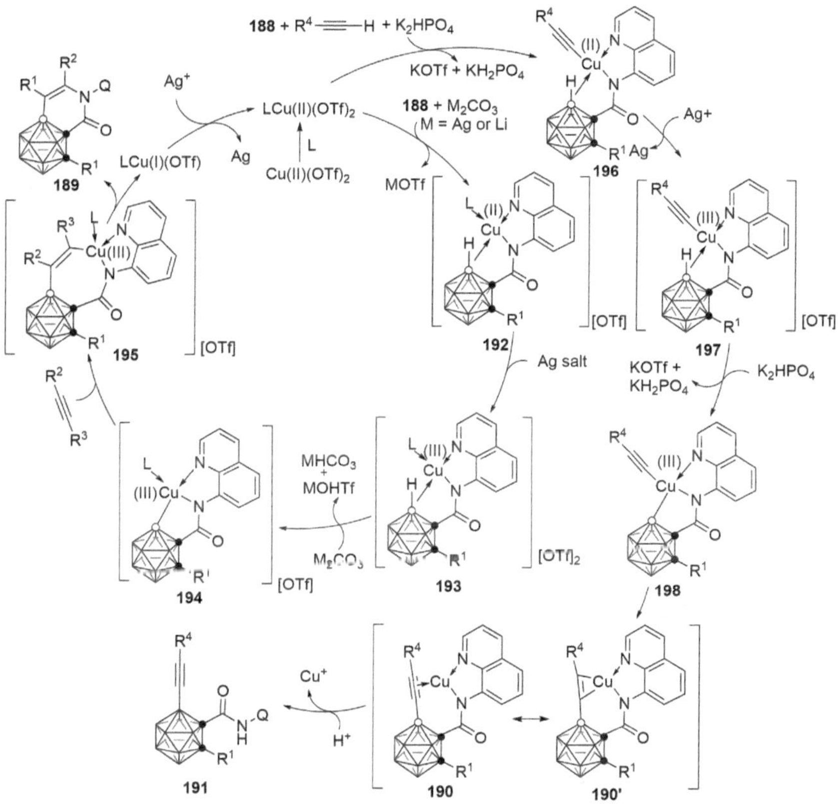

Scheme 42 Plausible reaction mechanism for Cu-catalyzed B(4)-alkenylation/annulation and B(4)-alkynylation.

a series of B(4,5)-disulfenylated *o*-carboranes **199** in good to excellent yield (Scheme 43) *(62)*. This disulfenylation reaction worked well for aryl disulfides such as phenyl, naphthalenyl, thiophenyl and furanyl disulfides, as well as alkyl disulfides. Both electron-donating and -withdrawing groups on the phenyl ring of aryl disulfides, such as alkyl, halo, trifluoromethyl, methoxyl and methylthio groups, were well compatible in this cross-coupling reaction. Interestingly, the bidentate directing group, 8-aminoquinoline amide, was *in-situ* removed after the reaction. A series of control experiments suggested that the disulfenylation involved a radical process. A plausible reaction mechanism involving a high valent Cu(III) species was proposed in Scheme 43. The catalytic cycle is initiated with bidentate chelation of **188** with Cu(II) species to generate complex **200**, which undergoes disproportionation or oxidation to give a Cu(III) complex **201**. Subsequent electrophilic substitution, reductive elimination and

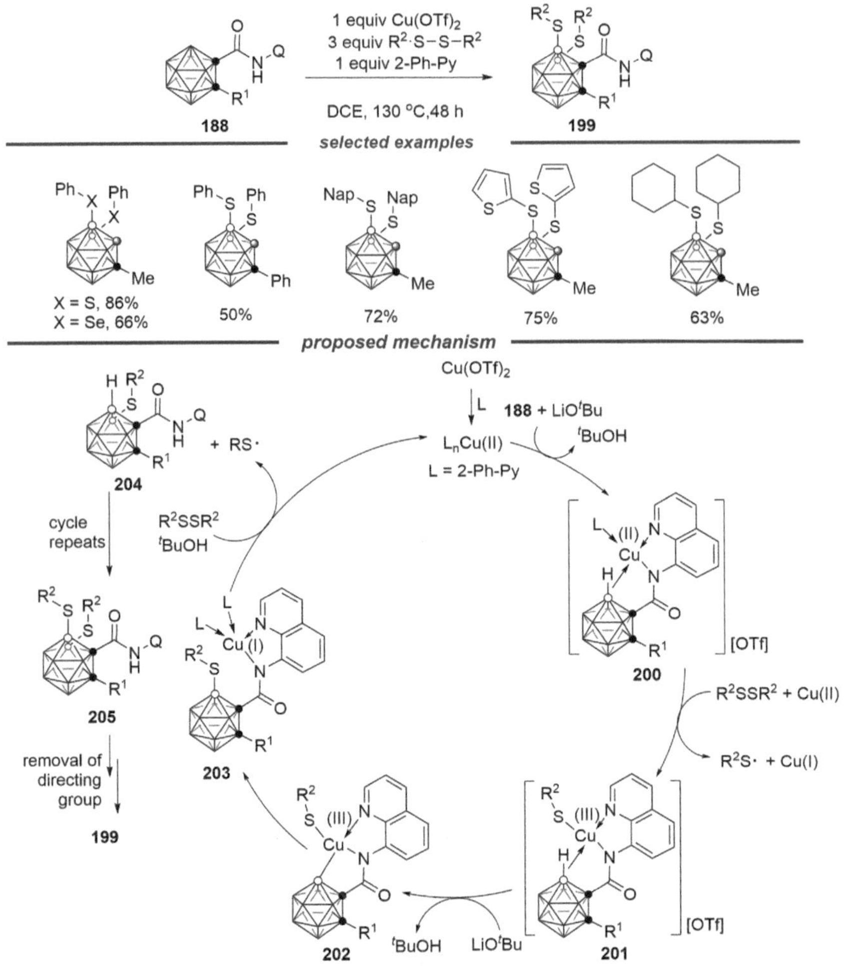

Scheme 43 Bidentate directing group guided Cu-catalyzed B(4,5)-disulfenylation of *o*-carboranes.

protonation with *t*-BuOH afford the mono-sulfenylated product **204**. Repeating of the catalytic cycle offers **205**, followed by removal of the directing group to yield the final products **199**.

4.2.3 Copper-catalyzed B(4,5)-dioxygenation

In comparison with thermal energy, electricity can serve as infinite electron pool and electron hole, and thus enable redox reactions without any external oxidants or reductants. Taking these points into consideration, the Xie group developed a strategy of merging base metal catalysis with electrochemistry for regioselective cage BH functionalization *(63)*. Subsequently, a

Cu-catalyzed electrochemical selective B—H oxygenation of *o*-carboranes was realized under constant electric current at room temperature, resulting in the facile synthesis of a series of B(4)-monooxygenated and B(4,5)-diphenolated o-carborane derivatives in an economic, sustainable and green manner.

In the presence of $Cu(OTf)_2$ catalyst, an 8-aminoquinoline-assisted B(4)-oxygenation of *o*-carboranyl amides **188** with lithium *tert*-butoxide took place smoothly under a constant current, giving access to B(4)-oxygenated *o*-carboranes **206** bearing alkyl and phenyl groups in high yields (Scheme 44). On the other hand, B(4,5)-dioxygenated *o*-carboranes **207** were prepared using lithium phenolates as coupling partners. Diverse

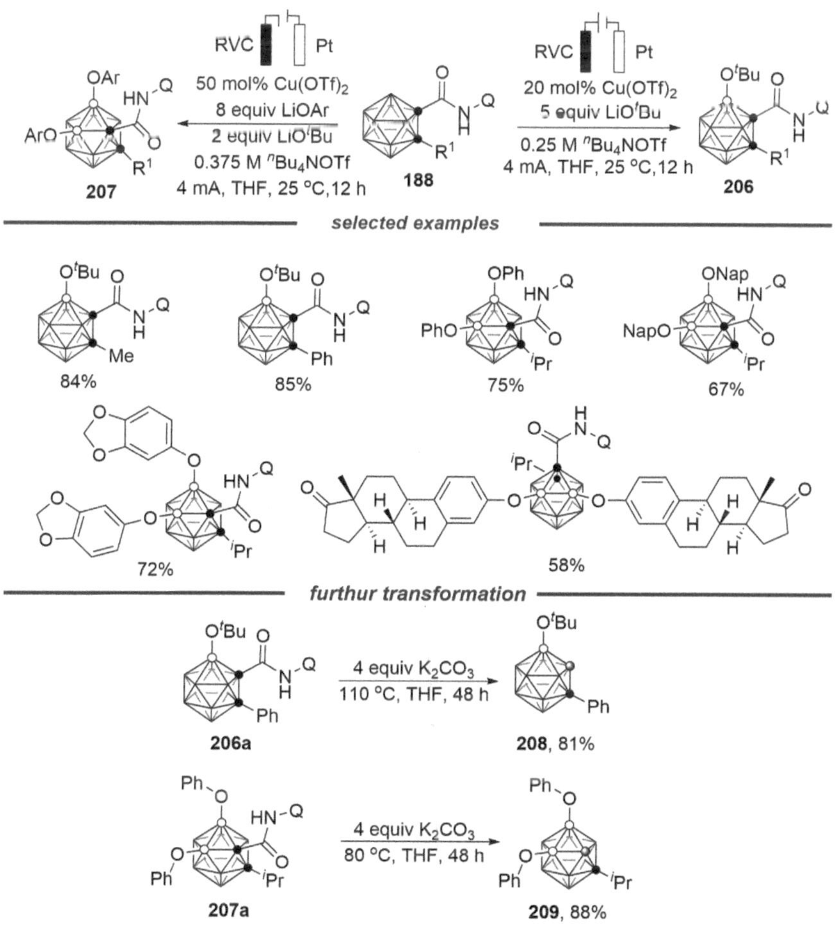

Scheme 44 Bidentate directing group guided Cu-catalyzed electrooxidative B(4,5)-oxygenation of *o*-carboranes.

functional groups ranging from halo, thiomethyl, methoxy to estrone were compatible. Furthermore, treatment of the resultant products with K_2CO_3 under heating resulted in efficient removal of the aminoquinoline auxiliary. It was speculated that the mono/di-selectivity was dependent on steric hindrance of the substituents. A plausible catalytic cycle was illustrated in Scheme 45. Coordination of *o*-carboranyl amides **188** with $Cu(OTf)_2$, followed by salt metathesis generates an intermediate **210**. Subsequent anodic oxidation of Cu(II) to Cu(III), electrophilic substitution of B(4)-H and reductive elimination proceed sequentially to afford B(4)-mono-oxygenated products **213**. Repeating of this cycle gives the B(4,5)-dioxygenated o-carboranes **217**.

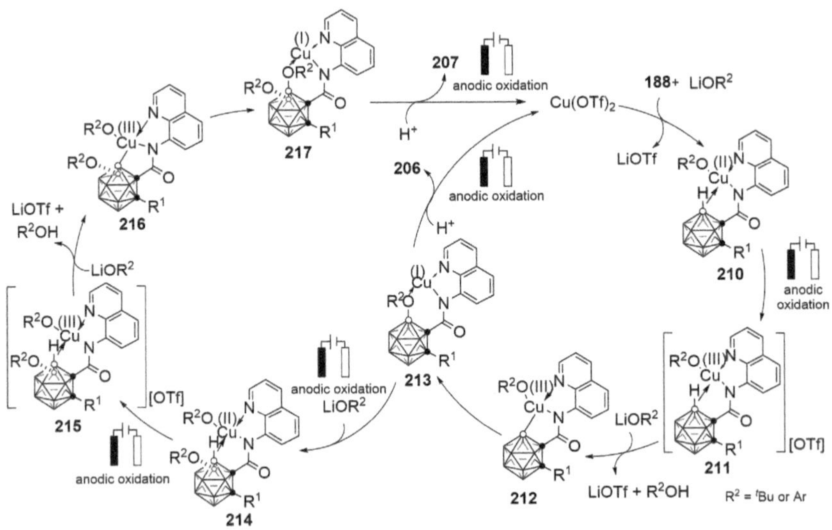

Scheme 45 Plausible reaction mechanism of Cu-catalyzed electrochemical B(4,5)-oxygenation.

4.2.4 Iron-catalyzed B(4)-arylation

The Xie group also realized a Fe-catalyzed intramolecular B(4)—H/C—H dehydrogenative coupling of *o*-carboranyl amides **188** containing an 8-aminoquinoline moiety which played dual roles as a directing group and a coupling partner (Scheme 46) *(64)*. This reaction demonstrated a good tolerance of diverse substituents, affording a class of C,B-substituted carborane-fused phenanthroline derivatives **218** in moderate to high yields. A proposed mechanism was shown in Scheme 46. Reaction of

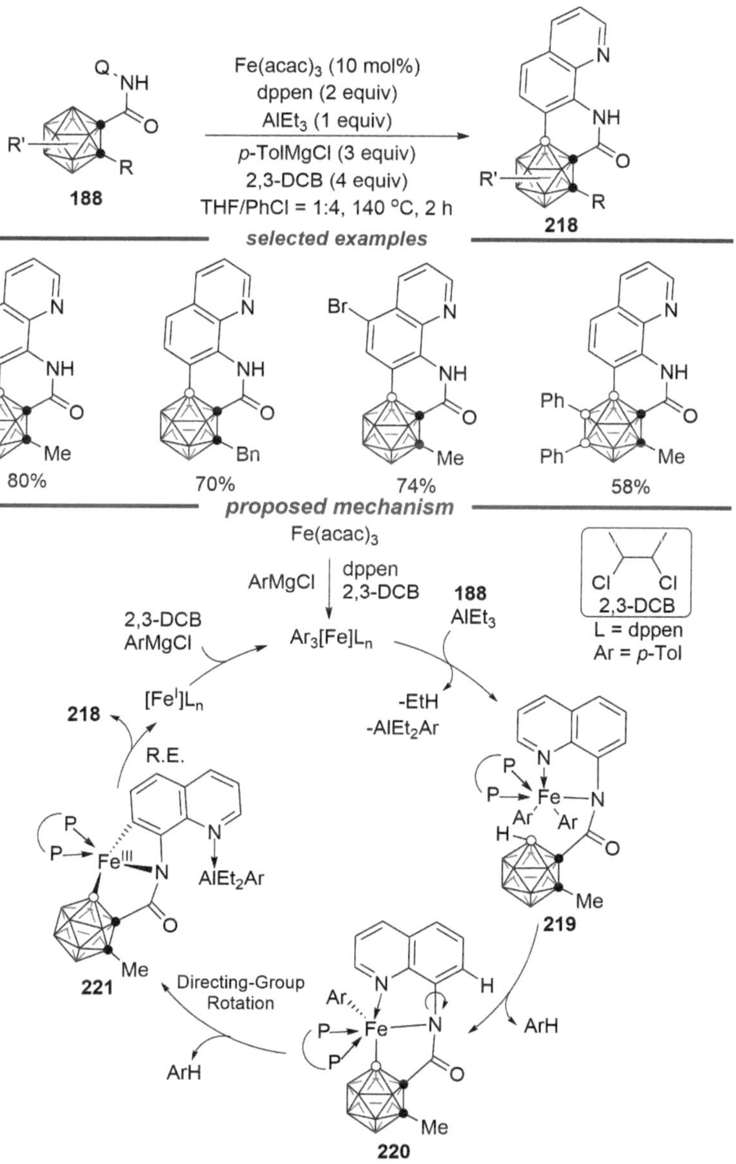

Scheme 46 Fe-catalyzed B(4)-arylation of o-carboranes.

$Fe(acac)_3$ with Grignard reagent generates an active aryl iron species, which chelates with bidentate **188** to afford intermediate **219**. Fe-promoted B(4)—H activation of **219** forms a bicyclic [B—Fe^{III}—N] species **220**, followed by rotation of the quinoline moiety to offer another [B—Fe^{III}—C] type of intermediate **221** due to the strong *trans*-effect of cage boron. Reductive elimination affords the annulation products **218**.

4.3 Cascade reaction for B(4,5) functionalization

A cascade reaction comprised of more than two consecutive reaction steps has shown superiority in atom economy and bond-forming efficiency. The Xie group reported an imine-directed Rh-catalyzed cascade cyclization of o-carboranyl N-arylimines **222** with α,β-unsaturated ketones, constructing three new C—C and one B—C bonds successively in a one-pot manner (Scheme 47) *(65)*. This cascade reaction provided a wide range of hardly accessible C,B-substituted carborane-fused N-polyheterocycles **223** in moderate to high yields with good tolerance of various functional groups. Substrates containing naphthalene and anthracene moieties worked well, producing C,B-substituted carborane-fused polycyclic aromatics, which might be potentially functional materials. In the absence of MesCOOH, the reaction of **222a** and methyl vinyl ketone at 90 °C for 0.5 h generated a C,B-substituted carborane-fused five-membered ring **227a**, which could be converted to the corresponding product under standard conditions. Accordingly, a plausible reaction mechanism involving an intermediate **227** was proposed in Scheme 47. The first catalytic cycle containing Rh-catalyzed B(4)—H activation, alkene insertion into the B—Rh bond and nucleophilic cyclization gives access to the intermediate **227**. Subsequent aryl C—H activation, nucleophilic cyclization towards C=O bond, dehydration and oxidative aromatization offer the final cyclization products **223**.

The Xie group also developed an Ir-catalyzed cascade reaction of carboranyl carboxylic acids **71** with benzamides via a sequence of regio-selective B—H/C—H and B—H/N—H dehydrogenative cross-couplings (Scheme 48) *(66)*. In this reaction, new B—C and B—N bonds in carboranoisoquinolinone derivatives **231** were constructed by successive activation of two B—H, one C—H and one N—H bonds. A wide array of benzamides bearing alkyl, alkoxy, halo and phenyl groups worked well. Under milder reaction conditions, only the B—C coupling took place to form α-carboranyl benzamides **232**, indicating the superiority of B—C coupling over the B—N one. Treatment of α-carboranyl benzamide **232a** with Cp*Ir(OAc)$_2$ gave access to an Ir(III) intermediate **233'**, which proved to be a key intermediate for B—H/N—H dehydrocoupling. A possible reaction mechanism involving Ir(III)—Ir(V)—Ir(III) catalytic cycle was

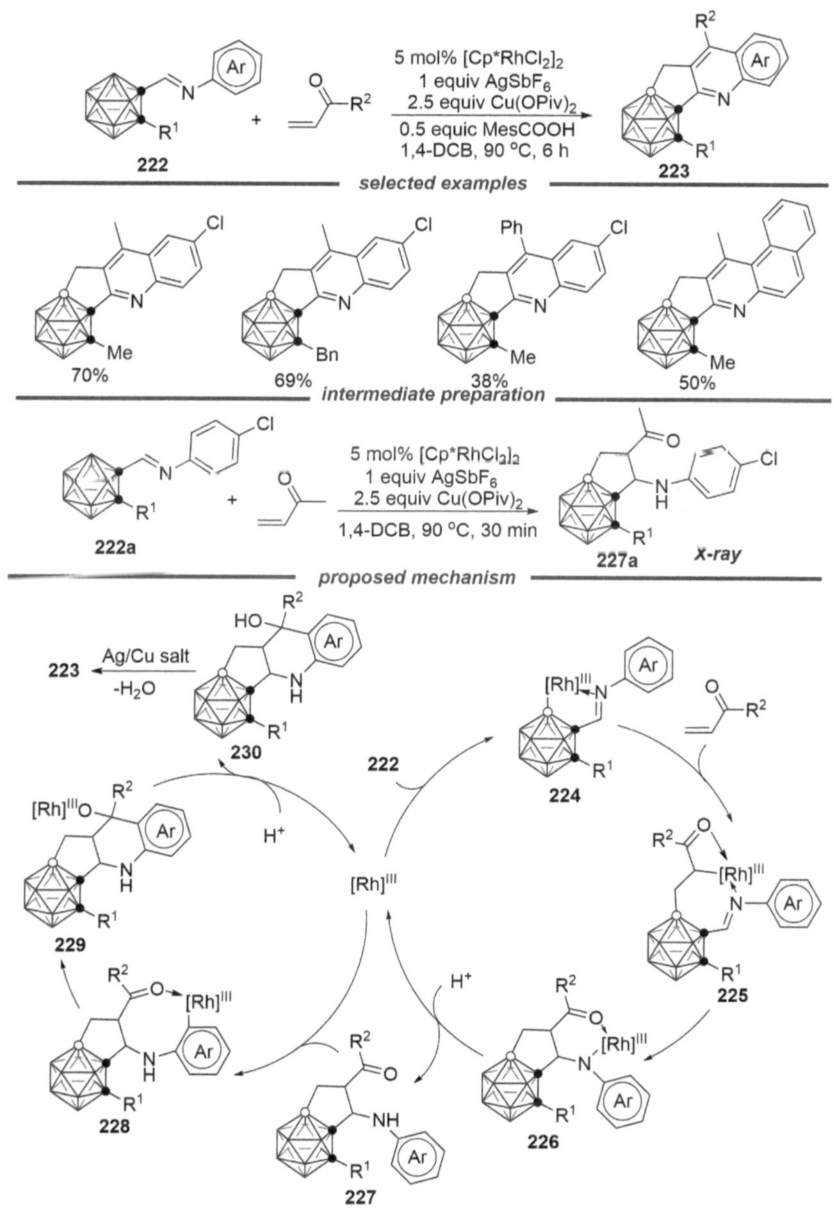

Scheme 47 Rh-catalyzed cascade cyclization.

described in Scheme 49. The first catalytic cycle was accounted for the formation of carboranyl benzamides **232**, involving a sequential of regioselective electrophilic B—H iridation, ligand exchange with benzamides, aryl C—H activation, and reductive elimination to build a

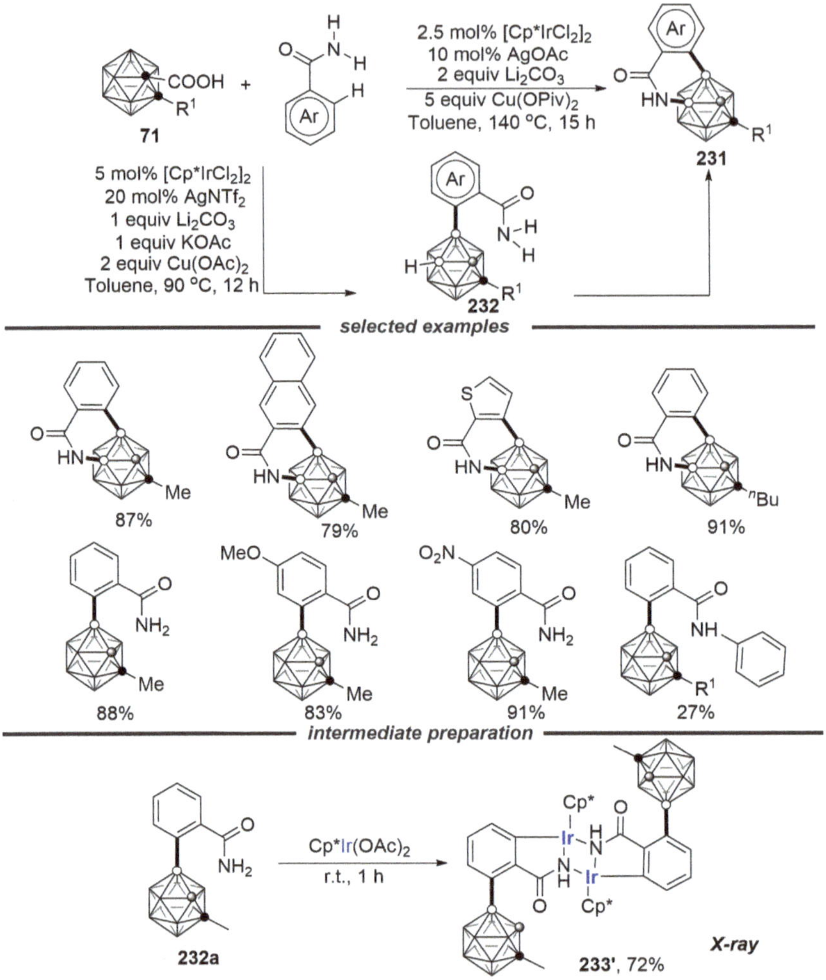

Scheme 48 Ir-catalyzed cascade dehydrogenative cross-coupling of *o*-carboranes with benzamides.

B—C bond. Then, reaction of **232** with Ir(III) species initiates the second catalytic cycle to give the cascade products **231** via an intermediate **233**.

Following the same strategy, an Ir-catalyzed cascade dehydrogenative B—H/C—H and B—H/O—H cross-coupling of carboranyl carboxylic acids **71** with benzoic acids was realized by the Xie group, leading to a batch of carboranocoumarins **242** in one-pot fashion (Scheme 50) *(67)*. Modulation of the reaction conditions made the reaction terminated after B—H/C—H dehydrogenative cross-coupling step, affording a class of α-carboranyl benzoic acids **243**. A variety of benzoic acids tethering with

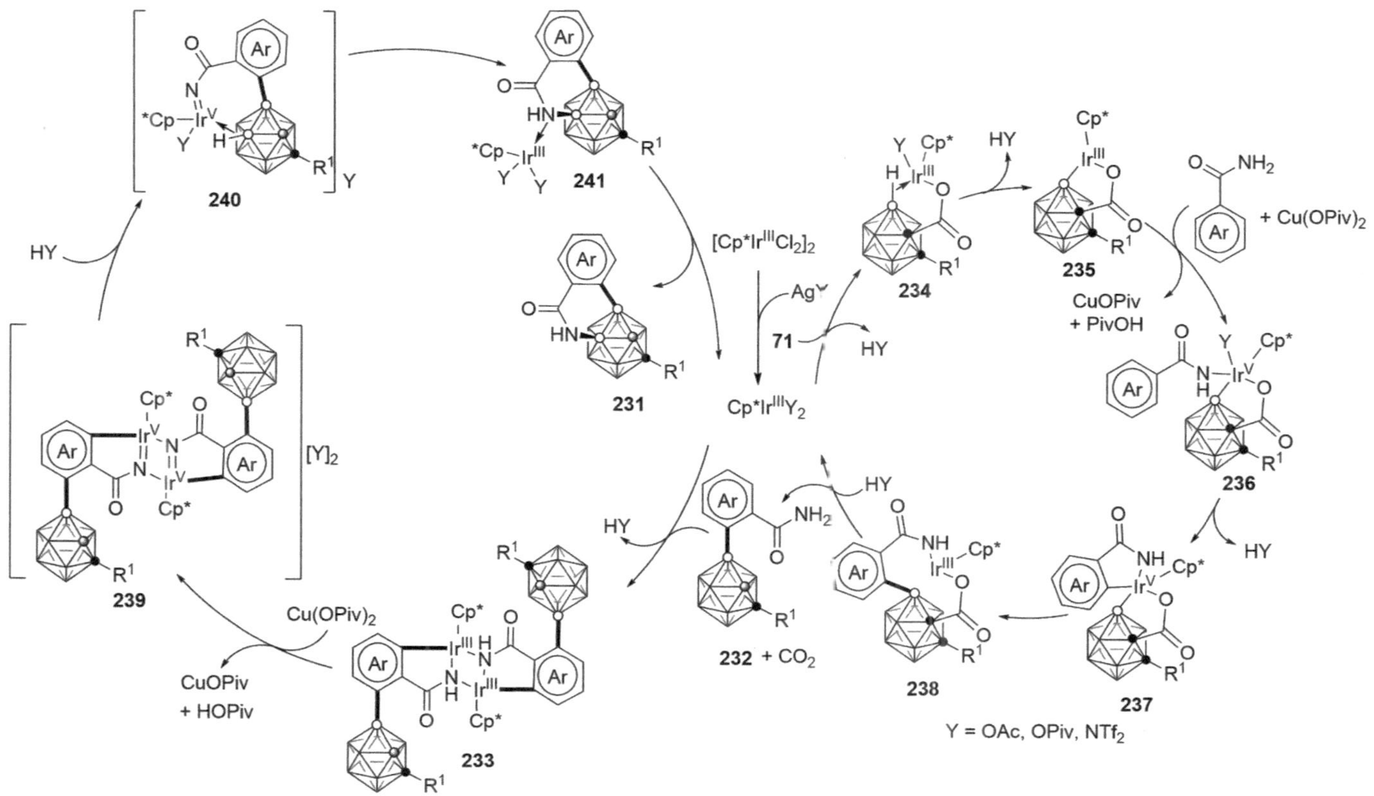

Scheme 49 Plausible reaction mechanism of Ir-catalyzed cascade dehydrocoupling with benzamides.

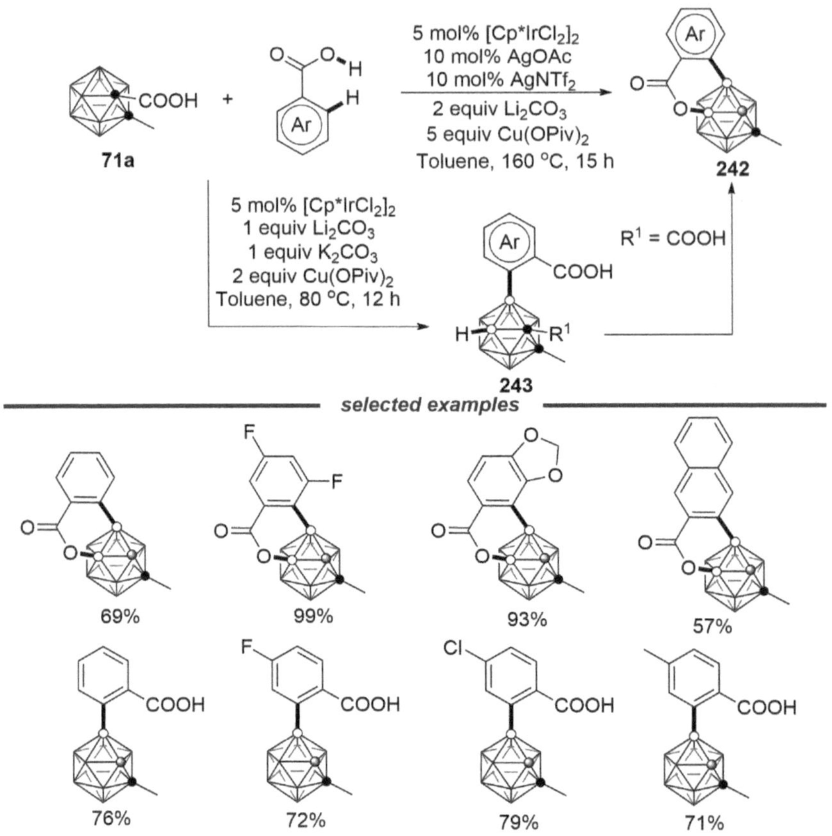

Scheme 50 Ir-catalyzed cascade dehydrogenative cross-coupling of *o*-carboranes with benzoic acids.

alkyl, aryl, nitro, ester, alkoxy and halo groups worked well in such a reaction. A plausible reaction pathway integrating two catalytic cycles was described in Scheme 51. The first B—C bond was built via a Ir(III)—Ir(V)—Ir(III) catalytic cycle similar to that illustrated in Scheme 49, forming an intermediate **249**. Subsequent cage C—COOH-enabled B—O formation took place through a successive sequence of electrophilic B(5)—H activation, oxidation of Ir(III) to Ir(V), ligand exchange and reductive elimination, generating the cyclized products **242** after a sequential of in-situ protonation and decarboxylation.

The Lee group reported a COOH-directed Ir-catalyzed cyclative indenylation and 1,3-dienylation of *o*-carboranes at B(4)-vertex from the sequential reactions of carboranyl carboxylic acids **71** with propargyl alcohols under different conditions (Scheme 52) *(68)*. When the reaction was conducted in TFE (2,2,2-trifluoroethanol) and toluene, previously

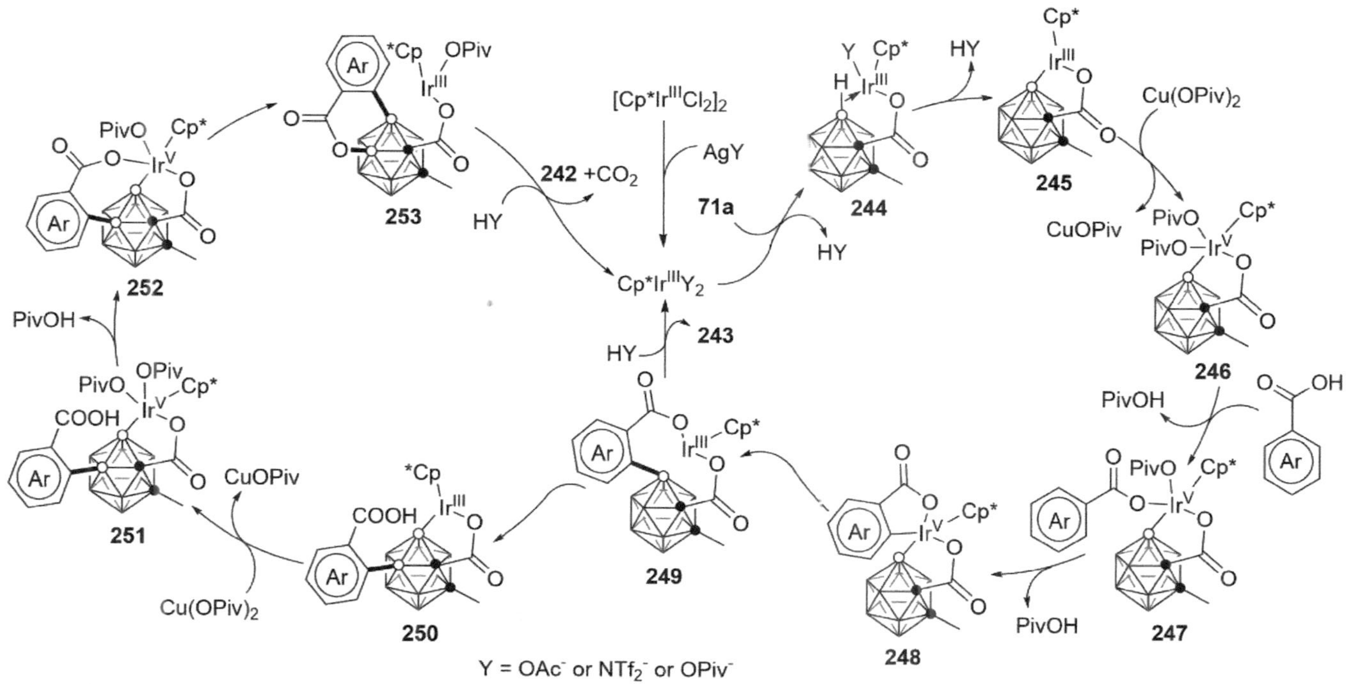

Scheme 51 Plausible reaction mechanism of Ir-catalyzed cascade dehydrogenative cross-coupling.

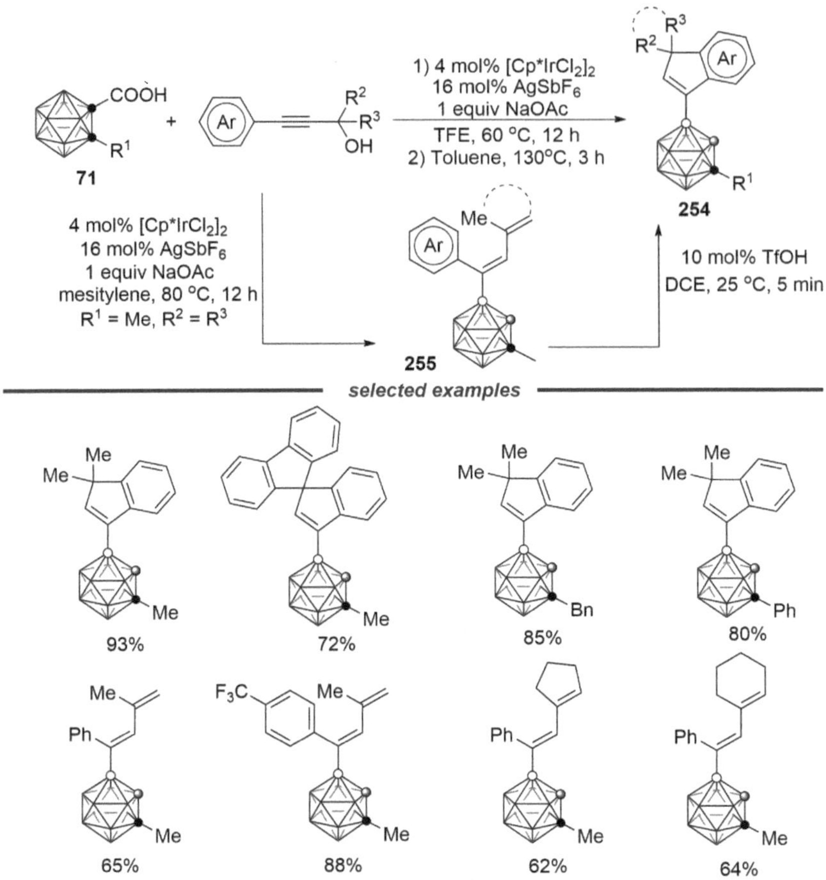

Scheme 52 Ir-catalyzed B(4)-indenylation and dienylation of *o*-carboranes with propargyl alcohols.

unaccessible B(4)-indenylated *o*-carboranes **254** were obtained in high yields with good tolerant of adamantyl, halo, ester and trifluoromethyl groups. In contrast, the reaction in mesitylene produced B(4)-dienylated *o*-carborane derivatives **255**, which could undergo TfOH-promoted cyclization to give cyclative products **254**. Several control experiments were conducted and a reaction mechanism was accordingly proposed in Scheme 53. Initially, a monomeric Ir(III) species is generated as active catalyst to form B(4)—Ir bond via electrophilic substitution. Subsequent alkyne insertion affords a seven-membered iridacycle **258**, followed by the release of OH^- group to form a carbocation intermediate **259**. Sequential intramolecular aromatic substitution, aromatization, protonation and decarboxylation give the cyclative indenylated products **254**. Alternatively, deprotonation of **259** results in the formation of the intermediate **262**

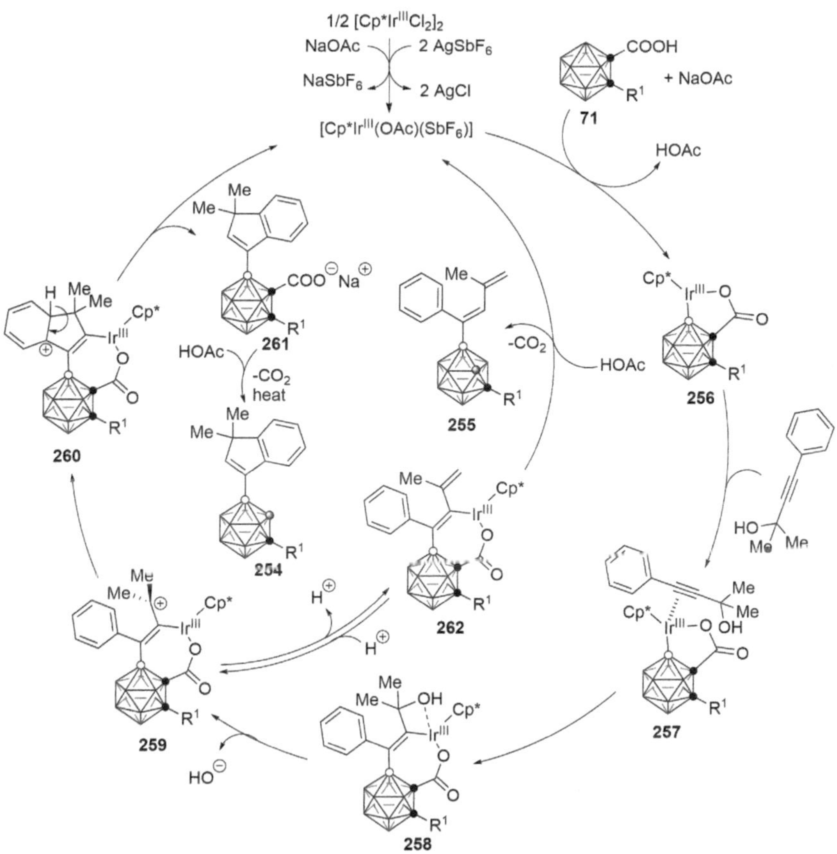

Scheme 53 Plausible reaction mechanism of Ir-catalyzed B(4)-indenylation and dienylation with propargyl alcohols.

containing dienyl moiety. Proto-deiridation and decarboxylation afford the dienylated *o*-carboranes **255**.

5. B(3,4,5,6)—H functionalization

The Xie group described a Pd-catalyzed selective B(3,4)-bifunctionalization of 3-iodo-*o*-carboranes **263** via a successive sequence of oxidative addition of B(3)-I to Pd(0), alkyne insertion into the B(3)-Pd bond, migration of Pd(II) to B(4)-vertex and cross-coupling with various coupling partners (Scheme 54) *(69)*. A series of 3,4-disubstituted *o*-carboranes **264** bearing alkyl, alkenyl, alkynyl, aryl, cyano and amido groups were prepared with excellent yields in a one-pot manner. It was proposed that the Pd-migration process might play a crucial role in such a bifunctionalization. To verify this hypothesis, B(4,5,7,8,9,10,11,12)-octadeuterated

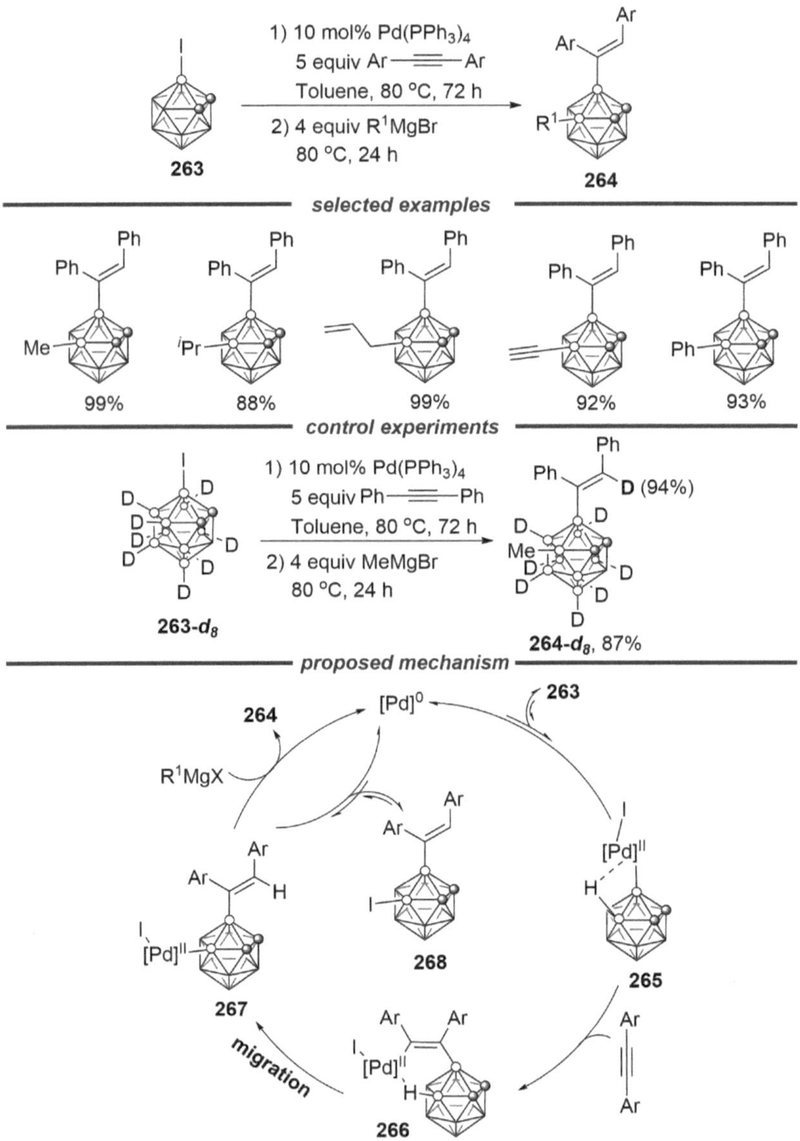

Scheme 54 Pd-Catalyzed B(3,4)-difunctionalization of 3-I-*o*-carboranes.

3-I-*o*-carborane **263**-d_8 was treated subsequently with PhC≡CPh and MeMgBr under standard reaction conditions, resulting in **264**-d_8 with 94% alkenyldeuteration. These results evidently confirmed the involvement of a concerted Pd/D-migration from **266** to **267** in the catalytic cycle illustrated in Scheme 54.

The Ackermann group disclosed an amide-assisted Pd-catalyzed B(3,4)-diarylation of carboranyl amides **269** (Scheme 55) *(70)*. A wide range

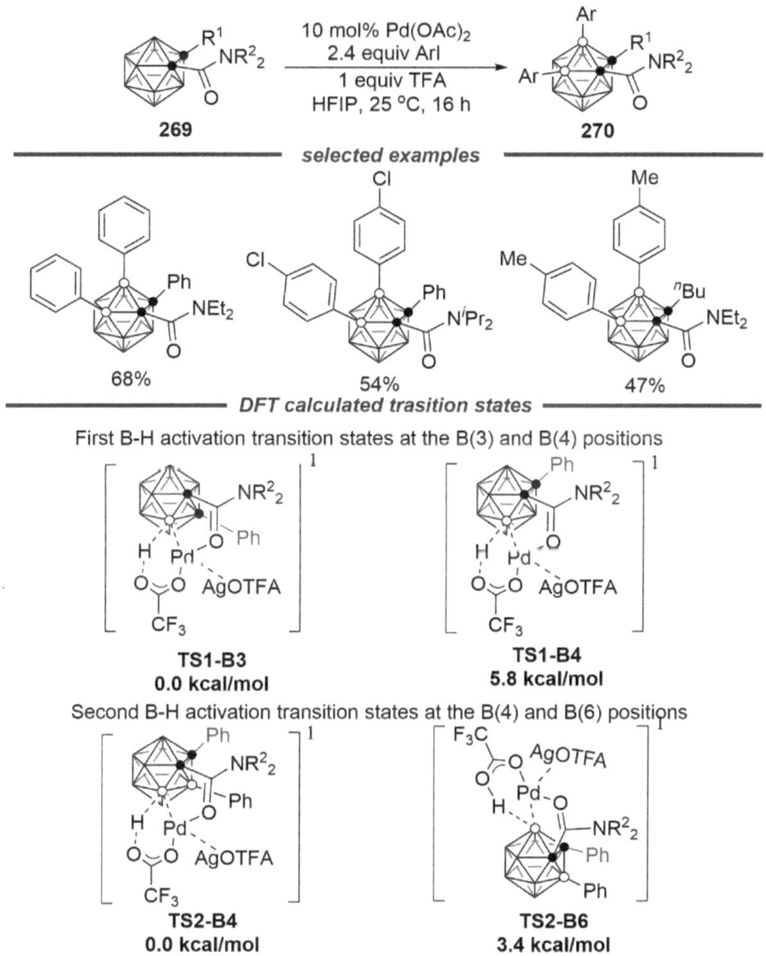

Scheme 55 Amide group guided Pd-catalyzed B(3,4)-diarylation of *o*-carboranes.

of aryl iodides substituted with alkyl, halo and acyl groups served as compatible coupling partners to afford B(3,4)-diarylated *o*-carboranes **270** in acceptable yields. Lowering the amount of substrate, catalyst and oxidant resulted in B(3)-monoarylated *o*-carboranes in fair yields. DFT calculations were carried out to figure out the origin of such unusual B(3,4)-selectivity. For the first B-H activation, B(3)-arylation was more favorable than B(4)-arylation by 5.8 kcal/mol. A catalytic cycle involving B—H palladation, oxidative addition and reductive elimination was proposed, which is similar to that shown in Scheme 22.

The Xie group reported a rhodium catalyzed one-pot B(3,4,5)-trifunctionalization of 1-carboxy-2-methyl-*o*-carboarne **71a** through consecutive substitution at B(5)-, B(4)-, B(3)-vertices (Scheme 56) *(71)*.

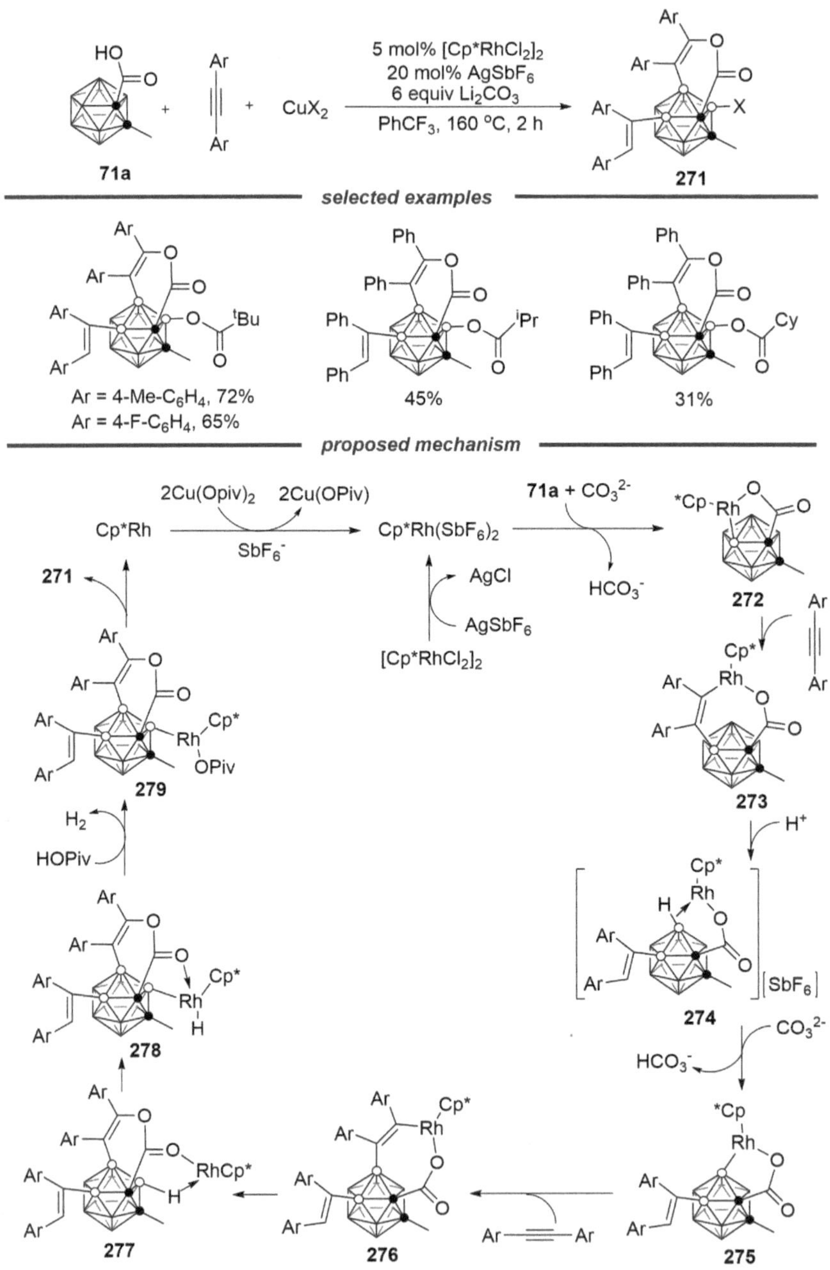

Scheme 56 Ru-catalyzed B(3,4,5)-trifunctionalization of *o*-carboranes.

With the assistance of carboxy directing group and Rh catalyst, **71a** reacted with ArC≡CAr and $Cu(OCOR)_2$ to furnish a batch of B(3,4,5)-trisubstituted *o*-carboranes **271** in moderate to good yields. It was supported by control experiments that B(3)-acyloxylation might proceed after the B(4,5)—H functionalization. Thus, a plausible mechanism was exhibited in Scheme 56. The catalytic cycle is initiated by B(5)-alkenylation, followed by B(4)-alkyne annulation and B(3)-acyloxylation to afford the trisubstituted products **271**.

The Yan group developed a Pd-catalyzed B(3,4,5,6)-tetraarylation of carboranyl aldehyde **107a** directed by in situ generated transient imine group, leading to B(3,4,5,6)-tetraarylated *o*-carboranes **280** in moderate to good yields (Scheme 57) *(72)*. When glycine was employed to form imine auxiliary, electron-donating substituents including alkyl, —OMe and —OTBS groups worked well, whereas the electron-withdrawing groups suffered from a problem of product purification by chromatography. It was proposed that this multi-arylation reaction shared a similar mechanism demonstrated in Scheme 23.

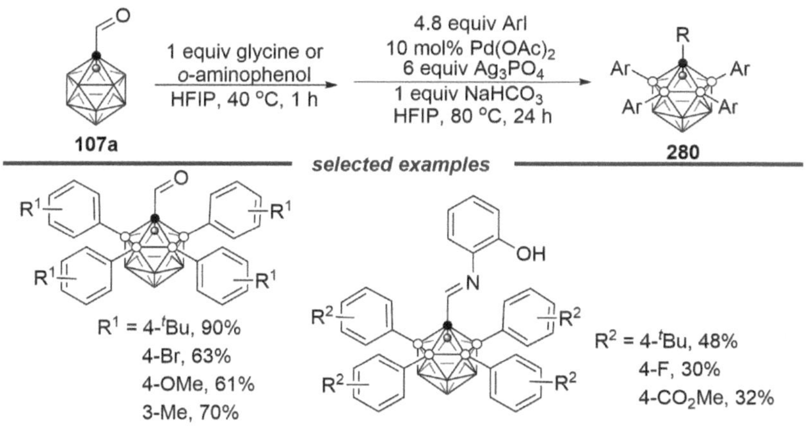

Scheme 57 Pd-catalyzed B(3,4,5,6)-tetra-arylation of *o*-carboranes.

6. B(3,4,6,11)/B(4,5,7,11)—H functionalization

Inspired by the Pd-catalyzed selective bifunctionalization of 3-I-*o*-carborane via Pd migration (Scheme 54), Pd-catalyzed regioselective one-pot tetrafunctionalization of 3,6- and 4,7-I_2-*o*-carboranes (**281** and **283**) via a similar Pd migration pathway was realized in the Xie group

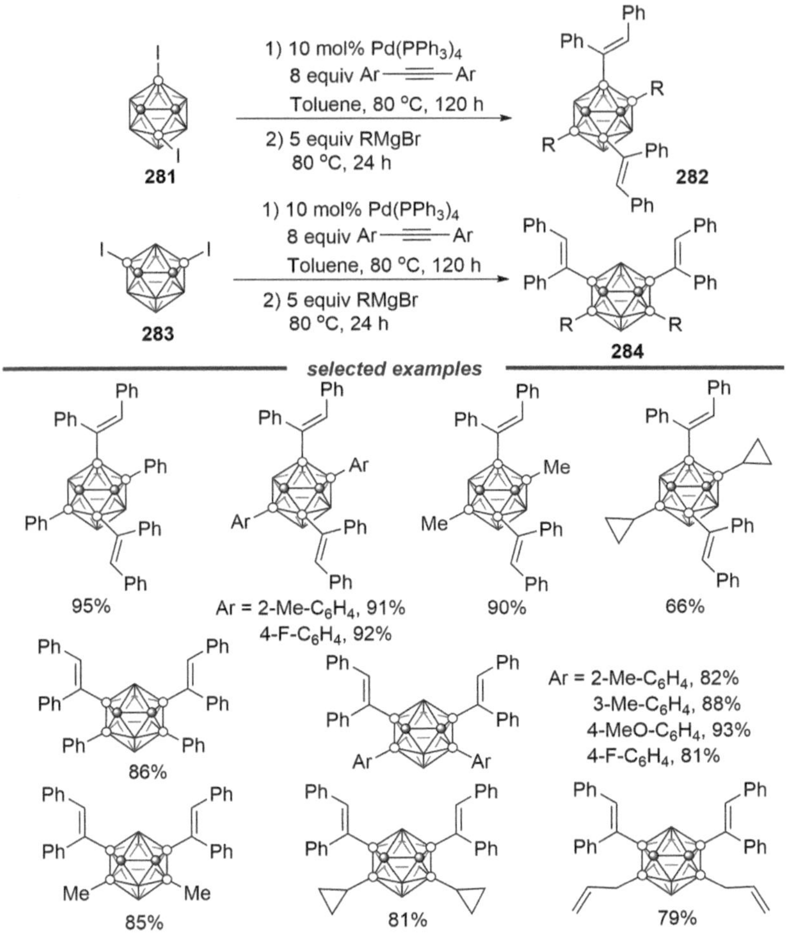

Scheme 58 Pd-catalyzed B(3,4,6,11)/B(4,5,7,11)-tetrafunctionalization of 3,6- or 4,7-I_2-*o*-carborane.

(Scheme 58) *(73)*. It was noted that no distal B(8/9)—H substituted products were detected, indicating that the Pd-migration was highly regioselective. Aryl, methyl and cyclopropyl Grignard reagents worked well to offer a class of tetrafunctionalized *o*-carboranes **282** and **284** in high yields. No obvious electronic effects were observed. This reaction proceeded in a similar mechanism described in Scheme 54.

7. Enantioselective cage B—H functionalization

Stereochemistry has been one of the most active research areas in modern chemistry. Despite a great variety of chiral structures incorporating

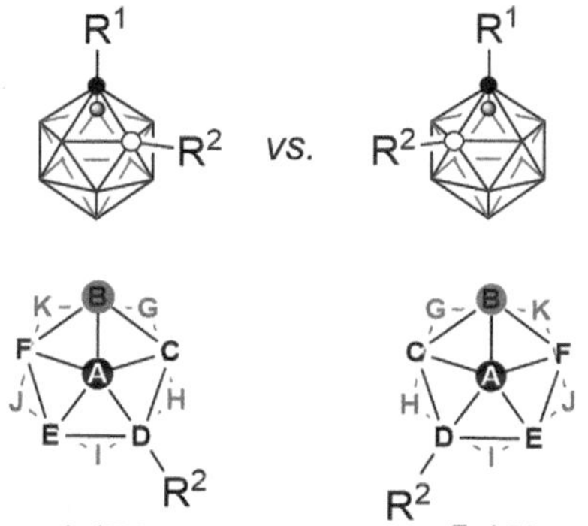

4-(*R*) 5-(*S*)

Fig. 3 Inherent cage B(4/5) chirality of *o*-carboranes.

central, axial, planar, and helical chirality achieved by catalytic asymmetric synthesis, to obtain the inherent chirality of three-dimensional cage compounds such as carboranes is extremely challenging and the currently available methods to access such an enantioenriched skeleton are still rather limited (Fig. 3) *(1,18)*. In 2018, the Xie group reported a proof-of-principle investigation on Pd-catalyzed enantioselective intramolecular B(5)-arylation of *o*-carboranes **285** with the assistance of a chiral phosphine ligand (Scheme 59) *(74)*. This work sets an excellent example for exploring the potential of asymmetric synthesis beyond conventional organic chemistry into the chemistry of chiral-at-cage *o*-carboranes.

In the presence of $Pd(OAc)_2$ catalyst and chiral (*R*)-BI-DIME ligand, substrates **285** underwent an enantioselective intramolecular B(5)-arylation, giving a broad range of B(5)-arylated *o*-carboranes (*S*)-**286** in good yields with very high ee values. Under the same reaction condition, replacement of (*R*)-BI-DIME with (*S*)-BI-DIME ligand afforded the corresponding B(4)-arylated *o*-carboranes (*R*)-**286**. Their absolute structures were determined by single crystal X-ray analyses. The resultant products were proved to be a pair of enantiomers by the mirror-like circular dichroism (CD) spectra. It was believed that the chiral ligand played an indispensable role in this enantioselective reaction. To understand the origin of such a high enantioselectivity, DFT calculations were performed. The results showed that the transition state **TS-*S*** is more energetically preferential than **TS-*R*** by 2.0 kcal/mol. The reason might be ascribed to the large steric hindrance between *t*-butyl group of (*R*)-BI-DIME and the phenyl group of the substrates in the transition state **TS-*R***. The reaction mechanism is similar to the one proposed in Pd-catalyzed intramolecular B-C coupling reaction *(75)*.

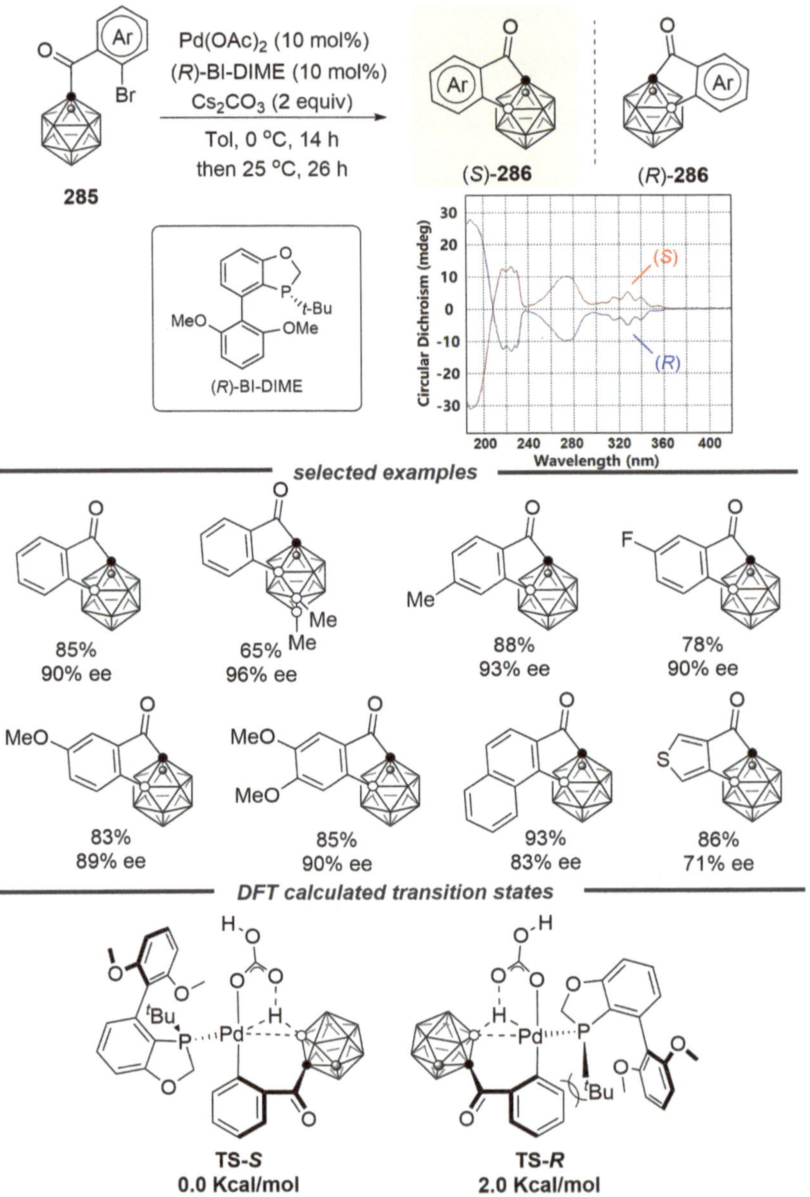

Scheme 59 Pd-catalyzed intramolecular asymmetric B(5)-arylation.

Following the similar protocol, an intermolecular asymmetric B(5)-alkenylation of 1-acylamino-*o*-carboranes **287** was realized via iridium catalysis with the aid of acylamino auxiliary and chiral phosphoramidite ligand, giving (*S*)-B(5)-alkenyl-*o*-carboranes **288** in excellent yields and

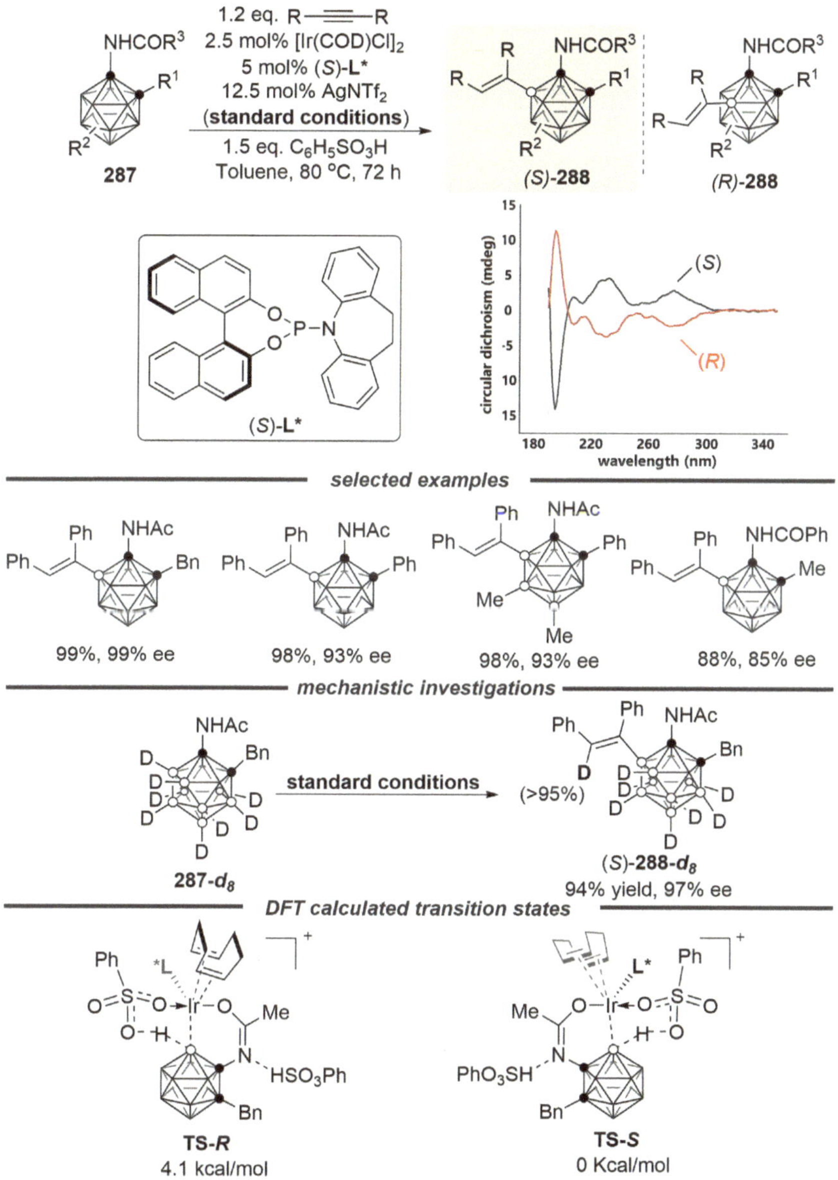

Scheme 60 Ir-catalyzed intermolecular asymmetric B(5)-alkenylation.

enantioselectivity (Scheme 60) *(76)*. The mirror-like circular dichroism (CD) spectra suggested that (*R*)-B(4)- and (*S*)-B(5)-alkenyl-*o*-carboranes **288** were a pair of enantiomers, whose absolute structures were confirmed by single crystal X-ray analyses. Deuterium labelling experiment resulted in the formation of (*S*)-**288**-d_8 with >95% D-incorporation in excellent yield and ee value. DFT calculations indicated that the transition state **TS-*S*** was energetically more stable than **TS-*R*** by 4.1 kcal/mol, originating from the additional C—H...π interactions in **TS-*S***. A plausible mechanism was accordingly proposed in Scheme 61. In the presence of chiral phosphoramidite ligand, $[Ir(COD)Cl]_2$ is oxidized by silver salts to generate a more active Ir(III) species. Subsequent electrophilic B(4/5)—H activation (dependent on the configuration of the chiral ligands) forms a six-membered iridacycle **290**, in which the acylamino directing group coordinates to the Ir(III) in an iminium form. Insertion of alkyne into the B—Ir bond gives an eight-membered iridacycle **291**, followed by protonation to afford the final products **288**.

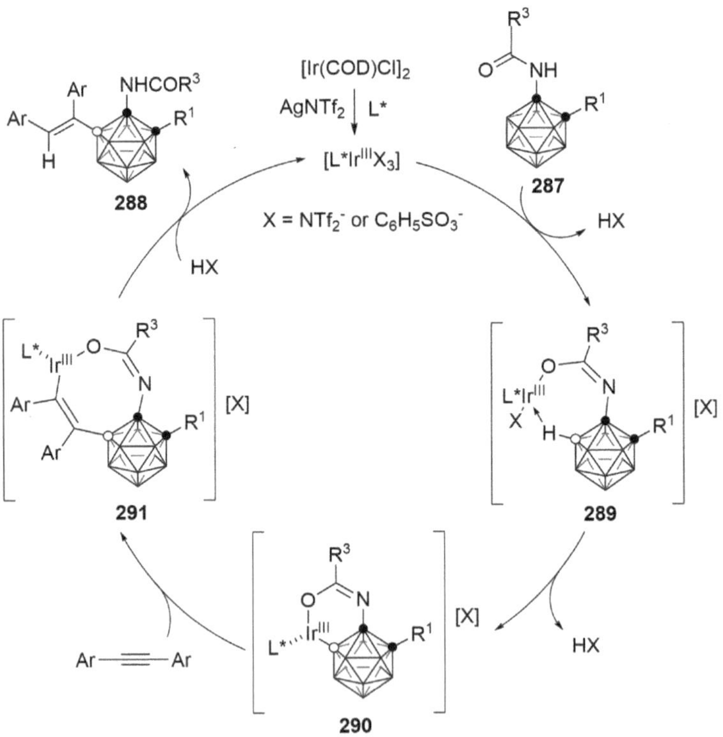

Scheme 61 Plausible reaction mechanism of Ir-catalyzed intermolecular asymmetric B(5)-alkenylation.

8. Conclusions and perspectives

This chapter summarizes recent advances in transition metal catalyzed regioselective and enantioselective functionalization of cage B—H bonds in *o*-carboranes. According to the different electron density of B—H vertices in *o*-carborane, a general strategy for selective cage BH functionalization has been proposed by the Xie group, which contains three principles: (1) electron-rich transition metal catalysts are good for the activation of the most electron-deficient B(3,6)—H bonds (connected to both cage CH vertices), (2) electron-deficient transition metal catalysts are good for the activation of relatively electron-rich B(8,9,10,12)—H bonds (with no bonding to either cage CH vertices), and (3) directing group assisted transition metal catalysis is appropriate for the activation of B(4,5,7,11)—H bonds (connected to only one cage CH vertex) whose vertex charge lays between that of B(3,6) and B(8,9,10,12) ones. Successful applications of such a strategy have resulted in the development of a series of methods for selective catalytic cage B—H functionalization, leading to the preparation of a large variety of cage B-functionalized carborane derivatives in a regioselective and catalytic manner.

With a view of atom-economy, step-economy and environmental friendliness, new synthetic methodologies have been developed, including dehydrogenative cross-couplings, merging base metal catalysis with electrochemistry, cascade catalysis, one-pot multi-functionalization of *o*-carboranes, resulting in a large pool of highly functionalized *o*-carborane derivatives with potential applications in various realms.

Though significant progress has been made in transition metal catalyzed selective cage B—H functionalization of *o*-carboranes over the past decade, this research area remains as a young developing realm in comparison to the well-developed C—H functionalization of organic molecules. Many issues need to be addressed in the future. For example, higher catalyst loadings are generally required in comparison to those for C—H modification of carbon-based substrates, probably due to the steric bulkiness of *o*-carboranes. Thus, more efficient catalyst systems are desired. Regioselective functionalization of cage B(8,10)—H bonds remains as a big challenge. Moreover, regioselective *o*-carborane functionalization via base-metal catalysis or even transition-metal-free methods need to be further developed. The development of new synthetic methodologies incorporating transition metal catalysis with photochemistry, or electrochemistry, or radical chemistry, or organic catalysis needs to be further explored. Enantioselective activation of cage B—H bonds via transition metal catalysis is another developing field of great importance. Multicomponent cascade reactions involving B—H or C—H vertices enabled by transition metal catalysts are yet to be developed. The applications of the resultant functionalized *o*-carboranes are waiting for exploration in the future.

Acknowledgment

Financial supports from Research Grants Council of Hong Kong Special Administration Region (Project Nos. 14305918, 14306519, 14305420 and SRFS2021-4S05) are gratefully acknowledged.

References

1. Grimes, R. N., Ed. *Carboranes*; 3rd ed.; Elsevier: Oxford, UK, 2016.
2. For selected reviews, see: (a) Hawthorne, M. F.; Maderna, A. *Chem. Rev.* **1999**, *99*, 3421 – 3434; (b) Scholz, M.; Hey-Hawkins, E. *Chem. Rev.* **2011**, *111*, 7035 – 7062; (c) Stockmann, P.; Gozzi, M.; Kuhnert, R.; Sarosi, M. B.; Hey-Hawkins, E. *Chem. Soc. Rev.* **2019**, *48*, 3497 – 3512; (d) Issa, F.; Kassiou, M.; Rendina, L. M. *Chem. Rev.* **2011**, *111*, 5701–5722.
3. For selected reviews, see: (a) Núñez, R.; Tarrés, M.; Ferrer-Ugalde, A.; de Biani, F. F.; Teixidor, F. *Chem. Rev.* **2016**, *116*, 14307 – 14378; (b) Fisher, S. P.; Tomich, A. W.; Lovera, S. O.; Kleinsasser, J. F.; Guo, J.; Asay, M. J.; Nelson, H. M.; Lavallo, V. *Chem. Rev.* **2019**, *119*, 8262 – 8290; (c) Mukherjee, S.; Thilagar, P. *Chem. Commun.* **2016**, *52*, 1070–1093.
4. For selected reviews, see: (a) Xie, Z. *Coord. Chem. Rev.* **2002**, *231*, 23 – 46; (b) Yao, Z.-J.; Jin, G.-X. *Coord. Chem. Rev.* **2013**, *257*, 2522 – 2535; (c) Fisher, S. P.; Tomich, A. W.; Guo, J.; Lavallo, V. *Chem. Commun.* **2019**, *55*, 1684 – 1701; (d) Hosmane, N. S.; Maguire, J. A. Comprehensive Organometallic Chemistry III; Crabtree, R. H., Mingos, D. M. P., Eds.; Elsevier: Oxford, 2007; Vol. 3, Chapter 5.
5. For selected reviews, see: (a) Qiu, Z.; Ren, S.; Xie, Z. *Acc. Chem. Res.* **2011**, *44*, 299 – 309; (b) Zhao, D.; Xie, Z. *Coord. Chem. Rev.* **2016**, *314*, 14 – 33; (c) Qiu, Z.; Xie, Z. *Chem. Soc. Rev.*, **2022**, *51*, 3164–3180.
6. For reviews, see: (a) Qiu, Z. *Tetrahedron Lett.* **2015**, *56*, 963 – 971; (b) Olid, D.; Núñez, R.; Viñas, C.; Teixidor, F. *Chem. Soc. Rev.* **2013**, *42*, 3318 – 3336; (c) Dziedzic, R. M.; Spokoyny, A. M. *Chem. Commun.* **2019**, *55*, 430 – 442.
7. Teixidor, F.; Barberà, G.; Vaca, A.; Kivekäs, R.; Sillanpää, R.; Oliva, J.; Viñas, C. *J. Am. Chem. Soc.* **2005**, *127*, 10158–10159.
8. (a) Potenza, J. A.; Lipscomb, W. N.; Vickers, G. D.; Schroeder, H. *J. Am. Chem. Soc.* **1966**, *88*, 628–629; (b) Zheng, Z.; Knobler, C. B.; Mortimer, M. D.; Kong, G.; Hawthorne, M. F. *Inorg. Chem.* **1996**, *35*, 1235–1243.
9. (a) Herzog, A.; Maderna, A.; Harakas, G. N.; Knobler, C. B.; Hawthorne, M. F. *Chem. Eur. J.* **1999**, *5*, 1212–1217; (b) Teixidor, F.; Barberà, G.; Viñas, C.; Sillanpää, R.; Kivekäs, R. *Inorg. Chem.* **2006**, *45*, 3496–3498.
10. Hawthorne, M. F.; Wegner, P. A. *J. Am. Chem. Soc.* **1968**, *90*, 896–901.
11. Ramachandran, B. M.; Knobler, C. B.; Hawthorne, M. F. *Inorg. Chem.* **2006**, *45*, 336–340.
12. Bondarev, O.; Sevryugina, Y. V.; Jalisatgi, S. S.; Hawthorne, M. F. *Inorg. Chem.* **2012**, *51*, 9935–9942.

13. Safronov, A. V.; Sevryugina, Y. V.; Jalisatgi, S. S.; Kennedy, R. D.; Barnes, C. L.; Hawthorne, M. F. *Inorg. Chem.* **2012**, *51*, 2629–2637.
14. (a) Yu, W.-B.; Cui, P.-F.; Gao, W.-X.; Jin, G.-X. *Coord. Chem. Rev.* **2017**, *350*, 300–319; (b) Zhang, X.; Yan, H. *Coord. Chem. Rev.* **2019**, *378*, 466–482.
15. (a) Zhang, R.; Zhu, L.; Liu, G.; Dai, H.; Lu, Z.; Zhao, J.; Yan, H. *J. Am. Chem. Soc.* **2012**, *134*, 10341–10344; (b) Wang, Z.; Ye, H.; Li, Y.; Li, Y.; Yan, H. *J. Am. Chem. Soc.* **2013**, *135*, 11289–11298.
16. Hoel, E. L.; Talebinasab-Savari, M.; Hawthorne, M. F. *J. Am. Chem. Soc.* **1977**, *99*, 4356–4367.
17. Mirabelli, M. G. L.; Sneddon, L. G. *J. Am. Chem. Soc.* **1988**, *110*, 449–453.
18. For reviews, see: (a) Quan, Y.; Qiu, Z.; Xie, Z. *Chem. Eur. J.* **2018**, *24*, 2795 – 2805; (b) Quan, Y.; Xie, Z. *Chem. Soc. Rev.* **2019**, *48*, 3660 – 3673; (c) Au, Y. K.; Xie, Z. *Bull. Chem. Soc. Jpn.* **2021**, *94*, 879 – 899; (d) Qiu, Z.; Xie, Z. *Acc. Chem. Res.* **2021**, *54*, 4065–4079.
19. Cheng, R.; Qiu, Z.; Xie, Z. *Nat. Commun.* **2017**, *8*, 14827.
20. Bai, H.; Xu, H.; Zhang, H.; Guo, Y.; Shan, J.; Wei, D.; Zhu, Y.; Zhang, S.; Zhang, W. *Catal. Sci. Technol.* **2018**, *8*, 5165–5177.
21. Li, C.-X.; Zhang, H.-Y.; Wong, T.-Y.; Cao, H.-J.; Yan, H.; Lu, C.-S. *Org. Lett.* **2017**, *19*, 5178–5181.
22. Cui, C.-X.; Zhang, J., Qiu, Z.; Xie, Z. *Dalton Trans.* **2020**, *49*, 1380–1383.
23. Xu, T.-T.; Cao, K.; Zhang, C.-Y.; Wu, J.; Ding, L.-F.; Yang, J. *Org. Lett.* **2019**, *21*, 9276–9279.
24. Zhang, Z.-Y.; Zhang, X.; Yuan, J.; Yue, C.-D.; Meng, S.; Chen, J.; Yu, G.-A.; Che, C.-M. *Chem. Eur. J.* **2020**, *26*, 5037–5050.
25. Cheng, R.; Qiu, Z.; Xie, Z. *Chem. Eur. J.* **2020**, *26*, 7212–7218.
26. Cheng, R.; Qiu, Z.; Xie, Z. *Chin. J. Chem.* **2020**, *38*, 1575–1578.
27. (a) Naito, H.; Morisaki, Y.; Chujo, Y. *Angew. Chem. Int. Ed.* **2015**, *54*, 5084–5087; (b) Naito, H.; Nishino, K.; Morisaki, Y.; Tanaka, K.; Chujo, Y. *Angew. Chem. Int. Ed.* **2017**, *56*, 254–259; (c) Yang, X.; Zhang, B.; Zhang, S.; Li, G.; Xu, L.; Wang, Z.; Li, P.; Zhang, Y.; Liu, Z.; He, G. *Org. Lett.* **2019**, *21*, 8285–8289; (d) Ochi, J.; Tanaka, K.; Chujo, Y. *Angew. Chem. Int. Ed.* **2020**, *59*, 9841–9855; (e) Guo, J.; Liu, D.; Zhang, J.; Zhang, J.; Miao, Q.; Xie, Z. *Chem. Commun.* **2015**, *51*, 12004–12007.
28. Au, Y. K.; Zhang, J.; Quan, Y.; Xie, Z. *J. Am. Chem. Soc.* **2021**, *143*, 4148–4153.
29. Qiu, Z.; Quan, Y.; Xie, Z. *J. Am. Chem. Soc.* **2013**, *135*, 12192–12195.
30. Xu, T.-T.; Zhang, C.-Y.; Cao, K.; Wu, J.; Jiang, L.; Li, J.; Li, B.; Yang, J. *ChemistrySelect* **2017**, *2*, 3396–3399.
31. Wu, J.; Cao, K.; Xu, T.-T.; Zhang, X.-J.; Jiang, L.; Yang, J.; Huang, Y. *RSC Adv.* **2015**, *5*, 91683–91685.
32. Cao, K.; Huang, Y.; Yang, J.; Wu, J. *Chem. Commun.* **2015**, *51*, 7257–7260.
33. Lyu, H.; Zhang, J.; Yang, J.; Quan, Y.; Xie, Z. *J. Am. Chem. Soc.* **2019**, *141*, 4219–4224.
34. (a) Guo, C.; Qiu, Z.; Xie, Z. *ACS Catal.* **2021**, *11*, 2134–2140; (b) Dziedzic, R. M.; Martin, J. L.; Axtell, J. C.; Saleh, L. M. A.; Ong, T.-C.; Yang, Y.-F.; Messina, M. S.; Rheingold, A. L.; Houk, K. N.; Spokoyny, A. M. *J. Am. Chem. Soc.* **2017**, *139*, 7729–7732; (c) Eleazer, B. J.; Smith, M. D.; Popov, A. A.; Peryshkov, D. V. *Chem. Sci.* **2017**, *8*, 5399–5407; (d) Liu, D.; Dang, L.; Sun, Y.; Chan, H.-S.; Lin, Z.; Xie, Z. *J. Am. Chem. Soc.* **2008**, *130*, 16103–16110.
35. (a) Cao, K.; Xu, T.-T.; Wu, J.; Jiang, L.; Yang, J. *Chem. Commun.* **2016**, *52*, 11446–11449; (b) Xu, T.-T.; Cao, K.; Wu, J.; Zhang, C.-Y.; Yang, J. *Inorg. Chem.* **2018**, *57*, 2925–2932.
36. Quan, Y.; Xie, Z. *J. Am. Chem. Soc.* **2014**, *136*, 15513–15516.
37. Lyu, H.; Quan, Y.; Xie, Z. *Angew. Chem. Int. Ed.* **2015**, *54*, 10623–10626.
38. Zhang, C.; Wang, Q.; Tian, S.; Zhang, J.; Li, J.; Zhou, L.; Lu, J. *Org. Biomol. Chem.* **2020**, *18*, 4723–4727.

39. Au, Y. K.; Quan, Y.; Xie, Z. *Chem. Asian J.* **2020**, *15*, 2170–2173.
40. Cheng, B.; Chen, Y.; Xie, Z. *J. Org. Chem.* **2021**, *86*, 12412–12418.
41. Quan, Y.; Tang, C.; Xie, Z. *Chem. Sci.* **2016**, 7, 5838–5845.
42. (a) Quan, Y.; Xie, Z. *Angew. Chem. Int. Ed.* **2016**, *55*, 1295–1298; (b) Mu, W.-H.; Liu, W.-Z.; Cheng, R.-J.; Fang, D.-C. *ACS Omega* **2019**, *4*, 465–474.
43. Zhang, X.; Zheng, H.; Li, J.; Xu, F.; Zhao, J.; Yan, H. *J. Am. Chem. Soc.* **2017**, *139*, 14511–14517.
44. Zhang, X.; Yan, H. *Chem. Sci.* **2018**, *9*, 3964–3969.
45. Xu, T.-T.; Cao, K.; Zhang, C.-Y.; Wu, J.; Jiang, L.; Yang, J. *Chem. Commun.* **2018**, *54*, 13603–13606.
46. Chen, Y.; Quan, Y.; Xie, Z. *Chem. Commun.* **2020**, *56*, 7001–7004.
47. Quan, Y.; Lyu, H.; Xie, Z. *Chem. Commun.* **2017**, *53*, 4818–4821.
48. Wang, Q.; Tian, S.; Zhang, C.; Li, J.; Wang, Z.; Du, Y.; Zhou, L.; Lu, J. *Org. Lett.* **2019**, *21*, 8018–8021.
49. Cao, K.; Zhang, C. Y.; Xu, T.-T.; Wu, J.; Ding, L.-F.; Jiang, L.; Yang, J. *J. Organomet. Chem.* **2019**, *902*, 120956.
50. Cao, H.-J.; Wei, X.; Sun, F.; Zhang, X.; Lu, C.; Yan, H. *Chem. Sci.* **2021**, *12*, 15563–15571.
51. Zhang, L.-B.; Xie, Z. *Org. Lett.* **2022**, *24*, 1318–1322.
52. Ko, G. H.; Um, K.; Noh, H. C.; Kim, J. Y.; Jeong, H.; Maeng, C.; Han, S. H.; Han, G. U.; Lee, P. H. *Org. Lett.* **2022**, *24*, 1604–1609.
53. Lyu, H.; Quan, Y.; Xie, Z. *Angew. Chem. Int. Ed.* **2016**, *55*, 11840–11844.
54. Lyu, H.; Quan, Y.; Xie, Z. *J. Am. Chem. Soc.* **2016**, *138*, 12727–12730.
55. Li, H.; Bai, F.; Yan, H.; Lu, C.; Bregadze, V. I. *Eur. J. Org. Chem.* **2017**, 1343–1352.
56. Baek, Y.; Kim, S.; Son, J.-Y.; Lee, K.; Kim, D.; Lee, P. H. *ACS Catal.* **2019**, *9*, 10418–10425.
57. Han, G. U.; Shin, S.; Baek, Y.; Kim, D.; Lee, K.; Kim, J. G.; Lee, P. H. *Org. Lett.* **2021**, *23*, 8622–8627.
58. Cao, K.; Wu, J.; Zhang, C.-Y.; Ding, L.-F.; Yang, J. *J. Organomet. Chem.* **2021**, *954–955*, 120956.
59. Lyu, H.; Quan, Y.; Xie, Z. *Chem. Eur. J.* **2017**, *23*, 14866–14871.
60. Wu, J.; Cao, K.; Zhang, C.-Y.; Xu, T.-T.; Ding, L.-F.; Li, B.; Yang, J. *Org. Lett.* **2019**, *21*, 5986–5989.
61. Chen, Y.; Au, Y. K.; Quan, Y.; Xie, Z. *Sci. China Chem.* **2019**, *62*, 74–79.
62. Chen, Y.; Quan, Y.; Xie, Z. *Chem. Commun.* **2020**, *56*, 12997–13000.
63. Au, Y. K.; Lyu, H.; Quan, Y.; Xie, Z. *J. Am. Chem. Soc.* **2020**, *142*, 6940–6945.
64. Chen, Y.; Lyu, H.; Quan, Y.; Xie, Z. *Org. Lett.* **2021**, *23*, 4163–4167.
65. Lyu, H.; Quan, Y.; Xie, Z. *Chem. Sci.* **2018**, *9*, 6390–6394.
66. Au, Y. K.; Lyu, H.; Quan, Y.; Xie, Z. *J. Am. Chem. Soc.* **2019**, *141*, 12855–12862.
67. Au, Y. K.; Lyu, H.; Quan, Y.; Xie, Z. *Chin. J. Chem.* **2020**, *38*, 383–388.
68. Baek, Y.; Cheong, K.; Ko, G. H.; Han, G. U.; Han, S. H.; Kim, D.; Lee, K.; Lee, P. H. *J. Am. Chem. Soc.* **2020**, *142*, 9890–9895.
69. Ge, Y.; Zhang, J.; Qiu, Z.; Xie, Z. *Angew. Chem. Int. Ed.* **2020**, *59*, 4851–4855.
70. Liang, Y.-F.; Yong, L.; Jei, B. B.; Kuniyil, R.; Ackermann, L. *Chem. Sci.* **2020**, *11*, 10764–10769.
71. Cheng, B.; Chen, Y.; Zhou, P.; Xie, Z. *Chem. Commun.* **2022**, *58*, 629–632.
72. Cao, H.-J.; Chen, M.; Sun, F.; Zhao, Y.; Lu, C.; Zhang, X.; Shi, Z.; Yan, H. *ACS Catal.* **2021**, *11*, 14047–14057.
73. Ge, Y.; Qiu, Z.; Xie, Z. *Chem. Commun.* **2021**, *57*, 8071–8074.
74. Cheng, R.; Li, B.; Wu, J.; Zhang, J.; Qiu, Z.; Tang, W.; You, S.-L.; Tang, Y.; Xie, Z. *J. Am. Chem. Soc.* **2018**, *140*, 4508–4511.
75. Quan, Y.; Xie, Z. *J. Am. Chem. Soc.* **2015**, *137*, 3502–3505.
76. Cheng, R.; Zhang, J.; Zhang, H.; Qiu, Z.; Xie, Z. *Nat. Commun.* **2021**, *12*, 7146.

About the authors

Jie Zhang received her B.Sc. and M.Sc. degrees from Lanzhou University in 2011 and 2014, respectively. Subsequently, she went to The Chinese University of Hong Kong (CUHK) and obtained her Ph.D. in chemistry under the supervision of Prof. Zuowei Xie in 2018. After 3 years of post-doctoral research in Prof. Xie's group, she has been appointed a Research Assistant Professor in Chemistry Department of CUHK since January 2022. Her research interests focus on functionalization of *o*-carboranes and mechanistic studies through DFT calculations.

Zuowei Xie earned his PhD in 1990 from the Shanghai Institute of Organic Chemistry (SIOC), the Chinese Academy of Sciences. After working in SIOC as a research associate and at University of Southern California as a Postdoctoral Fellow, he took up a chemistry faculty appointment as Assistant Professor at CUHK in 1995, where he is now a Choh-Ming Li Professor of Chemistry. Prof. Xie is a member of The Chinese Academy of Sciences and The Hong Kong Academy of Sciences. He is also a Fellow of Chinese Chemical Society and Royal Society of Chemistry. Prof. Xie's main research interests center around the chemistry of (super)carboranes, metallacarboranes, catalytic B-H bond functionalization and low-valent main group chemistry.

CHAPTER FOUR

Boron compounds for catalytic applications

Tanzeela Fazal[a], Fayaz Ali[a], Narayan S. Hosmane[b], and Yinghuai Zhu[c,*]

[a]Department of Chemistry, Abbottabad University of Science and Technology, Abbottabad, Pakistan
[b]Department of Chemistry and Biochemistry, Northern Illinois University, DeKalb, IL, United States
[c]Sunshine Lake Pharma Co., Ltd., Songshan Lake Industrial Park, Dongguan, China
*Corresponding author: e-mail address: zhuyinghuai@hec.cn

Dedicated to Francesc Teixidor and Clara Viñas in honoring their contribution to the field.

Contents

Abstract

Boron-based catalysts are playing key role in synthetic chemistry for variety of chemical transformations to produce products ranging from commodity chemicals to selective pharmaceuticals. In synthetic chemistry hydrocarbon processing e.g. oxidation, reduction, selective hydrogenation, dehydrogenation, isomerization and polymerization are very important. Nowadays, it is almost impossible to drive any industrial chemical process without catalysts. Cheaper earth-abundant catalysts urge to replace the inherited high-cost noble metals-based catalysts. Therefore, boron-based catalysts are potentially

Advances in Catalysis, Volume 71
ISSN 0360-0564
https://doi.org/10.1016/bs.acat.2022.04.005

cost-effective candidates to switch the conventional expansive catalysts. The primary focus of the current study is to get an overview of the contributions of boron-based compounds as catalysts. Chemically, boron is a strong Lewis acid having an empty p-orbital and is electrophilic, which forms coordinate covalent bonds with nucleophiles. Applications of boron-based compounds as Frustrated Lewis Pairs (FLPs) catalysts which are combination of Lewis acidic and basic p-block species and sterically incapable to neutralize each other's reactivity through adduct formation, have become an attractive strategy in the field of catalysis. This chapter briefly introduces the basics of catalysis with the focus on heterogeneous catalysis of boron-based compounds, which is dynamic toolbox in synthetic chemistry. Common synthetic strategies of boron-based catalysts are also discussed as they are being applied in a growing number of transformations by performing comparably or even better than the gold standard catalysts and have gained special attention in photo-, electro- and thermal catalysis.

1. Introduction

Currently, the catalysts are being utilized to drive most of the industrial chemical processes to raise the living standards *(1)*. Petroleum, the primary source of global fuel, contributes the emission of CO_x and NO_x into the atmosphere hence, causing significant impact over the world climate change. Besides numerous other applications, catalysts are employed on a large scale for the treatment of crude petroleum feedstock by employing zeolites and silica-alumina based solid acid catalysts. Catalytic converters of metallic or ceramic support, decorated with precious metals, such as palladium, platinum and rhodium are being used to convert environmentally hazardous gases, i.e., carbon monoxide and nitrogen oxides to carbon dioxide, nitrogen and water. Due to scarcity and inherent high cost of noble metals, significant research is being carried out to replace the expansive catalysts with cheaper and earth-abundant analogs, such as Fe, Ni, Co, B, etc., and among these metal free catalysts, particularly boron-based catalysts are getting much attention *(2)*. This chapter is aimed to analyze the catalytic behavior of boron-containing catalysts. The boron-containing catalysts are stable and capable of working under both alkaline and acidic media, having efficiency greater than the carbon materials and are in comparison with the Pt metal *(3)*.

Globally, catalysis is a matter for interest for multi-disciplinary fields including material, chemical and biological sciences. Catalysts increase the kinetics of the reactions without disturbing the thermodynamics by decreasing the activation energies via alternative mechanistic pathways for transforming into the target products. Therefore, important and challenging goal in catalyst designing is to identify the mechanistic pathways and locate the components of the catalysts, responsible for influencing the chemical transformations. Catalysis is usually classified into three main types namely; homogeneous, heterogeneous and biological catalysis *(4)*.

1. Homogeneous catalysts have the same phase as that of reactants. Usually, homogeneous catalysts are organometallic complexes with a transition metal center coordinated with organic ligands. Organic ligands bind with the Lewis acidic center (metal) through Lewis basic atoms such as N, P, O or S which act as chelating sites. Transition metal acts as the active site, while coordinating ligands manipulate the electronic structure and shape of the central metal hence modify the activity of the catalysts.
2. Heterogeneous catalysts have different phase than the reactants and are mostly solids. They are the most common type of catalyst used for industrial chemical processes and account for 90% of all the catalysts used in industry.
3. Biocatalysts are enzymatic catalysts within the living organisms and facilitate the chemical reactions to sustain the life. Chemical nature, active sites and shapes of the proteins influence the activity, selectivity as well as specificity of the enzymatic catalysts to convert the reactants into the products.

Homogeneous catalysts readily dissolve in solution which make them enable to a suite of solution-based spectroscopic characterization techniques, but still industrial processes are mainly dependent on heterogeneous due to two main reasons. First, high stability of the heterogeneous catalysts at high temperature as compared to homogeneous catalysts, which usually decompose at temperature greater than 200 °C. Secondly, ease to remove and recycle the catalyst from reaction solutions makes heterogeneous catalysts industrially more suitable. Catalytic boron-based heterogeneous transformations also offer the opportunities to progress in green chemistry by eliminating the health, environmental, and economic concerns, associated with metal-based catalysts.

Boron is a widely spread non-metal, however, it constitutes only about 3 ppm of the earth's crust, naturally found as orthoboric acid, alkali and alkaline earth metal borates *(5)*. Prominent sources of boron are kernite, rasorite, and sodium borate minerals *(6)*. Chemistry of the boron compounds is a matter of interest, because boron-based compounds are often employed as Lewis acids owing to their strong electrophilic nature due to availability of a vacant p-orbital having potential to readily accept the electrons from donor molecules. Boron-containing compounds have been extensively exploited owing to their characteristic electronic and refractory properties.

They have high structural diversity and rich applications in all fields, including electrocatalysis, oxygen reduction reaction, oxygen evolution reaction, hydrogen evolution reaction, carbon dioxide reduction reaction and nitrogen reduction reaction *(7)*. A few of the applications is shown in Fig. 1A. They have been successfully involved to activate the element–element bonds (C—C, H—H, Si—H, N—H, B—H and C—H), catalytically, and also for isomerization, activation and functionalization of heterocyclic as well as aromatic compounds *(8)*.

Boron-based catalysts in the form of homogeneous catalysts are discussed here. For example, alkynes are hydrogenated *Z*-selectively to alkene in moderate yields with the help of boron-based catalysts. The $HB(C_6F_5)_2$ showed a record catalytic activity for *Z*-selective hydrogenation of hexyne to hexene with selectivity greater than 87%. This is an example of homogeneous catalysis based on bridged B/N Lewis pairs under high hydrogen pressure *(9)*. Similarly, carbon dioxide is an abundant molecule and can be a sustainable resource for carbon-based societies. Despite the high thermodynamic stability of CO_2, homogeneous reduction of CO_2 with hydroborane under very mild conditions (T less than 100 °C, $P < 5$ atm), was attained with hydroborates or with metal hydride complex mediated by boranes via homogeneous catalysis. Catalytic reductions of CO_2 are usually achieved through boron-based homogeneous catalysis which is carried by following three ways *(10)*:

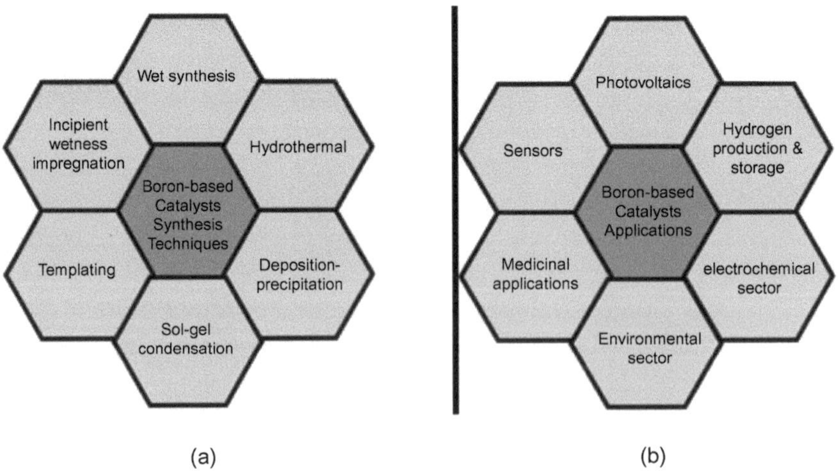

Fig. 1 (A) Synthetic techniques of boron based catalysts and (B) applications of boron-based catalysts.

(1) with hydroboranes: when the hydride transfer is catalyzed by a metal hydride complex, an amphiphilic compound, a strong Lewis base or a hydroborate.

(2) with hydrosilane: activated by strong Lewis acidic perfluorinated borane compounds.

(3) with diborane and silylborane by promoting the abstraction of one oxygen atom to afford CO.

Boron-based catalysts in the form of heterogeneous catalysts can be employed both in non-supported as well as supported forms. In non-supported catalysts, active component is a free particle while in supported catalyst, active component (mostly metals or its derivatives) are bound through inert material. Owing to high reactive and energetic surfaces, free particles aggregate in order to lower down their surface energies by forming larger particles, leading to decrease in the available active sites. To avoid this undesirable aggregation, the active component(s) are finely dispersed or embedded in any suitable inert support. To control the undesirable loss of the active component through leaching, embedded active sites are usually preferred over surface-confined active sites. Incorporation of the active component(s) into the inert support can be skilled by variety of ways and there are numerous strategies for synthesizing heterogeneous catalysts. Choice of synthetic strategy is dictated by different factors, including anticipated application of the catalyst, costs of the reagents as well as synthetic technique, final composition of the designed catalyst and desire of embedded active sites or surface-active sites. Boron and boron-based compounds are exploited in both forms, i.e., active catalysts as well as support.

2. Strategies for the synthesis of boron-based catalysts

Boron-based compounds are synthesized by a variety of techniques including simple wet synthesis *(11)*, hydrothermal *(12)*, deposition-precipitation *(13)*, sol–gel condensation *(14)*, templating *(15,16)*, incipient wetness impregnation *(17)*, etc. Some of the most common strategies for fabrication of boron-based heterogeneous catalysts are shown in Fig. 1A. Some of these strategies are explained here in detail.

(i) Wet synthetic technique is the most common, simple and facile method to produce boron-based nanoparticles. In this technique synthesis is carried out by precipitation from a soluble precursor. Particle formation takes place by nucleation from seed particle in the solution, which is then followed by subsequent particle growth and Ostwald ripening on the surface of soluble precursors present in the solution *(18)*. Solid particles are formed either by recrystallization, from a supersaturated solution containing solubilized precursor, or by simple mixing of two solutions, one containing solubilized catalyst precursor and the other solution containing a reducing or precipitating agent *(19)*. Synthesized phase can be solid (crystal growth), liquid (droplet condensation), or gas (bubble formation). Wet synthesis technique, although being the simple, but have little control over synthetic conditions leading to poor reproducibility in the catalysts production. Furthermore, as previously mentioned, freshly synthesized particles often contain highly energetic surfaces which are prone of rapid aggregation.

(ii) Deposition precipitation is a strategy to deposit active catalyst onto a support surface by precipitating the metal precursor in the presence of a solid support. This can be accomplished by suspending presynthesized support material, which acts as a nucleation site for the catalyst, in a solution containing dissolved metal precursor. Subsequently, precipitating the metal precursor as metal oxide by increasing the pH of the solution, or reducing it with a reducing agent. Chemical vapor deposition is a technique to deposit thin film-based boron catalysts, such as hexagonal boron nitride, diborane, carborane, and 4-methoxyphenylboronic acid *(20,21)*.

(iii) By employing arc discharge of graphene electrodes in the presence of diborane, H_3BO_3, B_2O_3 or B_4C are synthesized at a temperature greater than 2000 °C via solid state synthesis through top-down strategy. At lower temperatures, i.e., 900–1200 °C, more complex materials are formed ranging from substitutional boron to boronic and borinic esters, and boronic acids *(21)*. Carbon-doped boron nitride nanotubes are also reported to be synthesized through pyrolysis of melamine diborate *(3)*.

(iv) Plasma technology, an efficient up-to-down synthesis method, also proposes a feasible way for the design of functionalized and N-defects-rich catalysts *(22)*. Facile synthesis of boron and nitrogen dual-doped hollow mesoporous carbons for efficient reduction of 4-nitrophenol, and boron and nitrogen co-doped mesoporous carbon are reported by direct pyrolysis of phenylboronic acid/melamine for the reduction of 4-nitophenol *(23)*.

(v) Sol–gel synthesis is one of the most commonly used strategies for synthesizing heterogeneous catalysts owing to its potential to create a wide array of metal and ceramic oxide materials with controlled pore sizes from relatively inexpensive precursors under mild conditions. The metal alkoxides first react with water to form the corresponding metal hydroxide and alcohol, and the metal hydroxides subsequently condense to form the corresponding metal oxide solid and water *(24)*. Properties of the resultant products, such as morphology and pore size of the resulting sol–gel material, are easily tuned by varying the precursor concentrations.

(vi) Hydrothermal synthesis is a strategy through which crystalline materials are synthesized at elevated temperature and pressure *(25)*. Synthesis of boron carbon nitride (BCN) is reported by heat treatment of polyacrylonitrile in the presence of BCl_3 gas. Surface composition and catalytic activity of BCN catalysts is found to be dependent on the concentration of BCl_3 gas and the heat treatment during the preparation of catalysts *(26)*. Recent developments in boron-containing ternary catalytic materials are appreciable since the chemical functions of ternary materials can be easily tuned by controlling the configuration and composition of all three elements *(26)*.

Templating is practiced to design the nanostructure with diverse range of shapes including hollow spheres, crescents, rods, cubes, and porous materials either by a physical or chemical process. In physical process, adsorption of a species onto a template surface takes place, whereas chemical process involves grafting the species onto the template surface *(27)*. Shape of the nanoparticles plays an important role in dictating the applications of the designed catalyst. Catalyst having same constituents but with different shapes show different behavior.

Most of the literature found on catalyst, focuses primarily on the designing and optimizing the catalyst activity and selectivity to be industrially relevant. Catalyst longevity is also important for both environmental and economic purposes, as short lifetimes of catalyst lead to increased waste associated with disposing of inactive catalyst, and overall costs associated with synthesizing fresh catalyst. There are several mechanisms by which catalysts may deactivate, and among them three major factors involved are: poisoning, fouling, and degradation. The overall cost of the catalysts plays a crucial role in determining their role and place for industrial purposes.

In metal – supported catalysts, strong metal – support interactions with active oxide supports usually blocks metal-catalyzed surface reactions. During dry reforming of methane, weak oxidizing gases, such as CO_2 and H_2O, encapsulate the metal nanoparticles by boron oxide (BOx) layers in metal decorated hexagonal boron nitride nanosheets. BOx (i.e., B–O and

B–OH) surface encapsulation work synergistically with surface metal sites to promote the dry reforming of methane rather than blocking the surface reactions as observed in other supports, computational studies also verify the above findings *(3,28)*. Most of the boron-containing materials, such as boron carbide, titanium boride, nickel boride, cobalt boride, hafnium boride, and tungsten boride, as well as elemental boron, were used for oxidative dehydrogenation of alkanes *(29)*.

The composites of boron/polymers with powerful hydroxy radicals are highly efficient for catalytic decomposition of various aqueous micro-contaminants. Amorphous boron is exploited for peroxide activation and is exclusively reactive for per-oxy-mono-sulfate activation for degradation of benzenes, phenolics, dyes and antibiotics. Theoretical studies further confirm that both the amorphous boron clusters and crystal facets (100), (101), (110) directly dissociate polymer molecules to produce sulfate and hydroxyl radicals. Boron–oxygen coordination under the oxidative dehydrogenation of propane reaction conditions was found to be responsible for the selective reaction pathway to form olefins. The reactivity differences between elemental boron and boron-containing materials arise due to active-site density and electronic interactions between surface BOx species and the subsurface bulk structure *(30)*.

3. Frustrated Lewis pairs

In catalysis, Lewis acidity is one of the major properties employed for the activation of the compounds. Borane-based Lewis acids are very effective in several applications, including in Diels–Alder reactions, carbon dioxide reduction, and ammonia-borane dehydrogenation *(31)*. Revitalization in catalysis emerged with the discovery of frustrated Lewis pairs (FLPs). FLPs are combinations of sterically hindered Lewis acids and Lewis bases as shown in Fig. 2. The FLPs are capable of activating strong chemical bonds and exhibit quite interesting transformations, particularly small-molecule activation reactions (e.g., H_2, N_2O, CO_2, CO, SO_2, and lower alkenes, and alkynes) *(4)*. Boron relies solely on the vacant p orbital to activate unsaturated frameworks. In FLPs, the nucleophilic center can be a wide variety of atoms including hydrogen, carbon, nitrogen, oxygen, phosphorus or sulfur, while the Lewis acid component is usually limited to the first two rows of the p-block elements *(8)*. The boron-based FLPs led to numerous applications in the field of catalysis *(32)*. Typically, hard Lewis acids, such as boranes, alanes, borenium and carbenium cations, have been widely

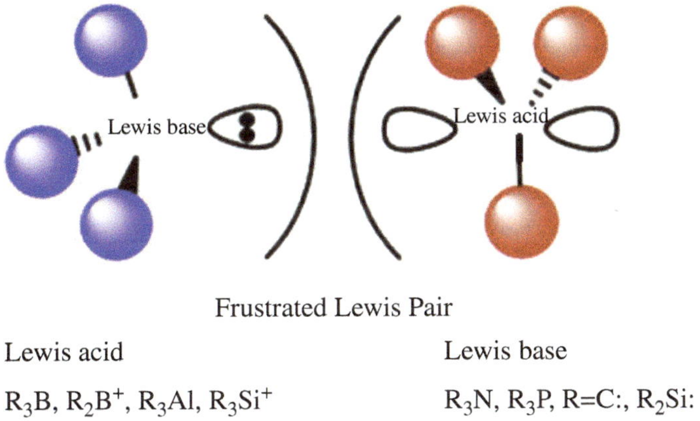

Fig. 2 Frustrated Lewis pairs.

employed. Selective hydrogenation as well as dehydrogenation and isomerizations are well reported to synthesize variety of components from food to explosives *(33)*. Along with these applications, boron-based FLP-mediate in synthetic chemistry in. hydroamination, hydroboration, cyclization, boration, asymmetric synthesis and polymerization as well *(34)*. Other reactivities associated with the use of strong boron Lewis acids involve the synthetically useful carboboration reactions as well as relevant cascade reactions of unsaturated C—C bonds. Interestingly, $BArF_4$ catalyzes the synthesis of phenoxy*alkyl*benzimidazoles with a chain length of four by dehydrogenation *(35)*, while tris(pentafluorophenyl)borane, $(B_9C_6F_5)_3$, and $BF_3.Et_2O$ catalyze the region-selective hydrothiolation of 1–3 dienes *(36)*.

Both BF_3 and BCl_3 are examples of Lewis acids in conjunction with water acting as a Lewis base and have the potential to initiate the cationic polymerization. Sustainable, controlled and selective polymerization reactions are crucial for the synthesis of important polymers or materials with specific structures or functions. The unsaturated olefin monomer generates protonated $(BF_3OH)^-$ counterion along with a carbenium ion to propagate the polymerization process. The $B(C_6F_5)_3$ has acidity comparable to BF_3 and is an ideal activator for Ziegler-Natta olefin polymerization *(6)*. Hence, in polymerization technique enabled by FLPs, i.e., $B(C_6F_5)_3$, catalyst activation generates metallacyclic cations paired with the hydridoborate anion $(HB(C_6F_5)_3)$ showing the catalytic polymerization of conjugated polar alkenes by an H-shuttling hydride between the cation and anion centers through the neutral borane *(37)*.

Involvement of boron-based organic LPs is found to mediate the ring opening polymerization of lactones and dimethyltrimethylene carbonate with reasonable control and activity of $(C_6F_5)_2PPhF)(B(C_6F_5)_4$ and $(C_6F_5)_3PF)(B(C_6F_5)_4$-assisted cationic polymerization of isobutylene which is promoted the ring opening polymerization of tetrahydrofuran. The mechanism responsible for the high regioselectivity in the BEt_3/LB-mediated copolymerization of CO_2 and epoxides has also been computationally investigated *(38–40)*.

The $B(C_6F_5)_3$-assisted hydrosilylation, and C=N bond reductions, reported in 2007 has broadened dramatically the metal-free reductions for range of substrates, such as C≡C, C=O, C≡N and C=C *(41)*. Development of the metal-free strategies for stereoselective hydrogenation of unsaturated substrates is very crucial in asymmetric synthesis. Although FLP are widely adapted for heterogeneous reductions, the most prestigious advancement of boron based FLP catalysts is their highly enantioselective hydrogenations for a variety of substrates *(41)*. The first example of chiral substrate induced asymmetric hydrogenation by mesityl borane $B(C_6F_5)_2(Mes)$. A highly diastereoselective catalytic hydrogenation of optically pure ketimines by $B(C_6F_5)_3$ has also been reported. Asymmetric hydrogenation with greater than 99% ee and asymmetric hydrosilylation with >87% have been achieved for the reaction of silyl enol ethers for the imine substrates by $B(C_6F_5)_3$ *(42)*. Highly enantioselective hydrosilylation of dicarbonyl compounds with 52–98% and 86–99% yields have been reported *(43)*. Alkenylborane produces highly enantioselective hydrogenation of ketones and enones in high yields with up to 95% *(44)*. Also, the FLP-catalyzed hydrogenation of variety of substrates, including nitrile, imine, enones, enamines, silyl enol ethers, oximes, N-heterocycles, aldehydes, ketones, etc., have already been reported *(45)*.

In addition, FLPs effectively activate molecular hydrogen by sequential H^+ and H^- transfer from the hydrogenated FLP to the substrate. The hydrogenation of alkene by ether/$B(C_6F_5)_3$ and of alkyne to cis-alkene conversion using ansa-aminohydroborane aniline and N-cyclohexyl ammonium salt using a $H_2/B(C_6F_5)_3$ pair and $Ph_2PC_6F_5/B(C_6F_5)_3$-assisted anthracene, tetracene, and tetraphene transformations are well reported *(46)*. Hydrogenation of imines, nitriles, aziridines, enamines, silyl enol ethers, olefins and alkynes has been reported via Lewis acidic borenium ions which have a great potential for hydrogenation of imine and enamine substrates. Highly enantioselective hydrogenation of imines has been reported via chiral borane catalyst *(6)*.

Since, catalytic hydrogenation is an important step involved in most of the industrial processes ranging from foods to fuels. Appreciable hydrogenation rates are achieved over defect-laden boron nitride for the hydrogenation of propene, cyclohexene, 1,1-diphenylethene, octadecene, and benzylideneacetophenone. Defects incorporation into boron nitride results into the electronic deficient and excess regions. Alkene and hydrogen-bond order are reduced over the specific defects of boron substitution for nitrogen, boron and nitrogen vacancies and Stone–Wales defects, particularly nitrogen vacancies, are the most efficient catalytically active sites *(47)*.

Because of high electrophilicity, boron-based Lewis acids with perfluorinated aryl rings induce transformations of unsaturated organic molecules by activating the π-bonds. Thus, the 1,1-carboboration of terminal alkynes by C – C cleavage lead to skeletal rearrangement of the alkyne by $B(C_6F_5)_3$ *(48)*. Alkyne hydrogenation is usually accompanied by (1) facilitating H   H heterolysis (2), stabilizing the hydride intermediate and (3) selectively reducing alkynes over alkenes. Incorporation of acidic and basic groups-based boron materials dictate the control over the activity and selectivity to activate a variety of small molecules *(49)*. By frustrated Lewis pairs-catalyst composed of $B(C_6F_5)_3$, ester substituted allenes have been reported to be successfully hydrogenated. Hydrogenation of electron-poor allenes with $B(C_6F_5)_3$ in combination with different Lewis bases has been reported *(50)*. By hydrogenation tetraphenylallene indene is reported to be synthesized by the addition of the central carbon of the allene, followed by Friedel-Crafts alkylation of one of the phenyl rings in the presence of catalytic amounts of $B(C_6F_5)_3$ *(51)*. Aromatic reduction of *N*-phenyl amines in the presence of $B(C_6F_5)_3$ and hydrogen yields the corresponding cyclohexylammonium derivatives. The arene reduction of *N*-phenyl amines is also reported *(52)*.

4. Classification of boron-containing compounds

4.1 Boranes

Trivalent boron compounds are prototypical Lewis acids with a very wide range of Lewis acidities. The triarylboranes are widely used as catalysts or co-catalysts and activators in organic transformations *(53)*. The trialkyl, trihalo, and triaryl-boranes (neutral boranes) have been extensively employed due to their strongly electrophilic nature exhibited by the vacant p-orbital. Lewis acidity is attenuated by variation of the substituents bound to boron *(54)*. Classical boranes BX_3 (X = H, F, Cl, Br and I) are reliable for cleavage of the C—O bond in ethers, acetals, and esters, and also for condensation reactions and Friedel–Crafts-type reactions *(4)*. Combination of $BF_3 \cdot Et_2O$ with organometallic reagents enhance the reactivity toward epoxides and oxabicyclic alkenes *(55)*. Intramolecular oxa-cyclization of epoxyalcoholsen to (+)-sorangicin A, hemibrevetoxin B is also reported to be accelerated *(54)*. The C—C bond formation is one of the most significant reactions in organic chemistry and in this regard, role of classical boranes is widely admitted throughout the synthetic community *(53)*. The tris(pentafluorophenyl)borane, owing to its good thermal as well as hydrolytic stability, high Lewis acidity and sterically hindered conformation at the boron center, has been widely utilized for olefin polymerization and is an excellent activator for metallocene Ziegler–Natta catalysis *(8)*. The hydrogenation, hydroboration, Friedel–Crafts C—H silylation, hydrosilylation, and dearomatization of N-heteroarenes are also reported to be catalyzed by $B(C_6F_5)_3$ *(8)*. Catalytic amount of $B(C_6F_5)_3$ promoted the allylation of a secondary benzylic alcohol *(56)*. Reactions catalyzed by $B(C_6F_5)_3$ for the construction of oxygen-containing bicycles from nitriles and ketones have also been reported. The highly *exo*-selective Diels–Alder reaction between multi-substituted acyclic dienes and unsaturated enals with a catalytic amount of $B(C_6F_5)_3$ has also been reported *(22)*.

The efficiently promoted Friedel–Crafts reactions between tertiary aliphatic fluorides and electron-rich arenesallylation of ketones or imines and unsaturated ketones by using $B(C_6F_5)_3 \cdot H_2O$ have been reported *(56)*. The chemo-and regio-selective CH insertion reaction was reported between phenol substrates and diazoesters. A catalytic amount of $B(C_6F_5)_3$ promoted the intramolecular cyclization of alkyne-containing carboxylic esters to produce dihydropyrones or isocoumarins *(57)*.

Boron trihalides act as both Lewis acids and nucleophiles to promote regioselective ring-opening reactions of vinylepoxides and vinylaziridines, in a regioselective fashion, to yield the vicinal halohydrins and haloamines, respectively. Also, selectively substituted aziridines undergo BF_3-catalyzed rearrangement (aza-pinacol rearrangement) to form the corresponding imines *(16)*, or isomerize to allylic amines *(17)*. Direct addition of N—H bonds to unsaturated organic compounds is an economic route to prepare valuable nitrogen-containing molecules. The frustrated Lewis pairs of $B(C_6F_5)_3$ react to yield zwitterionic vinyl phosphonium borate, and is also utilized for hydroelementation and hydroamination reactions of alkynes. The vanillic acid was converted to catechol with 90.3% yield using a catalytic system of bis(aminomethyl)-*nido*-$C_2B_9H_{10}$ *(58)*. More examples of catalytic amination of boron-based compounds are presented in Table 1.

Table 1 Catalytic amination assisted by boron-based compounds.

Catalyst	**Reaction**	**References**
Iodine assisted BF_3 catalysis	Decarboxylative amination of benzylic, aliphatic carboxylic acids	*(59)*
BF_3	Amination through C—H bond functionalization	*(60)*
BF_3	Reductive amination of amines/amides, cyano, carboxylic acid, carbonyl	*(61)*
Boron catalyst	Amination of carboxylic acids	*(62)*
$B(C_6F_5)_3$	Hydroaminations of trimethylsilylacetylene	*(6)*
$B(C_6F_5)_3$	Hydroamination of alkynes, aryl amines to enamines	*(63)*

The $B(C_6F_5)_3$ has been utilized to synthesize primary alcohols from aliphatic terminal epoxides *(64)* and is among few catalysts that are selective to prepare industrially relevant primary alcohol products starting from simple aliphatic terminal epoxides. The ring opening of 2-epoxyoctane with 2-propanol was used as model reaction and showed to be highly regioselective toward the formation of primary alkoxy alcohol *(64)*. The $BF_3 \cdot Et_2O$ catalyzes the alcoholysis of a large variety of aziridines *(65)* and azidolysis of vinyl aziridines as well *(66)*. Borane-catalyzed the stereoselective hydrogenation of pyridines and polyaromatic hydrocarbons. The $B(C_6F_5)_3$-methoxide bond is responsible for the hydrogenation of ketones and aldehydes yielding alcohols. The $B(C_6F_5)_3$ in ether and toluene, in combination with cyclodextrins, opened an avenue to H_2 activation and hydrogen-bonding. Hydrogenation of ketones and aldehydes using 1,4-dioxane/$B(C_6F_5)$, a key chemical combination in the hydroamination of terminal alkynes, has also been reported. Markovnikov hydroamination of alkynes, by a variety of secondary aryl amines is transferred to enamines products by $B(C_6F_5)_3$ catalysis *(63)*. The $B(C_6F_5)_3$ has also been used in a series of rearrangements of aryl-functionalized species to form complexes which are extensively fused polyaromatic systems, such as dibenzopentalenes and their derivatives.

Substituted aziridines undergo BF_3-catalyzed rearrangement to form the corresponding imines *(67)* or isomerize to allylic amines *(68)*. The borane activates the C—H bonds of heteroarenes and catalyze the borylation of furans, pyrroles, and electron-rich thiophenes *(69)*. Bromide and phenylboronic acid, along with Pd encapsulation, exhibit unique product selectivity *(70)*.

Boron Lewis acids are also extensively employed for catalytic borylation, silylation and functionalization of hydrocarbons *(53,71)*. Bifunctional boron catalysts are employed for the functionalization of heterocycles *(72)* particularly stressed heterocycles *(54)*, including N-heterocyclic, carbene-stabilized, o-carboranyl-substituted borenium ions, C—H bond activation and borylation of longer alkanes as well *(73)*. The frustrated Lewis pairs, constituted by weak coordinating anion and pyramidal boron Lewis super acids, have been used for the borylation and C—H bond activation of inactivated arenes and for their subsequent conversion to fluoroaromatics *(53)*.

For controlled ring opening polymerization of epoxides boron-based catalysts are widely employed *(74)*. The tris(pentafluorophenyl)borane $B(C_6F_5)_3$ is used as Lewis acid catalyst to lower the ring opening polymerization temperature of 1,3-benzoxazines *(75)*. Aryl assisted protonation of conjugated dienes produced allyl cationic intermediates which induced facile electrophilic aromatic substitution of aniline substrates in the presence of tris(pentafluorophenyl)borane *(61)*.

Since organoboron species are Lewis acids and trivalent boron substituents are electron-withdrawing groups by nature, the nucleophilic reactivity of the C—B bond is readily increased by combining the trigonal boron atom with Lewis bases, anions or nucleophiles. Triphenylborane catalyzes the polymerization of anhydrides and epoxides along with the block co-polymerization of anhydride/epoxides with epoxides/CO_2. Switching to the more Lewis acidic borane, $B(C_6F_5)_3$, the carbonate block of the polymer could be selectively depolymerized to yield the corresponding cyclic carbonate, thus opening the door toward chemical repurposing of the polymer *(76)*.

4.2 Organoboron acids

Organoboron acids, $R_2BOH/RB(OH)_2$, and their ester derivatives are being employed in catalysis since 1963. For the first time, organoboron acids were utilized for hydrolysis of chloroethanol by 8-quinolineboronic acid *(37)*. They are now being widely exploited in various reactions, including Diels–Alder cycloaddition, dipolar cycloaddition, aldol reactions, amidation, esterification, anhydride formation, and Friedel–Crafts alkylation reactions *(7,8,38)*.

Borenium ions are one kind of boron cations in the form of a three-coordinated boron center with a net positive charge and is served as a catalyst in variety of reactions, particularly for C—C bond formation.

Asymmetric reduction of achiral ketones by oxazaborolidine is also reported. Oxazaborolidinium cations have also been employed as catalysts for asymmetric Diels–Alder reactions, cycloaddition and cyanosilylation of aldehydes *(8)*.

4.3 Boric acid, borates, boronic acids and their derivatives

Boronic acids and esters have become very familiar among the synthetic organic chemistry community mainly due to their availability, easy handling, stability and their use in Suzuki-Miyaura type cross coupling reactions *(54)*.

Although oxygen having back donation reduces the Lewis acidity of the boron atom with respect to other trivalent boron derivatives still boric acid is extensively used in catalysis. Triphenyl borate in combination with vaulted biaryl ligand has been used for asymmetric aziridination and epoxidation *(77)*. Regio-and stereoselective ring opening of vinyl epoxides with alcohols based on palladium catalysts, in combination with organoboron derivatives, have been reported *(78)*.

The catalytic activation of C—H bonds with bifunctional boron catalysts *(71)* for the functionalization of heterocycles is well known *(72)*. N-heterocyclic carbene-stabilized *o*-carboranyl-substituted borenium ions are reported to activate C—H bond activation and borylation of alkanes *(73)*. The frustrated Lewis pairs are also employed borylation of inactivated arenes and their subsequent conversion to fluoroaromatics *(79)*. Conversion of carbon dioxide to cyclic carbonates *(80)* and to formates *(81)* using borylated catalysts is very well known. Highly challenging catalytic hydrogenation of amides to amines *(82)* and reductive cleavage of carbon monoxide *(83)* have also been reported with the help of botronic acids.

4.4 Boron-carbon based binary materials

Boron forms a strong covalent bond with carbon to produce variety of oxygen and moisture resistant boron/carbon-based compounds. These compounds have shown a wide range of applications in several technological fields, including photovoltaics, electrochemistry, sensors and catalysis *(21)*. The characteristic properties of organoboron compounds are attributed to the vacant p-orbital of the boron atom, which contributes to the Lewis acidity of the organoboron compounds. This nature not only allows them to act as a Lewis acid catalyst, but also allows the organic substituents to approach the boron atom as nucleophilic reagents *(84)*.

Among the photocatalysts, boron-based materials have gained the prestigious position due to their good performance and earth abundancy for water splitting, reduction of CO_2, acceptor less dehydrogenation, oxidation of C(sp^3)–H bonds, and C(sp^2)–H functionalization. Table 2 exhibits the

Table 2 Photocatalytic performance of boron-based catalysts.

Catalyst	Reaction	Reference
Boron nitride with TiO_2 doped with gadolinium	H_2 generation capabilities by water splitting under visible light	*(85)*
Boron nitride doped with sulfur atoms	H_2 generation efficiency under light	*(86)*
Co doped BN core-shell nanocomposite	H_2 generation efficiency	*(87)*
Boron nitride incorporated with TiO_2	79% decomposition rate of methyl orange, methylene blue, rhodamine B	*(88)*
Boron nitride nanofiber coupled with TiO_2	Photocatalytic degradation of crystal violet and Rhodamine B	*(89)*
SnO_2 coupled with boron nitride	Methyl orange degradation under visible light irradiation with decomposition rate 92%	*(90)*
Boron nitride on the surface of graphene-carbon nitride	Rhodamine B degradation efficiency up to 98.2%	*(91)*
Boron nitride on the surface of graphene-carbon nitride hybrids	Photocatalytic activity for hydrogen generation	*(92)*
Carbon nitride/boron nitride	Photocatalytic ability for H_2 generation	*(93)*
Ternary B–C–N alloy	Photo-redox catalyst to utilize solar energy for chemical conversion.	*(94)*
Boron carbon nitride containing cobalt oxide	Photocatalyst for oxygen evolution reaction and methyl orange and methyl blue dyes degradation under direct sunlight irradiation	*(95)*
Cobalt phosphide modified boron carbon nitride:	Photocatalyst for hydrogen generation	*(96)*
Boron-carbon nitride	Photocatalyst for hydrogen evolution and H_2O_2 production under visible light irradiation	*(97)*

photocatalytic performance of different boron-based catalysts. Boron-doped carbon nanotubes, synthesized through chemical vapor deposition, were evaluated as electrocatalysts. Electrocatalyst CeO_x-NiB is reported as highly efficient electrocatalyst for water splitting *(98)*. Potential of boron-doped graphene was also explored for the electrocatalytic oxygen reduction reaction. The boron-doped polymeric carbon nitride showed promising catalytic activity and high selectivity in the oxidation of benzylic aromatics *(99)*.

Boron carbide having rhombohedral structure, composed of 12-atom icosahedral clusters linked by covalent bonds *(100)*, is a p-type semiconductor with band gaps of approximately 3.0 eV. This semiconducting material is employed in a traditional boron carbide photocathodic cell as well as photocatalyst for H_2 evolution *(101)*. Carbon-doped boron nitride materials have potential to be used as energy storage materials *(102)* and in photocatalysis *(103)*. The $B_{4.3}C$ and $B_{13}C_2$ have contribution in photocatalysis by extending the absorption of light to the fully visible spectrum. The band gap of boron carbide catalysts is engineered by inductive defects for improving the properties as a photocatalysis. Boron-doped nanotubes are highly effective in the electrocatalytic oxygen reduction reaction, and the same reactivity was achieved using boron-doped graphene *(3)*.

4.5 Boron-phosphorus and boron-nitrogen based binary materials

Boron phosphide is relatively inert semiconducting material in which boron and phosphorus are coordinated tetrahedrally through covalent bonds. Boron phosphide was discovered in 1957 by the reaction of boron with red phosphorus in an evacuated silica tube at elevated temperature *(104)*. Later on it was synthesized by chemical vapor deposition, molten salt reactions and gas-phase reactions *(105,106)*. Boron phosphide is successfully used as a photocatalyst for H_2 evolution in formaldehyde solution under irradiation with visible light in the presence of H_2PtCl_6 as a co-catalyst *(3)*.

Boron nitride (BN) with 1:1 ratio of boron to nitrogen exists in three different forms, i.e., hexagonal, cubic and wurtzite crystalline, and has prestigious position in heterogeneous catalysis either as an advanced catalyst or a catalyst support *(28)*. The hexagonal-BN consists of two-dimensional stacked layers and possesses excellent chemical inertness, thermal conductivity, and mechanical properties *(107,108)* and also acts as a physical carrier for metal or oxide catalysts in harsh environments *(28)*. The catalytic behavior of pure boron nitride nanosheets in oxidative reactions is attributed to reaction-induced creation of B–O/B–OH active sites *(109)*. The Ni NPs, supported on commercial boron nitride nanosheets were employed for dry reforming of methane and transforming CH_4 and CO_2 to syngas *(28)*. Owing to its unique structure, boron nitride exhibits the ability to function as a thermal catalyst by chemical functionalization of the boron nitride edge by the addition of hydroxy, alkoxy, amino, alkyl, and other groups for optimization of the catalytic behavior.

Catalytic activity promoted by confinement effect is also reported with various metals loaded boron nitride nanostructures in electrochemical reaction, CO methanation, water-gas shift reaction, and CO oxidation. The confined space influences the adsorption and diffusion coefficient of gases depending on their properties. This provides an opportunity to tailor the reaction selectivity with metal loadings on hexagonal -BN nanostructures. A typical case is the photocatalytic CO_2 reduction reaction, in which a stunning effect on the selectivity happened when Pt was covered by BN that overlayers Pt loaded boron nitride by switching the reaction path from $CO_2 \rightarrow CO$ to $CO_2 \rightarrow CH_4$ *(88)*. As discussed earlier, the most stable structures, O-terminated groups (BO_x, B–OH) on the h-BN surface are inevitable during the synthesis, which can be used to immobilize active metals. Thus, the interfacials hexagonal -BN-O-metal linkage at the defects/vacancies are responsible for the stability of supported metals and modify their catalytic properties. The surface complexity of defect-rich boron nitride is responsible to understand metal/hexagonal -BN interfacial bonding. Especially, operando characterizations are necessary to determine the true metal/h-BN interfaces under steady working conditions in the case of interfacial evolution *(110)*.

Hexagonal BN is employed in oxidative dehydrogenation reaction of propane to propylene especially N_2-treated boron nitrides exhibited a remarkable activity, (26.0% propane conversion with 89.4% selectivity toward olefins at 520 °C *(22)*. The BN in the form of thin films has been used as a tunneling layer for photocatalysts owing to efficient resistance for the electron–hole recombination *(111)*. Hexagonal BN nanosheets were also used for the electrocatalytic oxygen reduction reaction *(112)*. By surface functionalization with Au *(111)*, the catalyst activity was found to be enhanced. Hexagonal BN is also employed as a thermal catalyst in the form of nanosheets, and nanotubes were used as catalysts for selective (79% selectivity for propene at 14% propane) oxidative dehydrogenation of propane to propene *(113)*. Boron nitride is also used for aerobic oxidative deep desulfurization to remove dibenzothiophene and showed good performance, i.e., 84.8–98.8% removal *(114)*. The first-principles density functional theory for electrocatalytic activities reveal that with the two-dimensional hexagonal BN monolayers on metal, led to selectivity control for electroreduction of CO_2. This shows the great potential of tuning the chemical reactivity of a metal surface by interfacing the hexagonal boron nitride monolayers on metal surface for selectivity control in electrocatalysis *(110)*.

For oxidative dehydrogenation of light alkanes to olefins, B_4C, Ti_2B, NiB, BN, Co_2B, Co_3B, HfB_2, WB have shown reasonable performance. Hexagonal boron nitride is selective in the transformation of C_2–C_4 alkanes *(115)*. Again here, generation of BO_x species is critical to be observed for catalytic behavior. And hydrogen abstraction of B at OH groups by molecular oxygen dynamically generates the active sites, which triggers propane dehydrogenation. Supported boron oxides are also highly selective catalysts for the oxidative dehydrogenation *(116)*. Studies of boron-supported alumina have shown that surface boron, particularly the tetrahedral BO_4 sites, have Bronsted acid properties *(33)*. Boron has vast applications in hydrogenation, oxidation–reduction reaction, oxidative dehydrogenation, electrocatalysis and photocatalysis. Table 3 discusses the applications of boron nitride as catalyst and is depicted that surface functionalization is a good strategy to enhance the catalytic behavior.

Selectivity of boron nitride toward olefins, namely ethylene and propylene, was found high (above 90%) throughout the tested conversion range. This high olefin selectivity was also observed during the oxidative hydrogenation of *n*-butane and isobutane. The BN showed moderate alkane conversion of ~14%. Furthermore, BN is selective in the transformation of C_2–C_4 alkanes by activating the oxygen and forms surface oxygen species that are responsible for selective alkane oxidation *(115)*. During cracking of ethane, ethylene and only sparing amounts of heavier olefins are produced by creating a gap for C3 and C4 olefins. From 1990s onwards, supported boron oxide catalysts for the selective oxidation of ethane, and propane to olefins and oxygenated compounds are employed.

4.6 Boron carbonitride materials

Boron, like carbon and nitrogen, has capability to build graphitic structure by opening the pave to introduce the carbon into boron nitride to transform this insulator material into functional semiconductor *(26)*. A hexagonal graphene unit can be replaced by a h-BN unit as an imbedded defect to serve as an electron-withdrawing group for optimization of the electronic and

Table 3 Applications of boron nitride as catalyst for different chemical transformations.

Catalyst	Reaction	References
Ru-Pd on two-dimensional boron nitride	Hydrogenation of furfural into furfural alcohol via water-mediated protonation and production of fuel	*(117)*
Palladium on porous boron nitride	Hydrogenation of 1-hexyne	*(118)*
Ruthenium-stabilized on vacancy-rich boron nitride	Hydrogenation of (esters) dimethyl oxalate to methyl glycol	*(119)*
Palladium on defective hexagonal boron nitride nanosheet	Chemo-selective-hydrogenation of cinnamaldehyde	*(120)*
Cobalt on defective two-dimensional boron nitride	Oxidation reduction reaction	*(121)*
Cobalt-doped two-dimensional boron nitride	Electrocatalytic and oxygen reduction reaction	*(122)*
Boron nitride	Oxidative dehydrogenation of propane	*(123)*
Carbon-doped hexagonal boron nitride	Oxygen electroreduction	*(124)*
Hexagonal boron nitride on activated graphene	Electrocatalysts of Li-CO_2 battery	*(125)*
OH- and NH_2-functionalized hexagonal boron nitride	Photocatalytic reduction of nitrate without additional hole scavengers	*(126)*

optical properties of the material by retaining its monolayered structure and characteristic properties. Boron Carbonitride in the form of nanosheets were successfully employed for the oxidative dehydrogenation of ethylbenzene to styrene *(127)*. Boron carbonitride, synthesized by using boron oxide, urea, and glucose, is used to catalyze the splitting of water and reduction of CO_2 *(94)*. They have been utilized for assisting the reactions for H_2 evolution with excellent working stability. By manipulating the band structure, boron carbonitride is also employed for O_2 evolution with a Ni—Co alloy hydroxide as a co-catalyst *(95)*. Nitrogen and boron co-doped carbon catalysts in the form of nanotubes showed appreciable oxidative reduction reaction results *(128)*. Boron, carbon, and nitrogen-containing materials were also evaluated for the catalytic activity of the samples by Knoevenagel reaction of benzaldehyde with ethyl cyanoacetate to produce ethyl cyanocinnamate *(26)*.

4.7 Boron-modified graphitic carbon materials

Boron-modified graphitic carbon family gained special attention for photocatalysis, electrocatalysis and thermal catalysis. For H_2 and O_2 evolution under visible light, they have been extensively employed *(128,129)*. Conjugated polymers with graphitic carbon nitride exhibited appreciable results for the evolution of H_2 and O_2 gases. The photocatalytic production of H_2 was performed under irradiation of visible light, along with 3 wt.% Pt as a co-catalyst with graphene-carbon nitride due to the presence of reduced boron band gap *(130)*. The additional boron functional groups attached to the surface of g-carbon nitride might act as Lewis acid sites and couple with the Lewis base site of g-carbon nitride, thereby creating a metal-free organo-catalyst containing pairs of Lewis acid and Lewis base sites *(131)*. Such pairs are known to enhance the surface reactivity and speed up photochemical transformations, particularly of photocatalytic water splitting *(130)*. Graphene incorporated boron nitride were evaluated for the oxidative reduction reaction *(132)*. Boron and nitrogen substituted co-doped nanotubes along with graphene facilitated the electrocatalytic oxidative reduction reaction.

4.8 Boron-containing metal compounds

The Pt nanoparticles, encapsulated in inert hexagonal boron nitride sheets under oxidative conditions, were employed for enhancing CO oxidation reaction *(133)*. Palladium, supported on defective hexagonal boron nitride nanosheets, used for chemoselective hydrogenation of cinnamaldehyde has also been reported *(120)*. Nitrogen-containing boron vacancy in boron nitride strongly confined the palladium atoms at the atomic level. The vanillic acid and syringic acid were converted to terephthalic acid with a yield of 58.7% and selectivity of 88.4% using catalysts $PdNiO_x/AC$ and $MoWBO_x/AC$ *(134)*. Ionic composite of palladium(II)/iron bis(dicarbollide) for catalytic oxidative carbonylation in the production of diphenyl carbonate has also been reported *(135)*.

Development of regio- and stereo-selective nucleophilic ring-opening substitutions is a center of interest *(136)*, especially ring opening of strained heterocycles, such as oxa- and aza-benzonorbornadiene systems and hetero-bicyclic systems for many synthetic chemists. Oxygenophilic boron-based catalyst ($BF_3 \cdot Et_2O$) also assists to trigger the nucleophilic ring-opening reaction *(137)*. Ketones are reluctant to undergo the homogeneous hydrogenation in the presence of simple transition-metal catalysts. For hydrogenation and isomerization of alkenes, as well as hydrogenolysis and hydrosilanolysis of alkenyl carbox, rhodacarborane catalytic systems are also quite active and robust in alkene hydrogenation *(138)*. In addition, aryl boronic acids with electron-withdrawing groups were prepared from the partial protonolysis of Me_3Al and $C_6F_5B(OH)_2$ and were reported as efficient cocatalysts *(139)*. Boron nitride along with MgO showed high alkane conversion with good selectivity of ~60% HC conversion and ~65% C_2–C_3 olefin selectivity *(140)*.

Cobalt containing boron compounds, such as CoB, Co_2B, and Co_3B, have been exploited for use in alkaline batteries *(141)*. The Co-Ni-Mo-B catalyst has been employed for hydrodeoxygenation of phenol *(142)*. Li et al. reported amorphous mesoporous Co—B catalyst for hydrogenation of cinnamaldehyde to cinnamyl alcohol *(143)*. The Fe-Ni-B catalysts are active for hydrodeoxygenation of acetophenone to ethyl benzene in super-critical EtOH *(4)*. Boron composite was found to be highly active for the oxidative carbonylation of phenol, while carborane-based nickel complexes showed modest activity for norbornene polymerization *(144)*. Chemical reactions for the formation of amide bonds are highly inefficient and are among the most commonly used transformations in organic chemistry *(144)*. Borate ester catalyst $B(OCH_2CF_3)_3$ presents significant improvements over other catalytic amidation methods in terms of efficiency and safety with an unprecedented substrates, including functionalized heterocycles and even unprotected amino acids *(145)*.

Boron-based catalysts were also employed for dehydrative Friedel-Crafts alkylation of indole/pyrrole and aniline derivatives with β-hydroxyl ketones and, accordingly, methyl vinyl ketone was converted to β-hydroxyl ketone with water as a byproduct *(146)*. Pyridine-boryl radical mediated C—H alkynylation reaction of inactivated alkanes showed high selectivity to olefins under reaction conditions of oxidative dehydrogenation of propane. While boron nitride materials have the great potential for such reactions, other B-containing compounds, such as elemental B, B_4C, TiB_2, NiB, CoB, HfB_2, WB, and few other such species also have similar capability *(30)*.

Carbon-supported and boron-promoted Ni nanocatalysts were employed for in situ self-reduction process of hybrid borate-intercalated Ni—Al layered double hydroxide/carbon nanocomposites. The Ni—B

support interactions resulting into a significant enhancement of hydrodechlorination activity. Heterogeneous Ni-based catalysts were employed for the dehalogenation of the carcinogenic and mutagenic aromatic chlorides *(147)*. Catalytic activity of Pt/SiO_2 for nonoxidative dehydrogenation of *n*-butane was increased up to 3 times with the incorporation of boron. The role of boron in enhancing catalytic performance was attributed to facilitating the migration of alkyl groups from platinum catalytic centers to tetrahedrally coordinated boron sites *(33)*.

4.9 Borate esters-based catalyst material

Pyridinium salt of iron bis(dicarbollide) has been applied for oxidation of alcohol. Catalyst composite of zerovalent palladium and cocatalyst iron bis(dicarbollide) were efficient for the oxidation of benzyl alcohol and lignin to produce aromatic aldehydes *(148)*. The *N,N*-dialkylaminomethylarylboronic acids, interacted with sugars and diols, broke the crystal packing of cellulose by reversible binding with the multiple hydroxyl groups, thereby improving solubility and catalytic activity *(149)*.

For hydrogen evolution electrocatalysis, highly dispersed rhodium nanoparticles on boron nanosheets showed catalytic activity comparable to that of commercial platinum catalysts. Theoretical calculations suggested that the structure–activity relationship between B *(104)* crystal plane and Rh *(111)* crystal plane is responsible for the release of hydrogen *(150)*.

Selective oxidation of ethylbenzene into styrene and benzaldehyde is environmentally friendly and economically feasible by alternative oxidation reactions. Boron nitride supported monomeric molybdenum oxide-based catalytic evaluation revealed that the catalyst is conducive to the generation of benzaldehyde and formation of less undesired styrene *(151)*. Conversion of propane to olefins by oxidative dehydrogenation is a promising route, but usually with less selectivity control for olefins due overoxidation reactions leading to generate undesired CO_2. Since, all boron-containing compounds are active and selective for the formation of propylene from propane due to formation of oxidized complex with a distribution of BO_x surface sites. Therefore, effective use of these materials may lead to efficient uses of abundant light alkane resources into other desired compounds by oxidation processes *(110)*. The BN catalyzes the dehydrogenation of propane to propylene with superior selectivity (80.2%) with the minimum formation of CO_2 during 20.6% conversion of propane *(109)*. Introduction of defects into BN sheets for simple olefins, such as propene, as well as complex olefins of benzylideneacetophenone showed that the olefinic bond in ethene and propene is weakened over four types of defect sites: boron substitution for nitrogen, nitrogen vacancies, boron vacancies, and Stone–Wales defects. Catalytic activity is mainly due to nitrogen vacancies defects *(47)*. Radical polymerization reaction of carborane monomer produced carboranyl-functionalized polystyrenes for selective catalytic oxidation of glycerol *(152)*.

Thus, the boron-based derivatives are important class of compounds with high structural diversity and have rich applications in all fields of chemical and material sciences. Lewis acidity is one of their major characteristic to be used for the activation of the compounds These electron-deficient catalytic systems have exhibited great potential for activation of small groups, bond transformations and functionalization *(60)* and have been extensively studied for their electronic, refractory, and catalytic properties.

5. Conclusions

This review evaluated the current status of boron-based catalysts along with their synthetic techniques, potential applications, and previous research findings. Boron-based catalyst family gained special attention for photocatalysis, electrocatalysis and thermal catalysis and their applications proved to bring significant advancements in synthetic chemistry and materials sciences to attract the incoming researchers in academia and industry. Current knowledge of electron-deficient boron-based catalytic systems have exhibited great potential to make estimates for developing new and more efficient catalysts for C—H functionalization, construction of C—C and C–X (X=N, O, S, B, Si, etc.) bonds. Evidently, the boron-based compounds and their derivatives are anticipated to have a bright future and a profound impact on the future progress in chemical catalysis and materials science. Greater efforts must be directed toward developing methodologies for producing nano-structured materials with precise control over their size, shape, composition and ultimately the resulting properties. Aside from that, it is important to investigate the scalability for producing the cost-effective and green materials at industrial level.

References

1. Gates, B. C.; Marks, T. J. *Defragmenting Catalysis*; Wiley Online Library, 2012; pp. 11644–11645. Shi, H.; Wu, Q.; Wu, Z.; Liu, Y.; Wang, X.; Huang, H.; Liu, Y.; Kang, Z. *Catal. Sci. Technol.* **2022**, *12* (6), 1837–1842.
2. Shi, H.; Wu, Q.; Wu, Z.; Liu, Y.; Wang, X.; Huang, H.; Liu, Y.; Kang, Z. *Catal. Sci. Technol.* **2022**, *12* (6), 1837–1842.
3. Fang, Y.; Wang, X. *Angew. Chem. Int. Ed.* **2017**, *56* (49), 15506–15518.

4. Kandziolka, M. V., III. *Synthesis, Characterization, and Reactivity of Heterogeneous Boron-Containing Cobalt Catalysts for Catalytic Transfer Hydrogenation*; PhD diss. University of Tennessee, 2019.
5. Ono, K.; Erhard, A. *Ullmann's Encyclopedia of Industrial Chemistry*; Wiley-VCH Verlag GmbH & Co. KGaA: Weinheim, 2011.
6. Mahdi, T. *Hydrogenation and Hydroamination Reactions Using Boron-Based Frustrated Lewis Pairs*; University of Toronto (Canada): PhD diss, 2015.
7. Rafiq, M.; Hu, X.; Ye, Z.; Qayum, A.; Xia, H.; Hu, L.; Lu, F.; Chu, P. K. *Nano Energy* **2022**, *91*, 106661.
8. Rao, B.; Kinjo, R. *Chem. An Asian J.* **2018**, *13* (10), 1279–1292.
9. Szeto, K. C.; Sahyoun, W.; Merle, N.; Castelbou, J. L.; Popoff, N.; Lefebvre, F.; Raynaud, J.; Godard, C.; Claver, C.; Delevoye, L.; Gauvin, R. M.; Taoufik, M. *Catal. Sci. Technol.* **2016**, *6* (3), 882–889.
10. Bontemps, S. *Coord. Chem. Rev.* **2016**, *308*, 117–130.
11. Ji, L.; Zhou, H. *Tetrahedron Lett.* **2021**, *82*, 153411.
12. Baykal, A.; Kizilkyalli, M.; Toprak, M.; Knicp, R. *Turk. J. Chem.* **2001**, *25* (4), 425–432.
13. Sun, W.; Meng, Y.; Fu, Q.; Wang, F.; Wang, G.; Gao, W.; Huang, X.; Lu, F. *ACS Appl. Mater. Interfaces* **2016**, *8* (15), 9881–9888.
14. Avcioglu, S.; Kaya, F.; Kaya, C. *Mater. Lett.* **2019**, *249*, 201–205.
15. Duggan, P. J.; Tyndall, E. M. *J. Chem. Soc. Perkin Trans.* **2002**, *1* (11), 1325–1339.
16. Cao, L.; Hahn, K.; Scheu, C.; Rühle, M.; Wang, Y.; Zhang, Z.; Gao, C.; Li, Y.; Zhang, X.; He, M.; Sun, L.; Wang, W.; Cao, L. M. *Appl. Phys. Lett.* **2002**, *80* (22), 4226–4228.
17. Ferdous, D.; Dalai, A. K.; Adjaye, J. *Appl. Catal. A General* **2004**, *260* (2), 137–151.
18. Thanh, N. T.; Maclean, N.; Mahiddine, S. *Chem. Rev.* **2014**, *114* (15), 7610–7630.
19. Finney, E.; Finke, R. *J. Colloid Interface Sci.* **2008**, *317*, 351–374.
20. Cun, H.; Miao, Z.; Hemmi, A.; Al-Hamdani, Y.; Iannuzzi, M.; Osterwalder, J.; Altman, M. S.; Greber, T. *ACS Nano* **2021**, *15* (1), 1351–1357.
21. Agnoli, S.; Favaro, M. *J. Mater. Chem. A* **2016**, *4* (14), 5002–5025.
22. Liu, Z.; Yan, B.; Meng, S.; Liu, R.; Lu, W.-D.; Sheng, J.; Yi, Y.; Lu, A.-H. *Angew. Chem. Int. Ed.* **2021**, *60*, 19691–19695.
23. Chen, X.; Xie, Y.; Shao, Y.; Shen, K.; Li, Y. *ACS Appl. Mater. Interfaces* **2021**, *13* (36), 42598–42604.
24. Narula, C. K.; Rokosz, M.; Allard, L. F.; Kudla, R. J.; Chattha, M. S. *Langmuir* **2000**, *16* (8), 3818–3822.
25. Varodi, C.; Pogăcean, F.; Coros, M.; Magerusan, L.; Stefan-van Staden, R. I.; Pruneanu, S. *Sensors* **2021**, *21* (19), 6630.
26. Horimoto, T.; Kannari, N.; Sato, K. *Appl. Surf. Sci.* **2021**, *558*, 149841.
27. Primo, A.; Navalón, S.; Asiri, A. M.; García, H. *Chemistry* **2015**, *21* (1), 324–330.
28. Dong, J.; Fu, Q.; Li, H.; Xiao, J.; Yang, B.; Zhang, B.; Bai, Y.; Song, T.; Zhang, R.; Gao, L.; Cai, J.; Zhang, H.; Liu, Z.; Bao, X. *J. Am. Chem. Soc.* **2020**, *142* (40), 17167–17174.
29. Grant, J. T.; McDermott, W. P.; Venegas, J. M.; Burt, S. P.; Micka, J.; Phivilay, S. P.; Carrero, C. A.; Hermans, I. *ChemCatChem* **2017**, *9*, 3623–3626.
30. Han, J.-B.; San, H. H.; Guo, A.; Wang, L.; Tang, X.-Y. *Adv. Synth. Catal.* **2021**, *363* (9), 2366–2370.
31. Schweighauser, L.; Wegner, H. A. *Chemistry* **2016**, *22* (40), 14094–14103.
32. Lam, J.; Szkop, K. M.; Mosaferi, E.; Stephan, D. W. *Chem. Soc. Rev.* **2019**, *48* (13), 3592–3612.
33. Byron, C.; Bai, S.; Celik, G.; Ferrandon, M. S.; Liu, C.; Ni, C.; Mehdad, A.; Delferro, M.; Lobo, R. F.; Teplyakov, A. V. *ACS Catal.* **2020**, *10*, 1500–1510.

34. Ma, Y.; Zhang, S.; Chang, C.; Huang, Z.; Ho, J. C.; Qu, Y. *Chem. Soc. Rev.* **2018**, *47* (15), 5541–5553.
35. Bhunya, S.; Malakar, T.; Ganguly, G.; Paul, A. *ACS Catal.* **2016**, *6* (11), 7907–7934.
36. Wilkins, L. C.; Günther, B. A.; Walther, M.; Lawson, J. R.; Wirth, T.; Melen, R. L. *Angew. Chem. Int. Ed.* **2016**, *55* (37), 11292–11295.
37. Chen, E. Y. *Top. Curr. Chem.* **2013**, *334*, 239–260.
38. Hong, M.; Chen, J.; Chen, E. Y.-X. *Chem. Rev.* **2018**, *118* (20), 10551–10616.
39. Zhao, W.; He, J.; Zhang, Y. *Sci. Bull.* **2019**, *64* (24), 1830–1840.
40. Hamza, A.; Sorochkina, K.; Kotai, B.; Chernichenko, K.; Berta, D.; Bolte, M.; Nieger, M.; Repo, T.; Papai, I. *ACS Catal.* **2020**, *10* (23), 14290–14301.
41. Stephan, D. W. *Chem* **2020**, *6* (7), 1520–1526.
42. Feng, X.; Du, H. *Tetrahedron Lett.* **2014**, *55* (51), 6959–6964.
43. Ren, X.; Du, H. *J. Am. Chem. Soc.* **2016**, *138* (3), 810–813.
44. Gao, B.; Feng, X.; Meng, W.; Du, H. *Angew. Chem. Int. Ed.* **2020**, *59* (11), 4498–4504.
45. Khan, I.; Manzotti, M.; Tizzard, G. J.; Coles, S. J.; Melen, R. L.; Morrill, L. C. *ACS Catal.* **2017**, 7 (11), 7748–7752.
46. Pal, R.; Ghara, M.; Chattaraj, P. K. *Catalysts* **2022**, *12* (2), 201.
47. Nash, D. J.; Restrepo, D. T.; Parra, N. S.; Giesler, K. E.; Penabade, R. A.; Aminpour, M.; Le, D.; Li, Z.; Farha, O. K.; Harper, J. K.; Rahman, T. S.; Blair, R. G. *ACS Omega* **2016**, *1* (6), 343–1354.
48. Patrick, E. A.; Piers, W. E. *Chem. Commun.* **2020**, *56* (6), 841–853.
49. Tseng, K. N.; Kampf, J. W.; Szymczak, N. K. *J. Am. Chem. Soc.* **2016**, *138* (33), 10378–10381.
50. Inés, B.; Palomas, D.; Holle, S.; Steinberg, S.; Nicasio, J. A.; Alcarazo, M. *Angew. Chem. Int. Ed.* **2012**, *51*, 12367–12369.
51. Gellrich, U. *Eur. J. Org. Chem.* **2021**, *2021* (33), 4707–4714.
52. Osi, A.; Mahaut, D.; Tumanov, N.; Fusaro, L.; Wouters, J.; Champagne, B.; Chardon, A.; Berionni, G. *Angew. Chem. Int. Ed.* **2022**, *61*, e202112342.
53. Berionni, G. *Chem. Synth.* **2021**, *1* (1), 10.
54. Pineschi, M. *Adv. Synth. Catal.* **2021**, *363* (9), 2325–2339.
55. Yamamoto, Y. *Angew. Chem. Int. Ed.* **1986**, *25* (11), 947–959.
56. Dryzhakov, M.; Moran, J. *ACS Catal.* **2016**, *6* (6), 3670–3673.
57. Dasgupta, A.; Stefkova, K.; Babaahmadi, R.; Gierlichs, L.; Ariafard, A.; Melen, R. L. *Angew. Chem. Int. Ed.* **2020**, *59* (36), 15492–15496.
58. Zhu, Y.; Bai, Z.; Phuan, W. C.; Ghaida, F. A.; Hosmane, N. S.; Ding, J. *J. Organomet. Chem.* **2018**, *865*, 58–64.
59. Narobe, R.; Murugesan, K.; Schmid, S.; Burkhard König, B. *ACS Catal.* **2022**, *12* (1), 809–817.
60. Ma, Y.; Lou, S.-J.; Hou, Z. *Chem. Soc. Rev.* **2021**, *50* (3), 1945–1967.
61. Kumar, G.; Qu, Z.-W.; Grimme, S.; Chatterjee, I. *Org. Lett.* **2021**, *23* (22), 8952–8957.
62. Morisawa, T.; Sawamura, M.; Shimizu, Y. *Org. Lett.* **2019**, *21* (18), 7466–7469.
63. Kumar, G.; Qu, Z.-W.; Ghosh, S.; Grimme, S.; Chatterjee, I. *ACS Catal.* **2019**, *9* (12), 11627–11633.
64. Bennett, C. K.; Bhagat, M. N.; Zhu, Y.; Yu, Y.; Raghuraman, A.; Belowich, M. E.; Nguyen, S. T.; Notestein, J. M.; Broadbelt, L. J. *ACS Catal.* **2019**, *9* (12), 11589–11602.
65. McGhee, A.; Cochran, B. M.; Stenmark, T. A.; Michael, F. E. *Chem. Commun.* **2013**, *49* (60), 6800–6802.
66. Righi, G.; Bovicelli, P.; Marucci, C.; Ilaria Tirotta, I. *Synthesis* **2012**, *44* (20), 3202–3208.
67. Sugihara, Y.; Iimura, S.; Nakayama, J. *Chem. Commun.* **2002**, 134–135.

68. Cabré, A.; Sciortino, G.; Ujaque, G.; Verdaguer, X.; Lledós, A.; Riera, A. *Org. Lett.* **2018**, *20* (18), 5747–5751.
69. Légaré, M. A.; Courtemanche, M. A.; Rochette, É.; Fontaine, F. G. *Science* **2015**, *349* (6247), 513–516.
70. Dai, C.; Du, K.; Song, C.; Guo, X. In *Advances in Catalysis*; Song, C., Ed.; Academic Press Inc., 2020; pp. 91–133.
71. Légaré, L. J.; Jayaraman, A.; Misal Castro, L. C.; Rochette, É.; Fontaine, F. G. *J. Am. Chem. Soc.* **2017**, *139* (41), 14714–14723.
72. Iqbal, S. A.; Pahl, J.; Yuan, K.; Ingleson, M. J. *Chem. Soc. Rev.* **2020**, *49*, 4564–4591.
73. Liu, Y.; Dong, W.; Li, Z. H.; Wang, H. *Chem* **2021**, 7 (7), 1843–1851.
74. Andrea, K. A.; Plommer, H.; Kerton, F. M. *Eur. Polym. J.* **2019**, *120*, 109202.
75. Arslan, M.; Kiskan, B.; Yagci, Y. *Polymer* **2018**, *10* (3), 239.
76. Andrea, K. A.; Wheeler, M. D.; Kerton, F. M. *Chem. Commun.* **2021**, *57* (59), 7320–7322.
77. Guan, Y.; Lu, Z.; Yin, X.; Mohammadlou, A.; Staples, R. J.; Wulff, W. D. *Synthesis* **2020**, *52* (14), 2073–2091.
78. Hirai, A.; Yu, X.-Q.; Tonooka, T.; Miyashita, M. *Chem. Commun.* **2003**, 2482–2483.
79. Chardon, A.; Osi, A.; Mahaut, D.; Doan, T.-H.; Tumanov, N.; Wouters, J.; Fusaro, L.; Champagne, B.; Berionni, G. *Angew. Chem. Int. Ed.* **2020**, *59*, 12402–12406.
80. Zhang, Y.-Y.; Yang, G.-W.; Xie, R.; Yang, L.; Li, B.; Wu, G.-P. *Angew. Chem. Int. Ed.* **2020**, *59*, 23291–23298.
81. Zhao, T.; Hu, X.; Wu, Y.; Zhang, Z. *Angew. Chem. Int. Ed.* **2019**, *58* (3), 722–726.
82. Sitte, N. A.; Bursch, M.; Grimme, S.; Paradies, J. *J. Am. Chem. Soc.* **2019**, *141* (1), 159–162.
83. Sun, Q.; Daniliuc, C. G.; Mück-Lichtenfeld, C.; Bergander, K.; Kehr, G.; Erker, G. *J. Am. Chem. Soc.* **2020**, *142* (41), 17260–17264.
84. Duan, X.; Li, W.; Ao, Z.; Kang, J.; Tian, W.; Zhang, H.; Ho, S.; Sun, H. *J. Mater. Chem. A* **2019**, 7 (41), 23904–23913.
85. Nada, A.; Bekheet, M.; Viter, R.; Miele, P.; Roualdes, S.; Mikhael Bechelany, M. *Appl. Catal. B Environ.* **2019**, *251*, 76–86.
86. Zhao, G.; Wang, A.; He, W.; Xing, Y.; Xu, X. *Adv. Mater. Interfaces* **2019**, *6* (7), 1900062.
87. Liu, Q.; Li, Z.; Liu, Q.; Cheng, C.; Song, M.; Huang, C. *Appl. Surf. Sci.* **2020**, *514*, 146096.
88. Hayat, A.; Sohail, M.; Hamdy, M.; Hemida, T.; AlSalem, H.; Alenad, A.; Amin, M.; Shah, R.; Palamanit, A.; Khan, J.; Nawawi, W. I.; Mane, S. K. *Surf. Interfaces* **2022**, *29*, 101725.
89. Liu, D.; Zhang, M.; Xie, W.; Sun, L.; Chen, Y.; Lei, W. *Appl. Catal. B Environ.* **2017**, *207*, 72–78.
90. Wang, X.; Qiao, G.; Jin, Z. *J. Am. Ceram. Soc.* **2004**, *87* (4), 565–570.
91. Gu, J.; Yan, J.; Chen, Z.; Ji, H.; Song, Y.; Fan, Y.; Xu, H.; Li, H. *Dalton Trans.* **2017**, *46* (34), 11250–11258.
92. Jiang, L.; Yuan, X.; Zeng, G.; Wu, Z.; Liang, J.; Chen, X.; Leng, L.; Wang, H.; Wang, H. *Appl. Catal. B Environ.* **2018**, *221*, 715–725.
93. He, Z.; Kim, C.; Lin, L.; Jeon, T. H.; Lin, S.; Wang, X.; Choi, W. *Nano Energy* **2017**, *42*, 58–68.
94. Huang, C.; Chen, C.; Zhang, M.; Lin, L.; Ye, X.; Lin, S.; Antonietti, M.; Wang, X. *Nat. Commun.* **2015**, *6*, 7698.
95. Balakrishnan, T.; Anis, M.; Arun, S.; Kumar, M.; Kumar, S. A.; Mayavan, S. *RSC Adv.* **2016**, *6*, 79448–79451.
96. Yi, S.-S.; Yan, J.-M.; Wulan, B.-R.; Li, S.-J.; Liu, K.-H.; Jiang, Q. *Appl. Catal. B Environ.* **2017**, *200*, 477–483.

97. Ou, H.; Yang, P.; Lin, L.; Anpo, M.; Wang, X. *Angew. Chem. Int. Ed.* **2017**, *56* (36), 10905–10910.
98. Wang, H.; Liu, H.; Feng, T.; Wang, L.; Yuan, W.; Huang, Q.; Guo, Y. *Dalton Trans.* **2022**, *51* (2), 675–684.
99. Wang, Y.; Li, H.; Yao, H.; Wang, X.; Antonietti, M. *Chem. Sci.* **2011**, *2* (3), 446–450.
100. Liu, J.; Wen, S.; Hou, Y.; Zuo, F.; Beran, G. J. O.; Feng, P. *Angew. Chem. Int. Ed.* **2013**, *52* (11), 3241–3245.
101. Schnabel, C.; Wörner, M.; González, B.; Olmo, I.; Braun, A. *Electrochim. Acta* **2001**, *47*, 719–727.
102. Liu, Y.; Ali, R.; Ma, J.; Jiao, W.; Yin, L.; Mu, C.; Jian, X. *ACS Appl. Energy Mater.* **2021**, *4* (4), 3861–3868.
103. Fan, Z.; Cui, D.; Zhang, Z.; Zhao, Z.; Chen, H.; Fan, Y.; Li, P.; Zhang, Z.; Xue, C.; Yan, S. *Nanomaterials* **2021**, *11*, 41.
104. Popper, P.; Ingles, T. *Nature* **1957**, *179*, 1075.
105. Chu, T.; Jackson, J.; Smeltzer, R. *J. Electrochem. Soc.* **1973**, *120* (6), 802–806.
106. Kumashiro, Y.; Okada, Y. *Appl. Phys. Lett.* **1985**, *47* (1), 64–66.
107. Cao, Y.; Maitarad, P.; Gao, M.; Taketsugu, T.; Li, H.; Yan, T.; Shi, L.; Zhang, D. *Appl. Catal. B Environ.* **2018**, *238*, 51–60.
108. Wu, J.; Wang, L.; Yang, X.; Lv, B.; Chen, J. *Ind. Eng. Chem. Res.* **2018**, *57* (8), 2805–2810.
109. Shi, L.; Wang, D.; Song, W.; Shao, D.; Zhang, W.-P.; Lu, A.-H. *ChemCatChem* **2017**, *9* (10), 1788–1793.
110. Dong, J.; Gao, L.; Fu, Q. *J. Phys. Chem. Lett.* **2021**, *12*, 9608–9619.
111. Britnell, L.; Gorbachev, R. V.; Jalil, R.; Belle, B. D.; Schedin, F.; Katsnelson, M. I.; Eaves, L.; Morozov, V.; Mayorov, A. S.; Peres, N. M. R.; Neto, A. H. C.; Leist, J.; Geim, A. K.; Ponomarenko, L. A.; Novoselov, K. S. *Nano Lett.* **2012**, *12* (3), 1707–1710.
112. Uosaki, K.; Elumalai, G.; Dinh, H. C.; Lyalin, A.; Taketsugu, T.; Noguchi, H. *Sci. Rep.* **2016**, *6*, 32217.
113. Grant, J. T.; Carrero, C. A.; Goeltl, F.; Venegas, J.; Mueller, P.; Burt, S. P.; Specht, S. E.; McDermott, W. P.; Chieregato, A.; Hermans, I. *Science* **2016**, *354* (6319), 1570–1573.
114. Wu, P.; Zhu, W.; Chao, Y.; Zhang, J.; Zhang, P.; Zhu, H.; Li, C.; Chen, Z.; Li, H.; Dai, S. *Chem. Commun.* **2016**, *52* (1), 144–147.
115. Venegas, J. M.; McDermott, W. P.; Hermans, I. *Acc. Chem. Res.* **2018**, *51* (10), 2556–2564.
116. Dorn, R. W.; Cendejas, M. C.; Chen, K.; Hung, I.; Altvater, N. R.; McDermott, W. P.; Gan, Z.; Hermans, I.; Rossini, A. J. *ACS Catal.* **2020**, *10* (23), 13852–13866.
117. Bharath, G.; Rambabu, K.; Hai, A.; Banat, F.; Rajendran, S.; Dionysiou, D. D.; Pau Loke Show, P. L. *Fuel* **2021**, *290*, 119826.
118. Büchele, S.; Chen, Z.; Fako, E.; Krumeich, F.; Hauert, R.; Safonova, O. V.; López, N.; Mitchell, S.; Pérez-Ramírez, J. *Angew. Chem. Int. Ed.* **2020**, *59* (44), 19639–19644.
119. Huang, L.; Liu, X.-F.; Zou, J.; Duan, X.; Chen, Z.-C.; Zhou, Z.-H.; Ye, L.; Liang, X.; Xie, S.-Y.; Yuan, Y. *J. Catal.* **2022**, *406*, 115–125.
120. Li, Z.; Wei, W.; Li, H.; Li, S.; Leng, L.; Zhang, M.; Horton, J. H.; Wang, D.; Sun, W.; Guo, C.; Wu, W.; Wang, J. *ACS Nano* **2021**, *15* (6), 10175–10184.
121. Deng, C.; He, R.; Shen, W.; Li, M.; Zhang, T. *Phys. Chem. Chem. Phys.* **2019**, *21* (13), 6900–6907.
122. Deng, C.; He, R.; Wen, D.; Shen, W.; Li, M. *Phys. Chem. Chem. Phys.* **2018**, *20* (15), 10240–10246.
123. Venegas, J. M.; Zhang, Z.; Agbi, T. O.; McDermott, W. P.; Alexandrova, A.; Hermans, I. *Angew. Chem. Int. Ed.* **2020**, *59* (38), 16527–16535.

124. Marbaniang, P.; Patil, I.; Lokanathan, M.; Parse, H.; Sesu, D. C.; Ingavale, S.; Kakade, B. *ACS Sustain. Chem. Eng.* **2018**, *6* (8), 11115–11122.
125. Sun, Z.; Wang, D.; Lin, L.; Liu, Y.; Yuan, M.; Nan, C.; Li, H.; Sun, G.; Yang, X. *Nano Res.* **2022**, *15* (2), 1171–1177.
126. Jiang, C.; An, H.; Dong, G.; Feng, J.; Zhang, M.; Ren, Y.; Ma, J. *Chem. Eng. J.* **2022**, *428*, 131054.
127. Guo, F.; Yang, P.; Pan, Z.; Cao, X.-N.; Xie, Z.; Wang, X. *Angew. Chem. Int. Ed.* **2017**, *56* (28), 8231–8235.
128. Yang, L.; Jiang, S.; Zhao, Y.; Zhu, L.; Chen, S.; Wang, X.; Wu, Q.; Ma, J.; Ma, Y.; Hu, Z. *Angew. Chem. Int. Ed.* **2011**, *50* (31), 7132–7135.
129. Lu, Y.; Wang, W.; Cheng, H.; Qiu, H.; Sun, W.; Fang, X.; Zhu, J.; Zheng, Y. *Int. J. Hydrog. Energy* **2022**, *47* (6), 3733–3740.
130. Zhao, G.; Yang, H.; Liu, M.; Xu, X. *Front. Chem.* **2018**, *6*, 551.
131. Sun, W.; Zhu, J.; Zheng, Y. *Int. J. Hydrog. Energy* **2021**, *46* (75), 37242–37267.
132. Tao, H.; Gao, Y.; Talreja, N.; Gao, F.; Texter, J.; Yan, C.; Sun, Z. *J. Mater. Chem. A* **2017**, *5* (16), 7257–7284.
133. Song, T.; Dong, J.; Li, R.; Xu, X.; Hiroaki, M.; Yang, B.; Zhang, R.; Bai, Y.; Xin, H.; Lin, L.; Mu, R. Q.; Bao, X. *J. Phys. Chem. Lett.* **2021**, *12* (17), 4187–4194.
134. Bai, Z.; Phuan, W. C.; Ding, J.; Heng, T. H.; Luo, J.; Zhu, Y. *ACS Catal.* **2016**, *6* (9), 6141–6145.
135. Biying, A. O.; Yuanting, K. T.; Hosmane, N. S.; Zhu, Y. *J. Organomet. Chem.* **2017**, *849-850*, 195–200.
136. da Silva, A. R.; dos Santos, D. A.; Paixão, M. W.; Corrêa, A. G. *Molecules* **2019**, *24*, 630.
137. Sajisha, V. S.; Anas, S.; John, J.; Radhakrishnan, K. V. *Synlett* **2009**, *2009* (18), 2885–2895.
138. Zhu, Y.; Carpenter, K.; Bun, C. C.; Bahnmueller, S.; Ke, C. P.; Srid, V. S.; Kee, L. W.; Hawthorne, M. F. *Angew. Chem. Int. Ed.* **2003**, *42* (32), 3792–3795.
139. Tanaka, R.; Hirose, T.; Nakayama, Y.; Shiono, T. *Polym. J.* **2016**, *48* (1), 67–71.
140. Loiland, J. A.; Zhao, Z.; Patel, A.; Hazin, P. *Ind. Eng. Chem. Res.* **2019**, *58* (6), 2170–2180.
141. Liu, Y.; Wang, Y.; Xiao, L.; Song, D.; Wang, Y.; Jiao, L.; Yuan, H. *Electrochim. Acta* **2008**, *53* (5), 2265–2271.
142. Wang, W.; Yang, Y.; Luo, H.; Hu, T.; Liu, W. *Catal. Commun.* **2011**, *12* (6), 436–440.
143. Li, H.; Yang, H.; Li, H. *J. Catal.* **2007**, *251* (1), 233–238.
144. Sabatini, M. T.; Boulton, L. T.; Sheppard, T. D. *Sci. Adv.* **2017**, *3* (9), e1701028.
145. Hu, G.; Wu, Z.; Sheng Dai, S.; Jiang, D. *ACS Appl. Mater. Interfaces* **2018**, *10* (7), 6694–6700.
146. San, H. H.; Huang, J.; Lei Aye, S.; Tang, X.-Y. *Adv. Synth. Catal.* **2021**, *363* (9), 2386–2391.
147. Lan, M.; Fan, G.; Wang, Y.; Yang, L.; Li, F. *J. Mater. Chem. A* **2014**, *2* (35), 14682–14689.
148. Zhu, Y.; Chuanzhao, L.; Sudarmadji, M.; Hui Min, N.; Biying, A. O.; Maguire, J. A.; Hosmane, N. S. *ChemistryOpen* **2012**, *1* (2), 67–70.
149. Zhu, Y.; Kong, Z. N.; Stubbs, L. P.; Lin, H.; Shen, S.; Anslyn, E.; Maguire, J. *ChemSusChem* **2010**, *3* (1), 67–70.
150. Chen, K.; Wang, Z.; Wang, L.; Wu, X.; Hu, B.; Liu, Z. W.; M. *Nano-Micro Lett.* **2021**, *13*, 138.
151. Sheng, J.; Li, W.-C.; Wang, Y.-R.; Lu, W.-D.; Yan, B.; Qiu, B.; Gao, X.-Q.; Cheng, S.-Q.; He, L.; Lu, A.-H. *J. Catal.* **2021**, *400*, 265–273.
152. Zhu, Y.; Zhao, W.; Hosmane, N. S. *J. Orgmet. Chem.* **2015**, *798*, 80–85.

About the authors

Dr. Tanzeela Fazal was born in Abbottabad, Pakistan. She completed her M.Ss. from Hazara University Mansehra, Pakistan. She got her PhD degree from COMSATS University Islamabad, Pakistan in 2019. She also worked with Prof. Dr. Graham Hutchings in Cardiff University as research associate. Currently, she is working in Abbottabad University of Science and Technology, Pakistan as an Assistant Professor. Her research focus is on photosensitive and polymer materials for energy generation and catalysis.

Dr. Fayaz Ali completed his Bachelor and Master at Islamia College University Peshawar. He obtained his M.Phil. degree in Physical Chemistry from NCEPC, Peshawar, and then completed his PhD degree from King Abdulaziz University, Saudi Arabia. He was awarded the Best Graduate Student Excellence Award of 2018 in PhD by the Deanship of Graduate Studies, KAU. He worked as Assistant Professor of Chemistry for one year at the Department of Chemistry, Abbottabad University of Science and Technology. He also worked for two years as a postdoctoral research fellow in the School of pharmacy at Macau University of Science and Technology, Macau. His research focus is on the design, synthesis, and applications of novel materials for drug delivery, wastewater treatment and energy generation.

Dr. Narayan S. Hosmane obtained a PhD degree in Inorganic Chemistry in 1974 from the University of Edinburgh. In 1979 he joined the faculty at the Virginia Polytechnic Institute and State University. In 1982 he joined the faculty at the Southern Methodist University, where he became Professor of Chemistry in 1989. In 1998 he moved to Northern Illinois University and is currently a Distinguished Faculty, Distinguished Research Professor and Inaugural Board of Trustees Professor. He has published over 380 papers in leading scientific journals and an author/editor of six books.

Dr. Yinghuai Zhu received his PhD degree (1997) in organometallic chemistry from Nankai University. He worked at Nankai University for three years as an Assistant and Associate Professor, respectively. Before joining the Institute of Chemical and Engineering Sciences, A*STAR in 2002 as senior research fellow, he did his boron chemistry research in Northern Illinois University as a post-doctoral research fellow. He started to work at School of Pharmacy, Macau University of Science and Technology in 2017 and then joined Sunshine Lake Pharma Co. Ltd in 2020. His researches focus on boron chemistry and catalysis. He has published 85 articles in the peer reviewed journals and 12 book chapters.

CHAPTER FIVE

Recent trends and tactics in facile functionalization of neutral icosahedral carboranes ($C_2B_{10}H_{12}$) and *nido*-carborane (7,8-$C_2B_9H_{12}^-$)

Chun-Xiao Li and Hong Yan*
School of Chemistry and Chemical Engineering, Nanjing University, Nanjing, Jiangsu, People's Republic of China
*Corresponding author: e-mail address: hyan1965@nju.edu.cn

Contents

Abstract

Icosahedral *o*−/*m*−/*p*-carborane ($C_2B_{10}H_{12}$) and *nido*-7,8-carborane ($C_2B_9H_{12}^-$) are boron clusters with three-dimensional aromaticity, which has been well applied in material and pharmaceutical science. Although the diversity of highly functionalized carborane derivatives is continually desired, only a few have been afforded in the past few decades. Connecting specified vertices of carboranes with an organic fragment, termed as carboranylation, has been well explored under the aegis of novel synthetic strategies including transition-metal catalysis, metal-free catalysis, electrocatalysis, visible-light-induced photocatalysis, mechanochemical ball-milling reaction, and asymmetric catalysis. This chapter details and contextualizes the advancements in molecule construction brought forth by aforementioned means. The carboranylation approaches summarized in this chapter are categorized as (1) functionalization via B—H bond activation; (2) transformation of pre-functionalized carborane; (3) functionalization on carbon vertex.

Advances in Catalysis, Volume 71
ISSN 0360-0564
https://doi.org/10.1016/bs.acat.2022.06.002

1. Introduction

Icosahedral boron clusters $[C_nB_{12-n}H_{12}]^{n-2}$ ($n = 0, 1, 2$), with 26 delocalized framework electrons, have long been recognized as three dimensional analogs of benzene *(1)*. Among them, neutral *o*−/*m*−/*p*-carboranes ($C_2B_{10}H_{12}$) and *nido*-carborane (7,8-$C_2B_9H_{12}^-$) are extensively studied ones possibly by virtue of their unique properties and commercial availability (Fig. 1).)

According to the 3D aromaticity *(2)*, large hinderance, unique electronic properties, non-classical interactions (dihydrogen bonds, hydrophobicity, B—H ••• π interaction etc.) as well as tunable solubility between *closo*- and *nido*- clusters *(1b,c,3)*, carboranes play an important role in luminous materials *(4)*, metal-organic frameworks *(5)*, uranium capture *(6)*, electricity

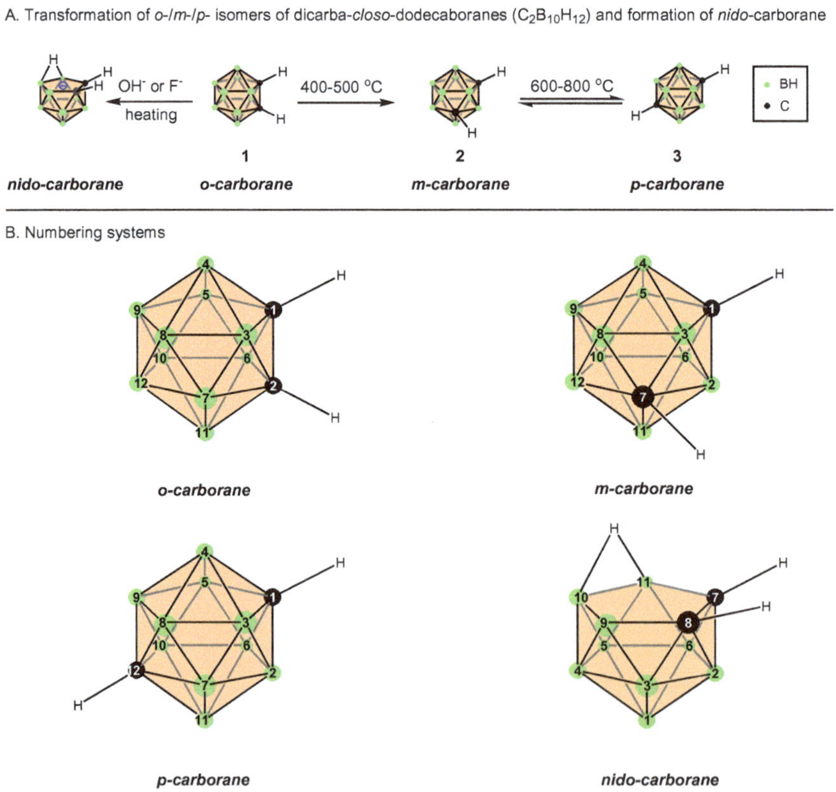

Fig. 1 Molecular geometry of *o*-, *m*-, *p*-dicarba-*closo*-dodecarboranes (**1–3**) and 7,8-*nido*-carborane. IUPAC numbering of the cage atoms and their transformations.

storage materials *(7)*, catalysis *(8)*, and so forth. At the same time, carborane derivatives are potential candidates for boron neutron capture therapy (BNCT) owing to its boron enrichment nature *(9)*.

On the other hand, the stability, large steric hinderance and hydrophobicity make carboranyl a useful scaffold in drug design. Hey-Hawkins and Leśnikowski have summarized previous investigations on applications of carborane clusters in small bio-active molecule design, respectively *(1c, 10)*. As far as weak interactions between carborane and drug target is concerned, the hydrophobicity and lipophilicity dominate the combination of carborane with receptor protein. According to X-ray structure analysis, however, the unconventional B—H ••• H—N dihydrogen bonds (unique interaction exists in polyhedral borane) play a vital role in some cases *(11)*.

The concept of "escape from flatland" are gradually prevalent in drug discovery during the last decade *(12)*. This guidance describes that enhanced potency, ideal metabolic liabilities, as well as better solubility can be acquired when flat phenyl moiety in a drug-like molecule is replaced by nonclassical bioisosteres (Fig. 2). These bioisosteres feature: (1) small compact sp^3-rich 3D structures or (2) presence of chiral centers. In 2017, Hirst and colleagues from GSK replaced a substituted phenyl motif in a $LpPLA_2$ inhibitor *Darapladib* with 1,3-disubstituted bicyclo[1,1,1]pentane skeleton *(13)*. The prepared anolog **4** maintained the original potency of *Darapladib* and showed improved physicochemical properties. Cubanyl *(14)*, adamantanyl (**5**) *(15)*, bicyclo[2,2,2]octanyl (**6**) *(16)*, bicyclo[2,1,1]hexanyl (**7**) *(17)* etc., in turn, have also been validated as bioisosteres of benzenes.

As a 3D analog of benzene, carboranes are competent to be bioisosteres of phenyl ring due to the following reasons: (1) different type of boron and carbon vertex in *o*−/*m*−/*p*-carborane has different electronic property, which results in protean combinations that can mimic mono-substituted or *o*−/*m*−/*p*-disubstituted phenyl ring (Fig. 2B); (2) the asymmetric substituted carborane is a chiral-at-cage molecule which can provide chiral element as a bioisostere; (3) it leaves a room for achieving novel patent-free molecules with improved bio-activity and physico-chemical profile. Recent studies have made important advances towards drawing upon carboranes as pharmacophores in drug development. For example, C-grafted *o*-carboranyl (abbreviated as *o*-carboranyl@C) has come to the fore among phenyl, cubanyl, adamantanyl, bicyclo[1,1,1]pentanyl, *m*-carboranyl@C and *p*-carboranyl@C *(18)*. Compound **8** exhibits the best and improved in vitro potency against *Plasmodium falciparum*. The Suga group incorporated L-carboranylalanine **9** (analog of L-phenylalanine) into peptide, which

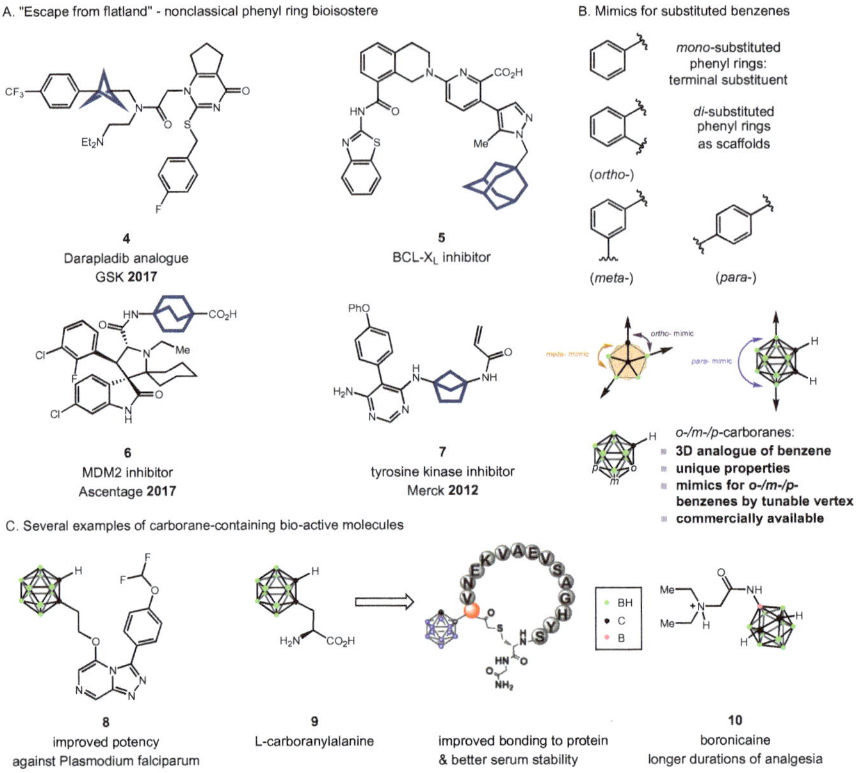

Fig. 2 Examples for non-classical phenyl ring bioisostere.

granted a unique pharmacophore for combining to the hydrophobic active surface of target protein and serum stability *(19)*. Despite C_{cage} tethered carboranyl has been well investigated, the B_{cage} tethered carboranyl falls far short of investigation. In 2015, Hawthorne and co-workers disclosed that the *o*-carboranyl@B(3) and 1,2-dimethyl-*m*-carboranyl@B(9) exhibited stronger effect and longer durations of analgesia than lidocaine and other analogs (adamantanyl, *m*-carboranyl@B(9) and *p*-carboranyl@B(2)) *(20)*. Recently, the Spokoyny group grafted *m*-carboranyl@B(9) on cysteine residues of peptides, which rendered some different properties compared to its phenyl counterpart *(21)*.

Although there are 10 boron atoms in a carborane cage, overwhelming synthetic methods focused on preparation of C-grafted molecules thus far. C-substituted carboranes are often synthesized by two facile approaches: (1) deprotonation of cage C—H bond (pK_a for *o*−/*m*−/*p*-carboranes ~23/28/30); or (2) coupling between alkynes and decaborane (Fig. 3B).

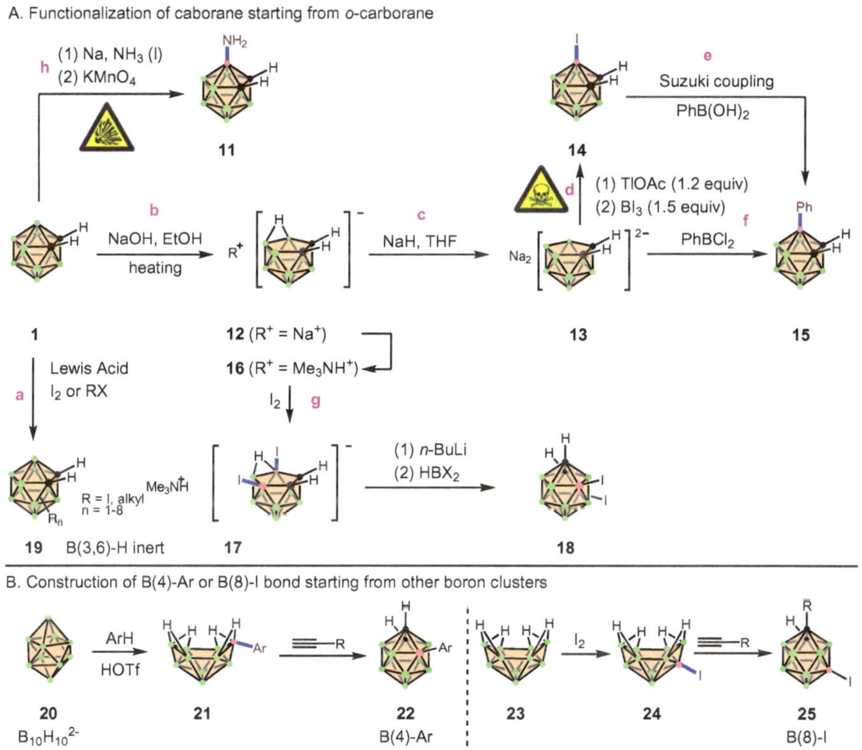

Fig. 3 Traditional strategies for functionalization of *o*-carborane.

The scare approaches for functionalization of boron vertices have been a drag on preparation of versatile poly-functionalized carboranes. In earlier days, the decoration of boron atoms in *o*-carboranes was generally rooted in reactions of carborane with electrophiles or nucleophiles (Fig. 3A). When Mulliken charge is considered, the electron densities order of *o*-carborane is B(9,12) > B(8,10) > B(4,5,7,11) > B(3,6) *(22)*, while the natural population analysis (NPA) charges revealed that B(8,10) > B(9,12) > B (4,5,7,11) > B(3,6) *(1c)*. Therefore the B(9) substituted *o*-carboranes can be achieved by electrophilic substitutions (route **a**). Particularly, 4,5,6,7,8,9,10, 11,12-X_8-*o*-carborane (X = Me or I) was synthesized under excess amount of Lewis acid and methyl or iodine precursor *(23)*. In contrast, the electron-deficient B(3) position of *o*-carborane is vulnerable to nucleophiles, which is consistent with consideration of the NPA charges and this property is well applied in *nido*-carborane synthesis. Strong Lewis bases including hydroxyl, alkoxides, and fluorides were discovered to deboronate *o*-carborane forming the corresponding anionic 7,8-*nido*-carborane (route **b**) *(24)*. After the deprotonation of *nido*-carborane **12** (route **c**) *(25)*, the capitation reaction of dicarbollide **13** with BI_3 or $PhBCl_2$ was then developed for forging B(3)-substituted *o*-carboranes (route **d-f**) *(25,26)*. By electrophilic diiodination of *nido*-carborane **16** and a following capitation, 4,7-I_2-*o*-carborane **18** was prepared (route **g**) *(27)*. In 1967, Zakharkin and Kalinin successfully introduced an amino group to B(3) of *o*-carborane through generation of *o*-carborane dianions in liquid ammonia followed by oxidation (route **h**) *(28)*. In addition, the decaborane intermediates were also well utilized (Fig. 3B). The carboranylation@B(4) arenes are obtained by acid-induced opening of *closo*-$B_{10}H_{10}^{2-}$ (**20**) via $[B_{10}H_{13}]^+$ intermediate (**21**) *(29)*. 8-I-*o*-carboranes (**25**) are achieved by iodination of decaborane (**23**) and a follow-up coupling with alkynes *(30)*.

Despite the wealth of literature on functionalization of carborane and corresponding organometallic investigations *(31)*, the development of facile and efficient decoration of carboranes is challenging and still in its infancy. Over the last half century, the advent and development in transition-metal catalysis, asymmetric synthesis, photoredox catalysis, ball milling, as well as eletrocatalysis have revolutionized the means of molecular construction, which also shed some light on the realm of carborane chemistry. In this chapter, we describe how the area of *functionalization of carborane* has flourished with illustration of recent developments benefitted from novel synthetic methods. We aim to show design principles for several catalytic approaches and highlight the accomplishments that are crucial for *functionalization of vertices on carborane cage* in the recent decade. We categorize the selected examples in three classes: (1) functionalization of carborane via direct B—H activations; (2) transformation of B_{cage} pre-functionalized carborane; (3) functionalization of cage carbon. Albeit equally important, the reactions of anionic icosahedral boron clusters ($B_{12}H_{12}^{2-}$, $CB_{11}H_{12}^-$) and the transformations via carboryne intermediate will not be discussed in this chapter.

2. Direct B—H functionalization

2.1 Directing group (DG) assisted B—H functionalization

Of the myriad reasons transition-metal catalysis has been widely adopted, perhaps most conceivable is the ability for novel activation modes of plentiful, often feedstock reagents. The organometallic investigations of carborane-based complexes, especially transition-metal induced B—H activations have been extensively studied since the 1970s *(32)*. In contrast, until very recently, they found only limited applications in preparation of carboranes. Nevertheless, transition-metal-enabled B—H cleavage and cage–functionalities forming mechanisms have allowed for versatile carboranylations of organic substrates in the past decade, providing fresh opportunities in synthetic and medicinal chemistry.

In Section 2.1, B—H functionalization assisted by directing groups are summarized. The type of DGs includes carboxyl group, imine (including transient imine intermediate), pyridyl, amide, hydroxyl, as well as phosphine. In addition, the intramolecular reactions between functionalities tethered on C_{cage} and B—H are also included in Section **2.1**.

2.1.1 Carboxyl group directed B—H functionalization

The early examples of heteroatom-induced B—H activation by iridium *(32,33)*, rhodium *(34)*, ruthenium *(35)*, palladium *(32)*, and cobalt *(31)* demonstrated the potential of heteroatom directed transition-metal catalyzed B—H functionalization. It was not until 2014 that the directing group (DG) assisted B—H functionalization was recognized as an independent field of synthetic research, which could provide an overarching and powerful strategy in derivatization of carborane.

A seminal work that capitalizes upon carboxyl assisted iridium promoted B—H activation in catalysis was reported by Xie in 2014 (Fig. 4A) *(36)*. The carboxy group was easily installed onto a C_{cage} when lithiated *o*-carborane is treated with CO_2. In the presence of 2.5 mol% $[Cp^*IrCl_2]_2$, 0.5 equiv. of $Cu(OAc)_2$ as well as 0.5 equiv. of AgOAc, carboranyl carboxylic acid **29** reacted well with 1.2 equiv. diaryl acetylenes that resulted in the achievement of regioselective B(4)-alkenylated products **30** in good to excellent yields. Yet it should be pointed out that **29** is a prechiral molecule, the use of achiral $[Cp^*IrCl_2]_2$ catalyst will give rise to formation of racemic mixtures, namely, B(4)-alkenylated and B(5)-alkenylated products. In order to simplify the description of such kind of racemate, we will use the B_{cage} with a lower ranking number to describe the reaction in the following context. The electron-poor diaryl acetylenes gave better yields, while alkyl substituted, or terminal alkynes did not react at all. The existence of free C_{cage}—H adjacent to C_{cage}—COOH did not lower the yield (**31**). According to the control experiment (Fig. 4, Eq. 1), a key intermediate **40** was acquired under room temperature, which indicates the B(4)—H is able to be activated even at 25 °C. Smooth decarboxylation of carboxylic acid **41** proceeded under the standard condition, which proved the carboxy

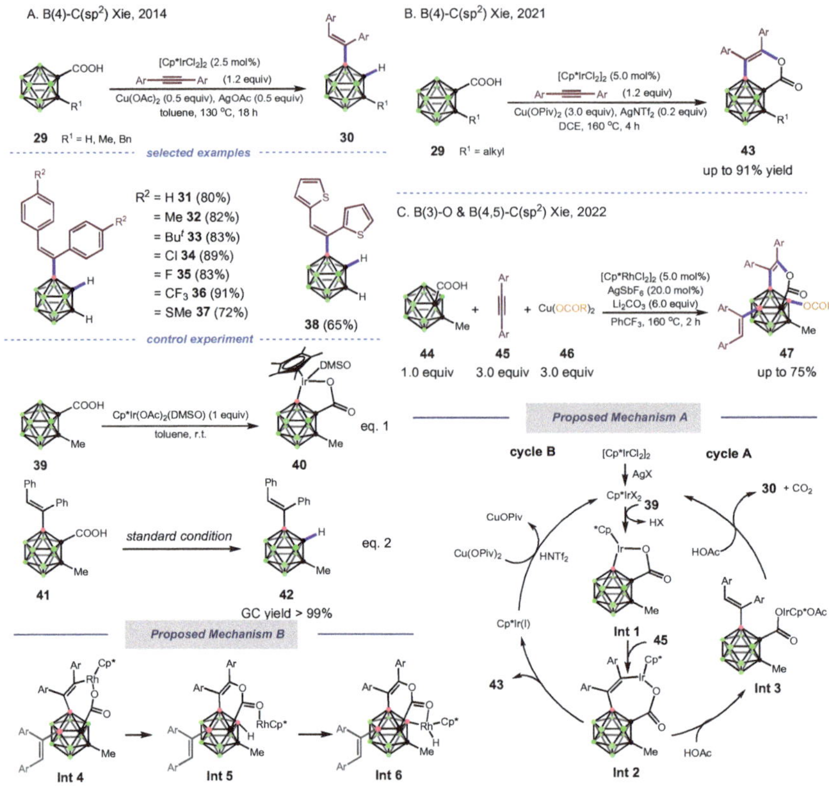

Fig. 4 Carboxyl directed catalytic cross-coupling between carborane and diaryl acetylenes.

group can serve as a traceless directing group. The employment of silver salt and high temperature were crucial to realize decarboxylation and monoselectivity. A plausible mechanism was proposed (Fig. 4, proposed mechanism A, cycle A). The **Int 2** undergoes protonation and decarboxylation to afford compound **30** and close the catalytic cycle. In fact, the **Int 2** is possible to undergo reductive elimination thus result in formation of carborane-isocoumarin derivatives (**43**) *(37)*. After optimizing the reaction conditions, the Xie group found this transformation can be facilitated with the addition of 5.0 mol% $[Cp^*IrCl_2]_2$, 20 mol% $AgNTf_2$, 3.0 equiv. of $Cu(OPiv)_2$, and 1.2 equiv. of diaryl acetylenes in 0.02 M 1,2-dichloroethane (DCE) at 160 °C (Fig. 4B). C_{cage} without DG was blocked by alkyls in this case. Since rhodium and iridium are in the same lane in periodic table, they have similar reactivities. However, the shorter radium makes rhodium often more reactive than its iridium counterpart.

When $[Cp^*RhCl_2]_2$ was alternated as the catalyst, an unexpected B(5)-alkenylation, B(4)-alkyne annulation and B(3)-acyloxylation of *o*-carborane was accomplished in one-pot (Fig. 4C). The copper salt is not only an oxidant for the reaction but also serves as the doner of acyloxy group at the same time. It is conceivable that the mechanism of B(5)-alkenylation is similar to cycle A shown in Fig. 4; the B(4)-alkyne annulation step is similar to cycle B exhibited in the same figure. When the **Int 5** was generated after the reductive elimination of **Int 4**, the carbonyl group served as a directing group, which mediated the consequently rhodium catalyzed B(3)-acyloxylation.

Concurrently, the Lee group disclosed a solvent-adjusted iridium-catalyzed B(4) cyclative indenylation or 1,3-dienylation between *o*-carboranes and propargyl alcohols (Fig. 5) *(38)*. When the model reaction was carried out in DCE, toluene, chlorobenzene and mesitylene, the 1,3-dienylated *o*-carborane (**51**) was yielded up to 65% yield. In sharp contrast, sodium salt (**49**) was achieved quantitatively at a lower temperature in trifluoroethanol (TFE). The carboxyl anion could be removed under heating condition, which is in consistent with Xie's work (Fig. 4, Eq. 2).

Following the mechanistic pattern depicted in Fig. 6, the generation of **Int 8** is very similar as formation of **Int 2** in Xie's B(4)-alkenylation,

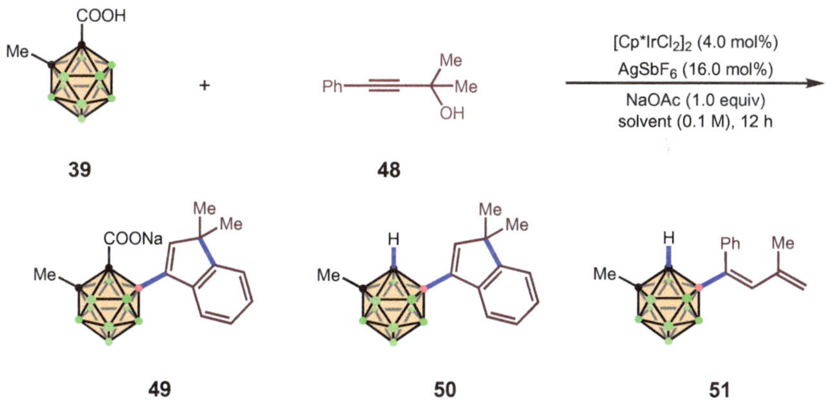

	Yield(%)		
condition	**49**	**50**	**51**
mesitylene, 80 °C	0	0	65
TFE, 60 °C	97	0	0
TFE, 60 °C; then toluene (0.1 M), 130 °C, 3 h	0	93	0

Fig. 5 Iridium-catalyzed cyclative B(4) indenylation or dienylation.

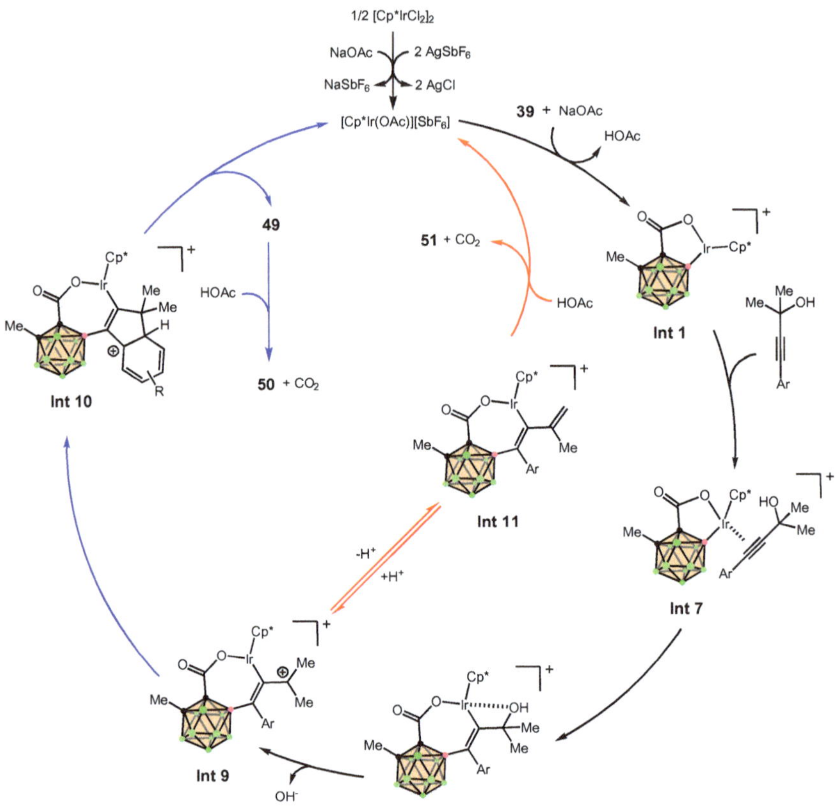

Fig. 6 Plausible mechanism of B(4) cyclative indenylation and 1,3-dienylation.

followed by dihydroxylation. The beneficial effect of TFE is broadly acknowledged in transition-metal catalysis, according to its protic but non-nucleophilic feature, high hydrogen bonding donor character, extreme polarity, capability to stabilize cation species, and high ionization power. TFE stabilized cation **Int 9** and guaranteed the following intramolecular aromatic substitution to form **Int 10** and consequent formation of indenylated *o*-carborane **49**. In another dienylation route, dehydration from **Int 9** is faster that leads to formation of **Int 11**, and diene **51** is achieved after proto-deiridation and decarboxylation steps. The propargyl alcohol bearing an electron-donating group (EDG) on aryl ring tends to forge B(4)-indenyl bond and propargyl adorned with an electron-withdrawing group (EWG) is favorable for B(4) dienylation. This result seems to be of a piece with the proposed mechanism. By the way, **Int 9** and **Int 11** are identified in equilibrium with each other by deuterium experiment. The cage carbon without DG is blocked by alkyl or aryl in this work.

A palladium catalyzed B(4) mono-alkynylation of *o*-carborane was unveiled by the Xie group in 2016 (Fig. 7) *(39)*. Either alkynyl bromides or terminal alkynes provide the alkynyl group, however the latter had a comparably broader substrate scope. According to the mechanism

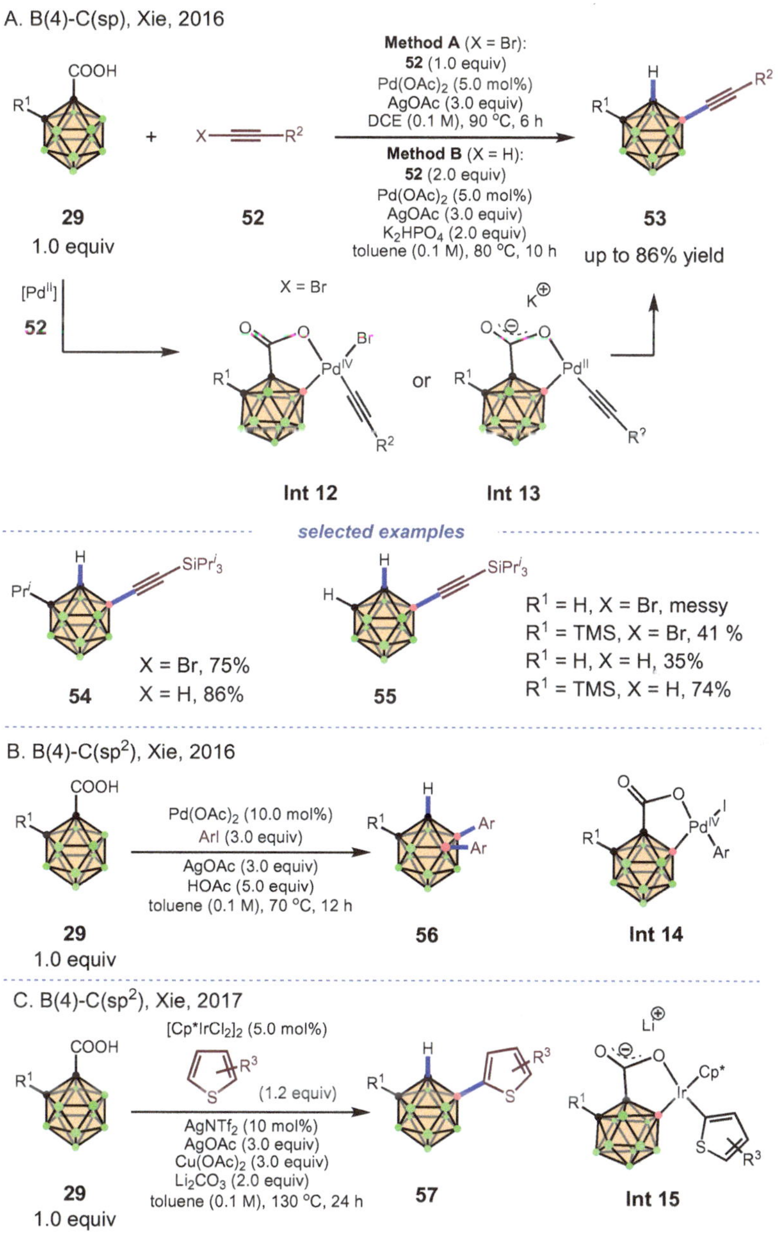

Fig. 7 Carboxy group assisted B(4) alkynylation and arylation.

investigation, both catalytic cycles initiated from the formation of the five-membered cyclic Pd(II) intermediate. Oxidative addition of $R^2C{\equiv}CBr$ affords a Pd(IV) intermediate (**Int 12**), while the ligand exchange by acetylide provides the Pd(II) intermediate (**Int 13**). When the free C_{cage}—H exists, the target products **55** was obtained in low yield ($R^2C{\equiv}CH$ as a substrate) or inseparable mixtures. Nevertheless, **55** could be acquired in a good yield when C_{cage} was blocked by trimethylsilyl (TMS). Akin to the mechanism of alkynyl bromides participated B(4) alkynylation, the B(4) arylation and B(4,5)-diarylation took place when aryl iodides were recruited as the coupling partners *(40)*. **Int 14** is proposed to be the key intermediate in this transformation. The ratio of diarylated products enhanced a lot when the temperature decreased to 70 °C. Silver acetate proved to be essential to realize the introduction of the second aryl group and thereof decarboxylation. The starting material with a free C_{cage}—H only afforded 41% corresponding diarylated product. While the same product was obtained in 69% yield when the cage carbon was protected by TMS. Following a kindred catalytic mode as terminal alkyne participated B(4)–alkynyl coupling, an Ir-catalyzed oxidative B(4)—H/C—H dehydrogenative coupling was discovered by the Xie group *(41)*. The thienyl α C—H activation accompanied with a following ligand exchange process gives key intermediate **Int 15**.

The carboxyl group guided B(4) functionalization could also expand to B(4,5)-dialkenylation (Fig. 8) *(42)*. In this case, a series of B(4,5)-dialkenylated *o*-carboranes with *E* configuration were achieved in presence of 10 mol% $Pd(OAc)_2$ catalyst and 4.0 equiv. of AgOAc. **Int 16** is a plausible intermediate in this transformation, followed by a *β*-H elimination to finish catalytic cycle. This transformation is sensitive to steric hinderance in aryl group. For example, employment of 1-methyl-4-vinylbenzene and 1-methyl-3-vinylbenzene gave rise to 76% and 72% yield, respectively. Nonetheless, the target molecule was not achieved when 1-methyl-2-vinylbenzene was investigated. It is possible that the hinderance of methyl group is not conductive to formation of **Int 16**. The yield decreased to 66% when C_{cage} was not blocked. Analogous to the Pd-enabled B(4,5)-dialkenylation, a rhodium catalyzed B(4,5)-dialkylation was developed by the Lu group *(43)*. The reaction condition is akin to the aforementioned B(4,5)-dialkenylation, allylic alcohols were selected as coupling partner. Enol generated after *β*-H elimination of rhodium intermediate, which underwent isomerization to produce the target molecules. Trace amount of product was detected when cage carbon was not blocked.

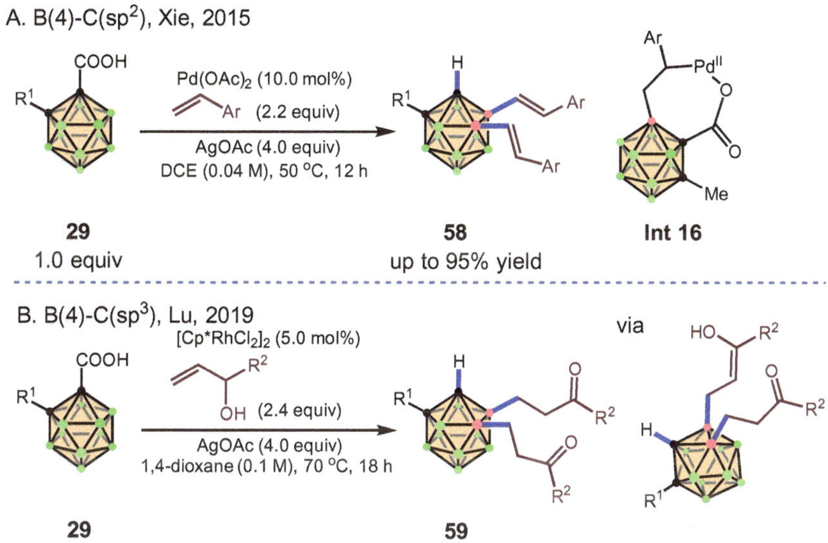

Fig. 8 Carboxyl group-guided cross-coupling of *o*-carboranes and alkenes.

The cascade dehydrocoupling reaction involved two different DG-tethered substrates is anticipated to possess the characters of both protocols. While it is challenging and is still underdeveloped. Xie and co-workers further extended this state-of-the-art strategy for constructing complex carborane derivatives (Fig. 9). The Ir-mediated preparation of carboran-oisoquinolinone derivatives **61** via a tandem B(4)-aryl and B(5)—N bond formation between **29** and benzamides was discovered (Fig. 9A) *(44)*. By adjusting silver salt and copper salt, the B(4)-aryl coupling products **62** were prepared. When **63** was treated with stoichiometric $Cp^*Ir(OAc)_2$, intermediate **64** was obtained. Control experiments exhibited that $Cu(OPiv)_2$ is very significant for the forge of B(5)—N bond in the second catalytic cycle. A plausible mechanism for B(5)—N formation was proposed at the end of Fig. 9. Leveraging the same mixed directing group approach, products **66** and **73** were successfully synthesized when **29** is treated with 2-pyridyl arene **65** or aryl carboxylic acid **72** by Ir catalysis *(45)*. Compound **71** was isolated and revealed that C—H activation is earlier than B—H activation in reaction.

It is well known that high bond energy is characteristic of B—O bond. The hydroxylation of B(4) was realized by the Xie group (Fig. 10A) *(46)*. The ^{18}O-labled experiment proved that the hydroxyl group in resultant products stem from oxygen gas. The structure of **Int 20**, formed by insertion

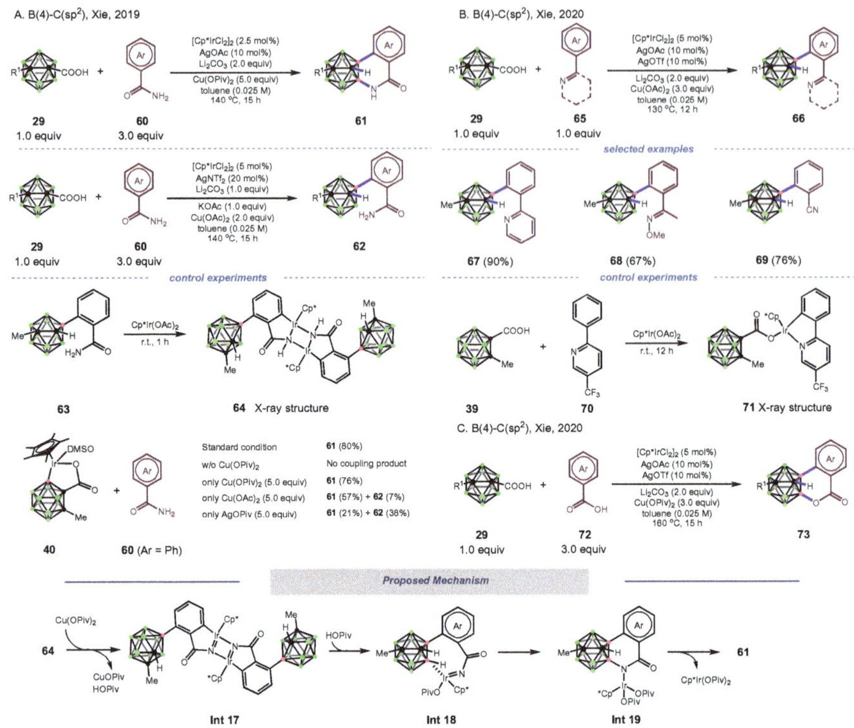

Fig. 9 Ir-catalyzed oxidative cascade dehydrogenative cross-coupling.

of O_2 to B(4)—Rh bond was proposed. In 2021, the Lee group reported a mono −/di-selectivity tunable Pd-catalyzed B(4) or B(4,5) acetoxylation of *o*-carborane acids *(47)*. The acetoxyl group was donated by phenyliodonium diacetate. A Pd(IV) intermediate **Int 21** was proposed. Product **85** was achieved in a moderate yield despite the existence of unblocked C_{cage}–H. Notably, a 1.2: 1.9: 1.0 stereoisomeric ratio of **88** was detected, which indicates B(4) and B(5) monosubstituted enantiomer generated from the achiral catalytic system. In principle, the ratio of B(4)-B(4′): B(4)-B(5′) & B(5)-B(4′): B(5)-B(5′) should be 1: 2: 1, which matches the experimental data.

In 2016, the Xie group came up with two methods for preparing B(4)-aminated *o*-carborane (Fig. 11A) *(48)*. Merging **29**, *O*-benzoyl hydroxylamines and palladium catalyst in one-pot, target product **90** is afforded. The five membered metallacycles **Int 22** is plausible the key intermediate. One the other hand, a ruthenium catalyzed B(4)—N coupling was disclosed. At almost the same time, the Yan group reported an Ir-catalyzed B(4)-amination *(49)*. Aryl azide, alkyl azide, as well as sulfonyl azide worked

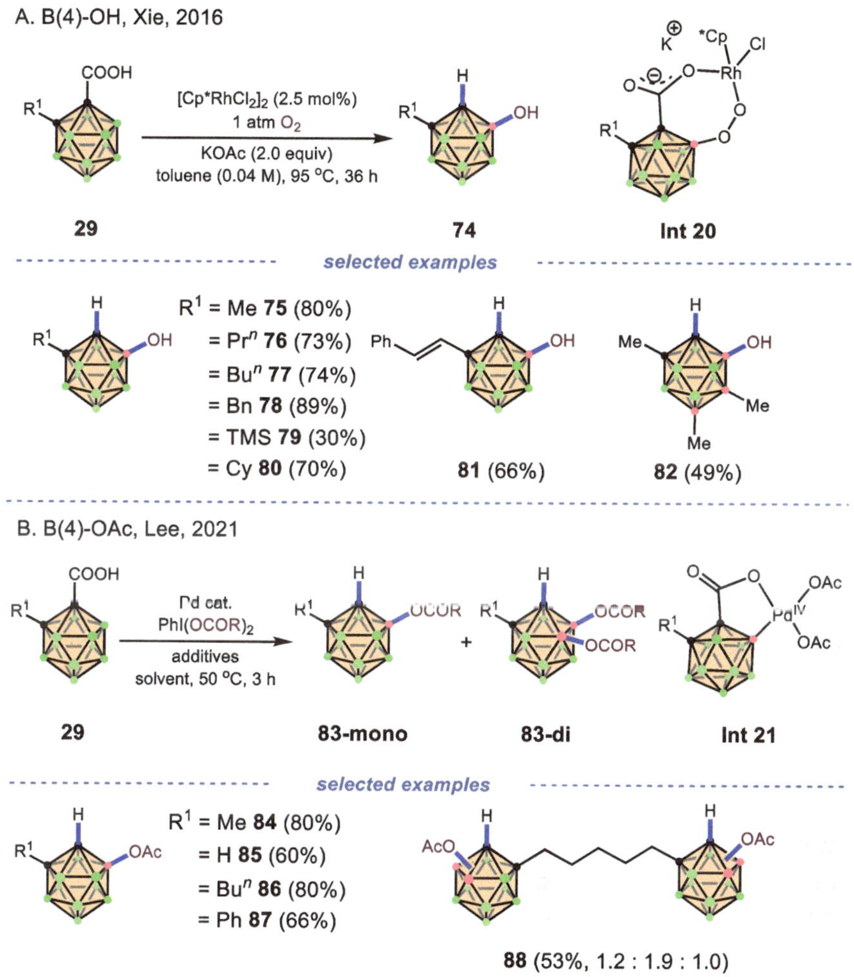

Fig. 10 Metal catalyzed B(4)—O construction.

as the precursor of amine group. The intermediate **Int 23** was detected by high-resolution mass spectrum (Fig. 11B). To further exploit the donor of amine group, the Lee group developed a rhodium or iridium-catalyzed B(4) amidation in a wide range of *o*-carboranes with alkyl-, and aryl-substituted dioxazolones (Fig. 11C) *(50)*. It is noteworthy that the Ir-enabled cross-coupling of *o*-carborane **29** and dioxazolones **93** has been modified by Lee and co-workers thus the reaction can be facilitated under a solvent-free ball milling condition (Fig. 11C, Eq. 1, method B) *(51)*. It's the first example of mechanochemistry, which offers brilliant atom efficiency and reduces

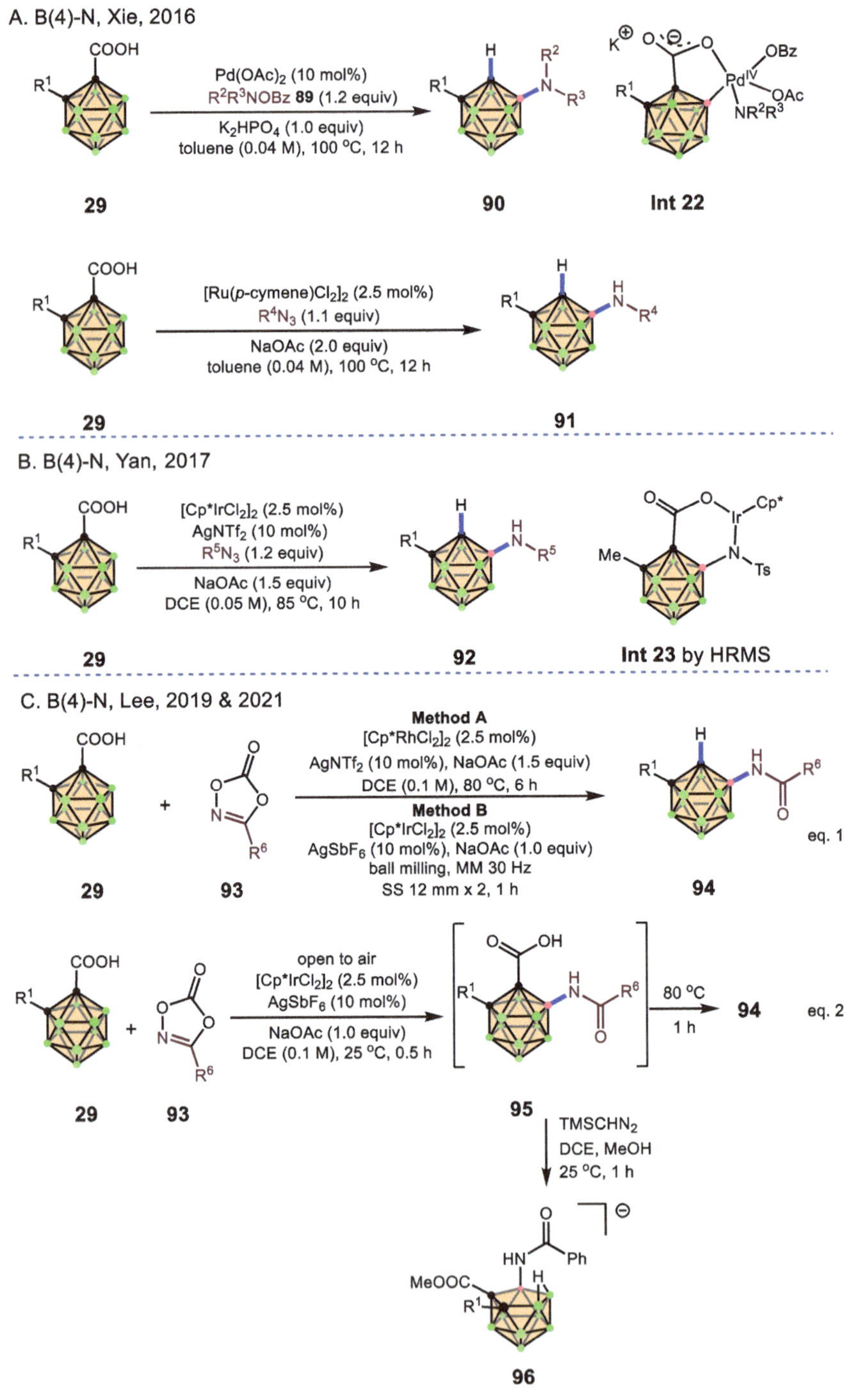

Fig. 11 Transition-metal catalyzed B(4)—N formation.

energy consumption under solve-free conditions, being applied into functionalization of carboranes. The reaction of **29** (solid, 1.0 equiv.) and **93** (solid, 1.2 equiv.) in the presence of $[Cp^*IrCl_2]_2$ (2.5 mol%), $AgSbF_6$ (10.0 mol%), NaOAc (1.0 equiv.) and milling balls (12 mm × 2) performed at 30 Hz for 1 h and gave the yields up to 99%. According to their report, NaOAc is crucial for this transformation and NaCl should be added as a solid auxiliary when phenyldioxazolone bearing a strong EWG was employed. This strategy also performs well for C(2) unsubstituted *o*-carborane acid **29** (R^1=H), gave the isolatable B(3)- and B(4)-amidated mixtures up to 82% yield. In addition to ball milling condition, the Ir-catalyzed B(4) amination can also be set up in solvent phase at ambient condition (room temperature, open to air; Fig. 11C, Eq. 2). The directing carboxyl group can be removed when the temperature lifted to 80 °C. *nido*-Carborane derivatives (**96**) was acquired when the carboxylic product **95** was esterificated by $TMSCHN_2$ reagent.

With the support of carboxyl directed Ir-catalyzed B(4) halogenation, a series of previously unreported halogenated carboranes were prepared in a good yield *(52)*. NIS, NBS as well as $FeCl_3$ donate the halogen atoms in this reaction. In addition, the preparation of B(4,5)-diiodinated products was enabled by Pd catalysis (Fig. 12).

2.1.2 Imine directed B—H functionalization

The carboxyl group represents a kind of weakly coordinating directing group, while strongly coordinating *N*-containing directing groups are also utilized frequently in C—H functionalization.

In 2017, the first imine-assisted transition-metal catalyzed functionalization of boron vertex was reported by the Yan group (Fig. 13A) *(53)*. This report detailed the Pd-catalyzed B(4) or B(4,5) arylation via transient generated imine

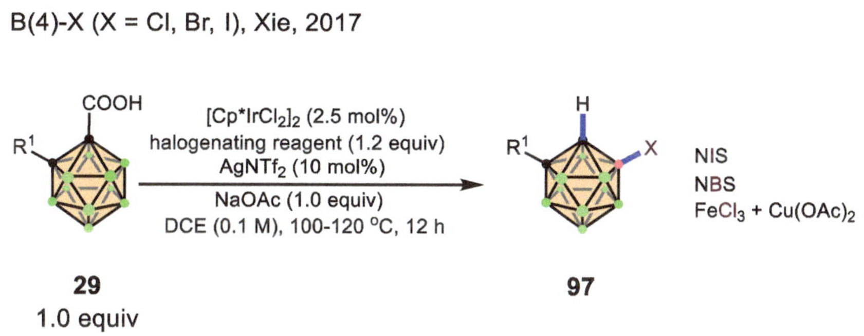

Fig. 12 Iridium catalyst promoted B(4)-halogenation.

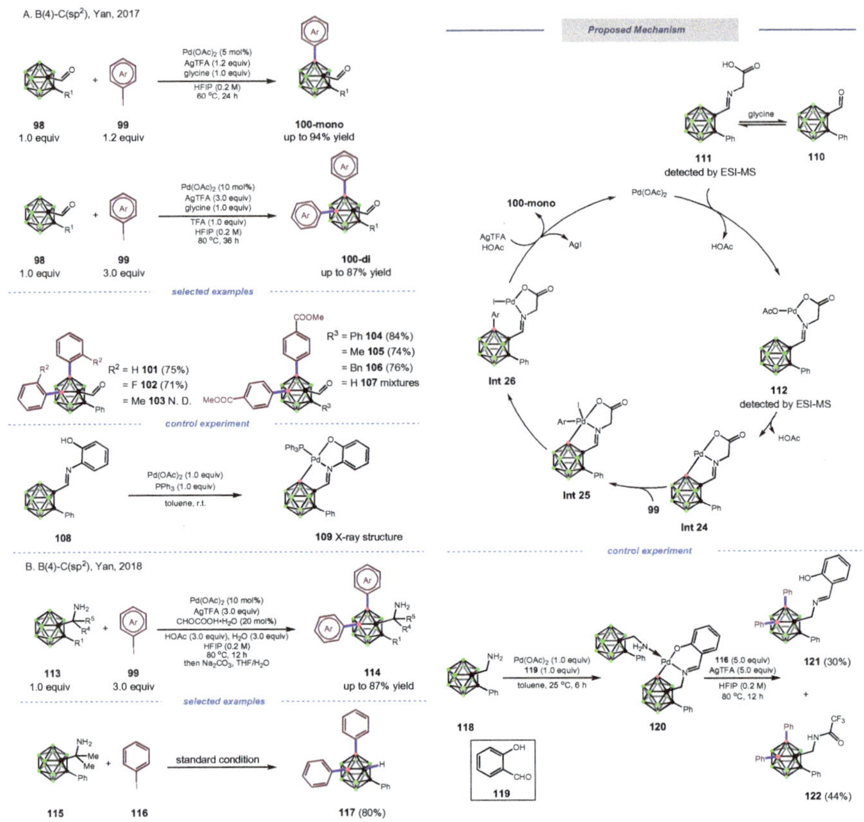

Fig. 13 B(4)-arylation or B(4,5)-diarylation enabled by palladium-catalyzed transient directing strategy.

species **111**, likewise the Pd-catalyzed C(sp^3)—Ar cross-coupling of aliphatic aldehydes and aryl iodide *(54)*. Leveraging the formation of imine directing group by condensation between an aldehyde with an amine group in situ, Yan and co-workers found that 1-CHO-2-R-*o*-$C_2B_{10}H_{10}$ (**98**) can be expediently converted to corresponding B(4) or B(4,5) arylated *o*-carboranes (**100**).

The proposed imine intermediates **111** and **112** were detected by high-resolution mass spectrometry (HRMS). In addition, the X-ray structure of **109**, an analog of **Int 24** was obtained by mixing imine **108** with stoichiometric amount of $Pd(OAc)_2$ and PPh_3. In complex **109**, a five/five-membered bicyclic palladacycle was formed via the chelating coordination of the N^O ligand and direct activation of B(4) vertex at palladium center. In addition, compound **109** has been shown capable to catalyze this transformation. Mono- and di- selectivity is controlled by adjusting the

catalyst loading, as well as the amount of aryl iodides and AgOAc. This reaction is sensitive to steric hinderance. Compound **103** with a bulky *ortho*-methyl group was not detected, which is evidence for the existence of five-coordinate intermediate **Int 25**. Complex mixtures were observed when the cage carbon was not blocked, which allows the development of palladium catalyzed poly-arylation of *o*−/*m*−/*p*-carboranes by Yan's team *(55)*. In light of the pioneering work by Ge and co-workers *(56)*, Yan targeted the Pd-catalyzed B(4,5)-diarylation of 1-CH_2NH_2–2-R-*o*-$C_2B_{10}H_{10}$ (**113**) via a transient bidentate imine directing group *(57)*. Many functionalities such as halo, OMe, CF_3, ketone, and ester were tolerated in this transformation. Compound **120**, which underwent B(4,5)-diarylation in the presence of 5.0 equiv. of **116**, indicates the structure of the key intermediate. The formation of **120** also demonstrated that bare NH_2 group may poisoned Pd center and the in situ generated imine intermediate excluded this possibility. According to substrate scope investigation, NH_2CMe_2 group turned out to be a traceless directing group under the standard condition (**117**).

The substrate scope of *o*-carboranes shown in Fig. 13 suggests that the excellent B(4) or B(4,5) selectivity is clearly tied to block of C_{cage} (without DG). After scrutinizing the component of mixtures in **107**, several kinds of mono-, di- or even triarylated *o*-carboranes were identified by gas chromatograph-mass spectrometry (GC–MS), which opens up prospects of preparation of poly-arylated *o*−/*m*−/*p*- carboranes. Condensation of **123** with *o*-aminophenol led to *p*-carboranyl imine **124**, which contains a bidentate N^O ligand. The palladium complex **125** was acquired when **124** was treated with equivalent amount of $Pd(OAc)_2$ and PPh_3. The structure of **125** was established by X-ray measurement. Complex **125** is readily reacted with 7.5 equiv. of aryl iodide **126** and Ag_3PO_4 in hexafluoroisopropanol (HFIP) at 80 °C and corresponding B(2,3,4,5,6)-pentaarylated compound **127** was separated in 39% yield. Compare with **124**, the glycine-derived DG was more effective for catalytic transformations, presumably due to the higher reactivity of corresponding pincer palladium intermediate. To evaluate the poly-arylation system, Pd-catalyzed reactions of **123** with x equivalent (x = 1, 2, 3, 4) of **126** and x equivalent of Ag_3PO_4 in the presence of 1.0 equiv. of glycine were carried out in HFIP at 60 °C. The same sort of arylated *p*-carborane derivatives with an exhaustive distribution of yield contingent upon the loading amount of the arylation reagent as well as silver salt was observed. B(2)-mono-arylated aldehyde **128**, B(2,4)-di-arylated aldehyde **129**, and B(2,4,5)-tri-arylated aldehyde **130** were

monitored under a low loading amount (x = 1, 2) of **126** and silver salt. When higher amount of **126** and silver salt were loaded (x = 3,4), B(2,3,4,5)-tetra-arylated aldehyde **131** and B(2,3,4,5,6)-penta-arylated imine **132** can be observed, yet in low yield (3–17%). The structures from **128** to **132** were confirmed by multinuclear NMR spectroscopy, HRMS, and X-ray analysis. The fivefold B—H activation was therefore proposed to proceeds in the sequence from vertex B(2) to B(4), then to B(5), and eventually to B(3) and B(6). In 2018, the Duttwyler group described preparation of B(2,3,4,5,6)-penta-arylated mono-carborane anion through a Pd-mediated consecutive activation of the adjacent B—H bonds *(58)*. The sharp distinction between B—H activation sequence in these two works is attributable to steric bulk that arises from the combination of bidentate DG and newly installed aryl group(s). When the third stage B—H activation was studied, a three-coordinated monocyclic Pd(II) center with a T-shaped geometry (**133**) was discovered. The bicyclic transition-metal containing pincer structures such as **109**, **120**, and **125** are always considered as the key intermediate when bidentate DGs are employed. The mono-cyclic coordination mode in presence of a bidentate DG is rarely reported in the past, let alone a switchable coordinative/catalytic mode tuned by steric hinderance. Treatment of **133** with 1.2 equiv. of **126** and silver salt at 80 °C gave the tri-arylated **134** in 30% yield. By employing aniline to form a transient mono-dentate imine in situ, the reaction of **129** with **126** afforded the tri-arylated aldehyde **130** in 19% yield, which indicates the existence of the O atom is not essential at the late stage of poly-arylations. After optimizing the catalytic conditions, a series of poly-arylated *o*−/*m*−/*p*-carboranes (**135–137**) were synthesized. This reaction condition tolerates a lot of functionalities, including drug fragments such as ibuprofen, estrone, and empagliflozin. The DGs in products can be removed or transferred to other functional groups readily. In addition, the poly-arylated carboranes with four or five aryl blades are sterically congested, as a result (P) and (M)-enantiomers existed equally in single-crystal structure of **136a** (Fig. 14).

A Rh-catalyzed cascade annulation including B(4)—H and aryl C—H activation was disclosed by the Xie group *(59)*. A novel C,B-substituted carborane-fused *N*-polyheterocycles **140** were constructed when carboranyl imine **138** (R^1 = alkyl and benzyl) reacted with vinyl ketones **139** in presence of $[Cp^*RhCl_2]_2$ and other additives. Imine directed regioselective formation of Rh(III)—B(4) bond, then alkene insertion and nucleophilic cyclization happened consecutively in the first catalytic cycle that produced intermediate **141**. Then ketone directed aryl C—H activation, nucleophilic

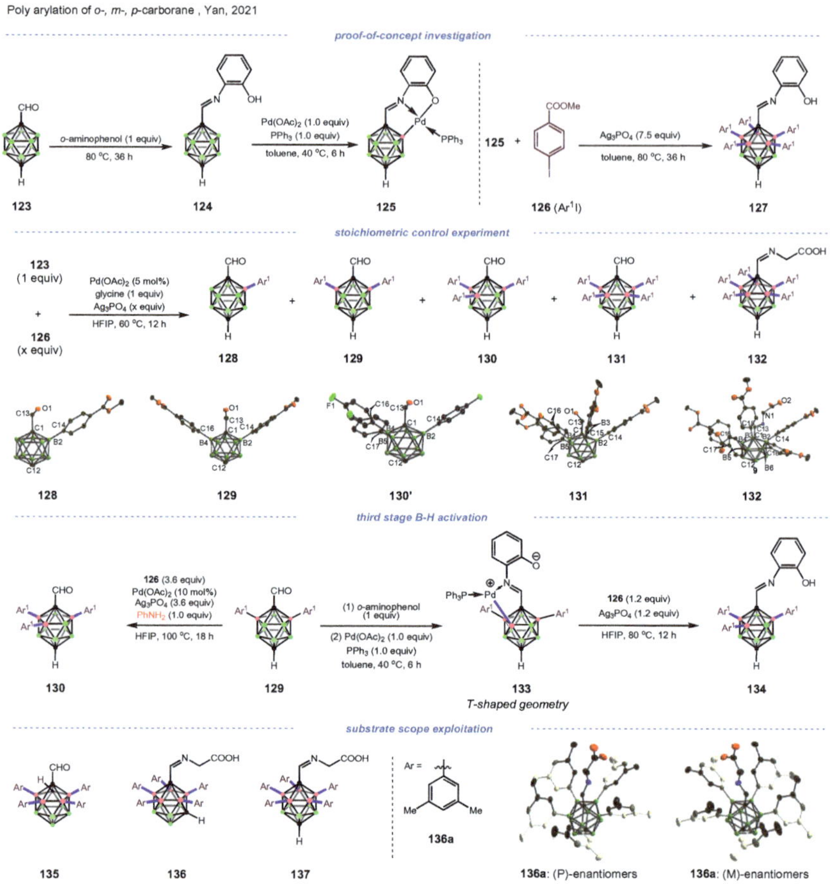

Fig. 14 Variable metal chelation modes and activation sequence in Pd-catalyzed poly-arylation of carboranes.

cyclization and dehydration steps were occurred sequentially. The intermediate **142** was oxidized to the final product **140** at last (Fig. 15).

Despite the state-of-the-art imine directed Pd-catalyzed arylation of carboranes were revealed by Yan and co-workers exhaustively, a little B—I coupled byproducts were detected in some cases, owing to large among of iodide anion left in the system after arylation. On the other hand, the range of substrates is limited to aryl iodides so that alkenyls or alkyls cannot be grafted to carborane by these approaches. To overcome these drawbacks, a Rh-catalyzed decarbonylative cross-coupling between *o*-carboranes and twisted amides was discovered by the Yan group in 2021 *(60)*. This reaction features precise B(3) or B(3,6) selectivity, additive free, and it can be even applied for

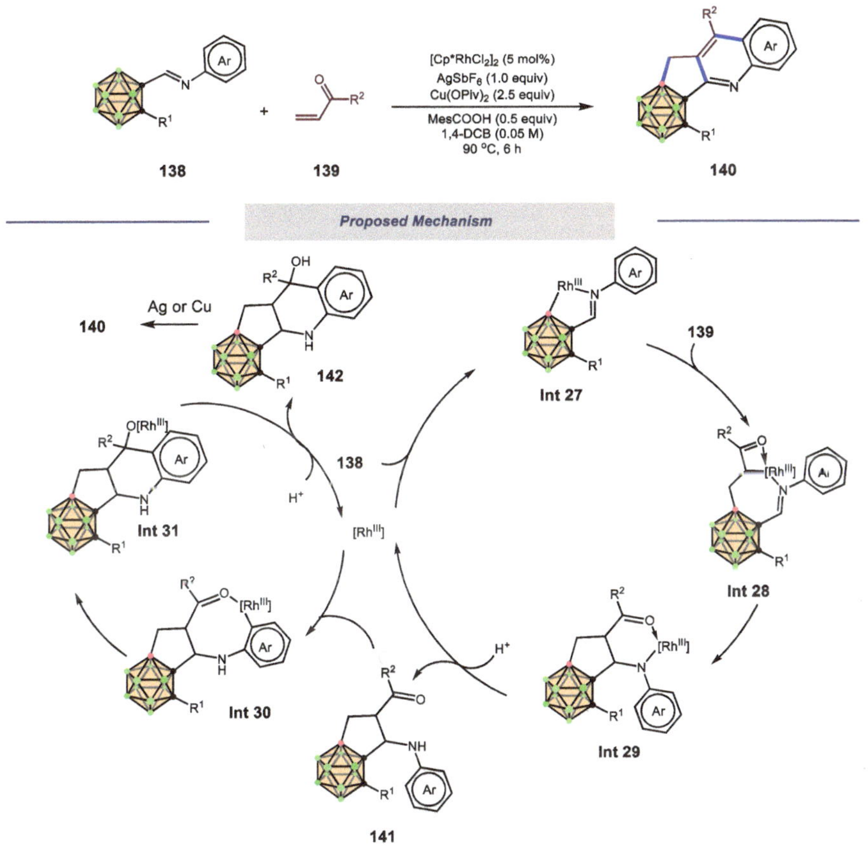

Fig. 15 Rh-mediated cascade annulation including B(4)—H and C—H activation.

late-stage carboranylation for versatile drug-like molecules (Fig. 16). After screening various of coupling partners, the *N*-acyl-glutarimides (**144**) exhibits the best reactivity. This kind of twisted amide was developed by Szostak et al. in 2015, featuring perpendicular twist of the amide bond (Winkler–Dunitz distortion parameters, $\tau = 85.7°$; $\chi_N = 5.6°$ cf. planar amide) thus renders high reactivity of *N*-acyl-glutarimides *(61)*. More than 40 products were prepared when 1.0 equiv. of **143**, 2.2 equiv. of **144**, and 5 mol% [Rh(*cod*)Cl]$_2$ was mixed. Halogens, CF_3, OCF_3, OMe, heteroaromatic rings, alkenyl, alkyl, galactose, as well as several drug fragments (probenecid, diacerein, fenofibrate, difufenican, estrone, indomethacin) were all tolerated. Moderate to good yields were obtained when H or EWGs were adorned at *para-* position of aryls. The yield declined a little when EDGs locate at *para-* positions of aryls (**151–153**). The yields decreased in presence of steric hinderance (**157–159**, **160–162**). Apart from

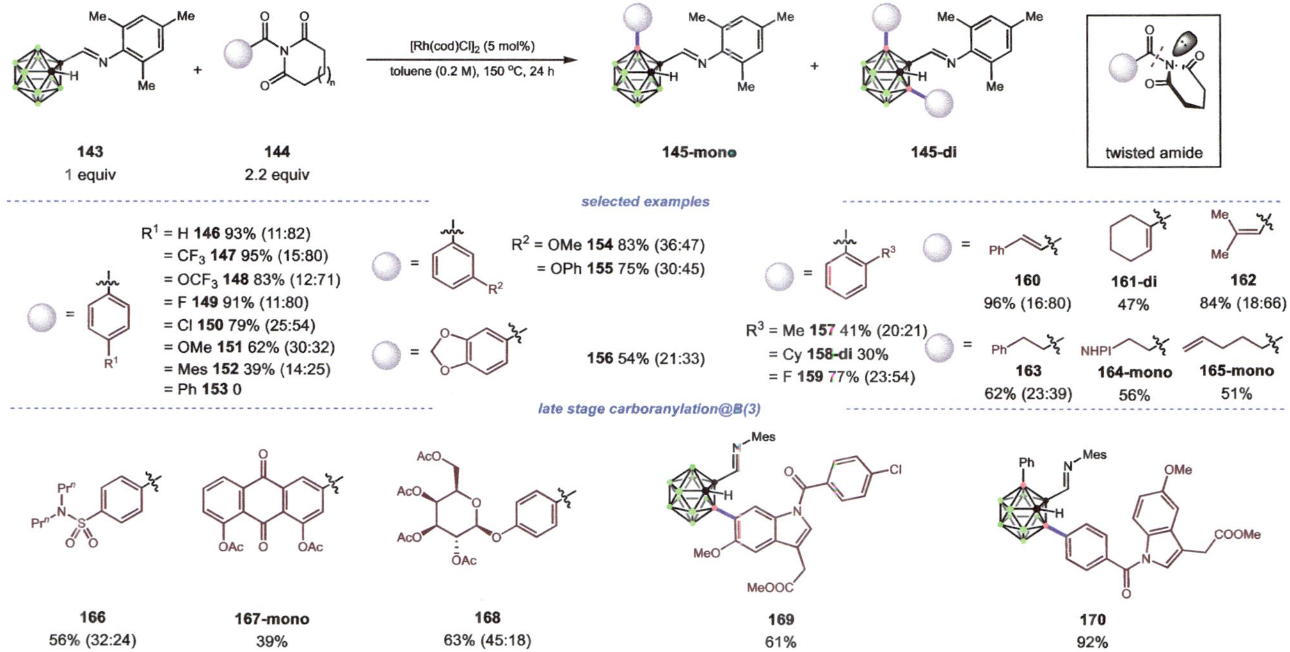

Fig. 16 Rh-catalyzed decarbonylative cross-coupling between *o*-carboranes and twisted amides.

2,4,6-trimethylphenyl, 4-fluorophenyl or 4-chlorophenyl based imine directing group also worked very well. No reaction occurred when another carbon vertex was blocked by methyl.

Density functional theory (DFT) studies clarified that B(3)–H activation takes place prior to C—N cleavage (Fig. 17). After the formation of **Int 34**, a ligand exchange between 1-benzoylpiperidine-2,6-dione and 1,5-cyclooctadiene (cod) occurs, followed by oxidative addition with a rate-determining 38.7 kcal mol^{-1} barrier. According to calculation, B(4) activated counterparts are found energetically unfavorable. Afterward, HCl exchanges with glutarimide ligand, followed by decarbonylation and reductive elimination to finish the cross-coupling. Pathways without decarbonylation are found to require a much higher barrier. To end up the reaction, cod re-enters the coordination sphere to expel CO, affording **Int 32-Ph** as the resting state of this system.

2.1.3 Pyridyl directed B—H functionalization

Pyridyl represents a kind of strongly coordinative directing group, and it is broadly explored in transition-metal catalysis realm. The pyridyl is easily installed on cage carbon by the reaction between 2-chloro-pyridine and *o*-carborane, treated with potassium bis(trimethylsilyl)amide (KHMSD).

In 2017, the Yan group reported the first pyridyl-guided B(3) or B(3,6) acyloxylation (Fig. 18A) *(62)*. In the presence of 5 mol% amount of $[Cp^*RhCl_2]_2$ and 2 equiv. of oxone, the reaction of **171** with 3 equiv. copper carboxylate in DCE at 120 °C for 24 h gave B(3)-acyloxylated or B(3,6)-di-acyloxylated *o*-carboranes in a high yield. No reaction took place in absence of rhodium catalyst. Addition of copper carboxylate is crucial for the reaction, although the role of copper ion is unclear. The combination of 3.0 equiv. of $Cu(OH)_2$ and 6.0 equiv. of carboxylic acid **172** are employed when the corresponding copper carboxylate is not commercially available. The B(3) selectivity is robust, no matter the cage carbon is blocked or not. This method accommodates either aryl carboxylic acids or alkyl carboxylic acids with various type of substitutions. In addition, the late-stage carboranylation@B(3) of indomethacin was achieved (**178**). However, the oxidative condition and high temperature impedes the application of carboxylic acids containing reductive functionalities (**179–182**).

A palladium-catalyzed tunable B(3)-acyloxylation and B(3,4)-di-acyloxylation was reported by Lu and co-workers recently (Fig. 18B) *(63)*. The reaction can be set up under room temperature, reductive functionalities are compatible even if ammonium cerium nitrate was employed as

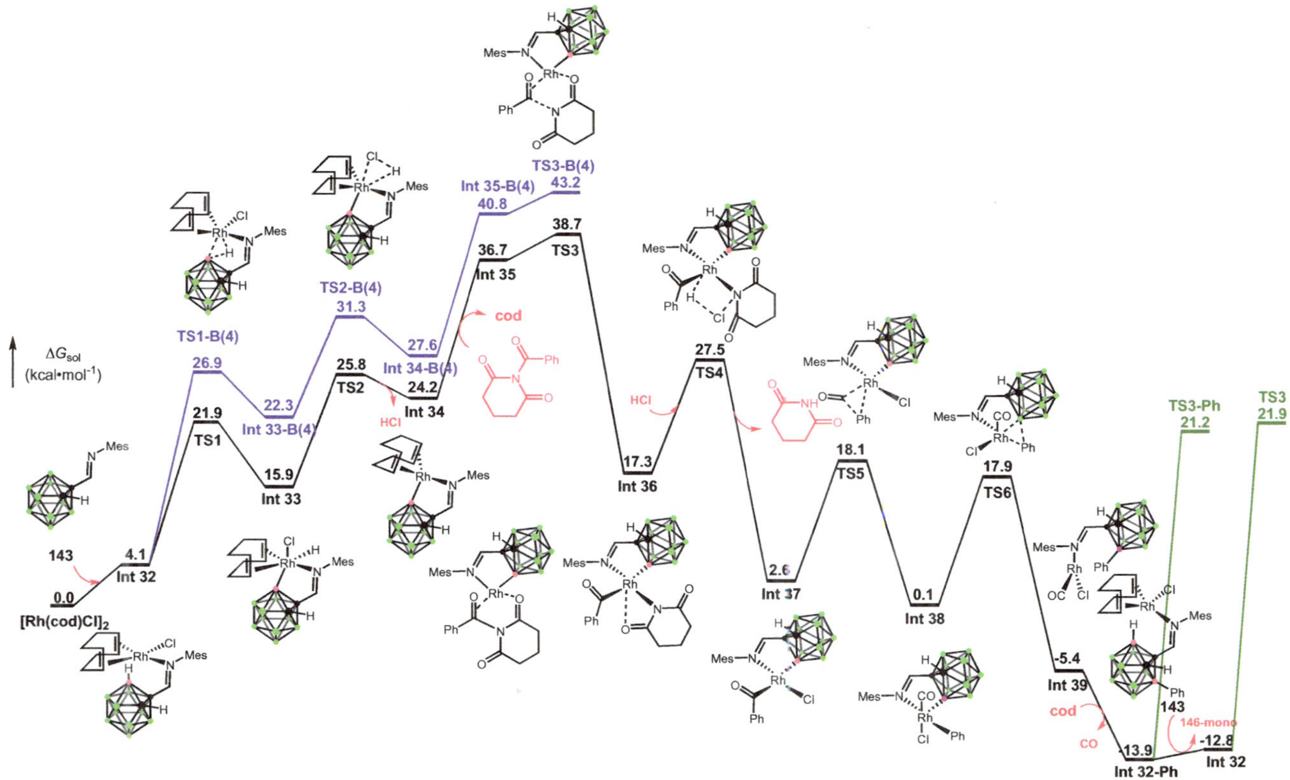

Fig. 17 DFT-computed pathways for the Rh-catalyzed reaction between substrates **143** and 1-benzoylpiperidine-2,6-dione. The structures of B(4) activated intermediates and transition states are omitted.

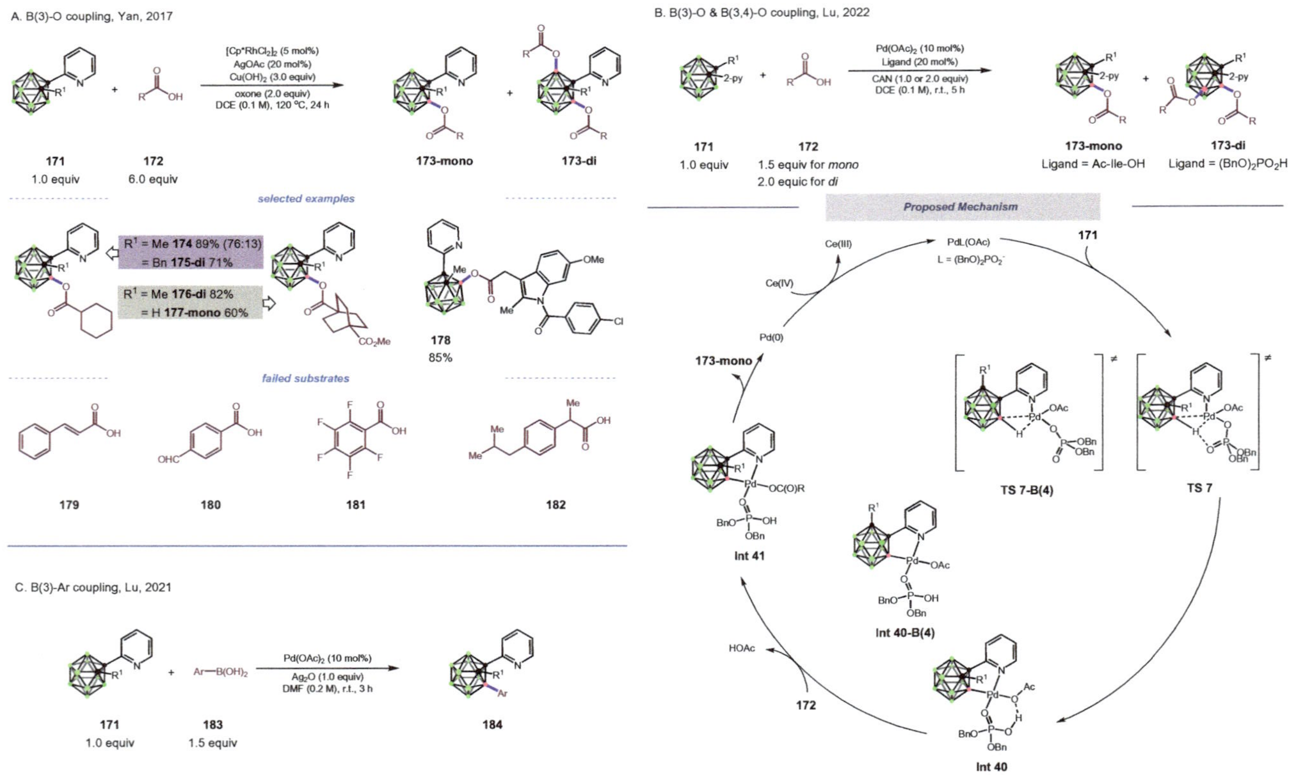

Fig. 18 Pyridyl guided functionalization of B(3) position.

the oxidant. In addition, they found generation of mono-B(3)-substitution can be guaranteed by using 20 mol% *N*-Acetyl-L-isoleucine (Ac-Ile-OH) as the ligand and di- B(3,4)-substitution can be realized by adding 20 mol% $(BnO)_2PO_2H$ as the ligand. DFT calculations suggested in the first step B(3)—H bond rather than B(4)—H bond activation is more feasible. Furthermore, B(4)—H activation is more favorable than B(6)—H activation in the second B—H activation step according to its lower energetic barrier. In a related study, B(3) mono-arylation of 1-R^1–2-pyridyl-*o*-$C_2B_{10}H_{10}$ was achieved by palladium catalysis under room temperature (Fig. 18C) *(64)*.

Leveraging pyridyl directed B(3)—H activation, bis(*o*-carborane)s connected with B(3)–B(3′) and B(3)–B(6′) bonds were prepared by the Cao group for the first time *(65)*. $Pd(MeCN)_4(BF_4)_2$ performed much better than other Pd catalysts. As B(3) and B(6) are hard to be distinguished from each other, B(3)–B(3′) coupling product **186** and B(3)–B(6′) coupling product **187** were formed simultaneously. It is reasonable to supposed that the reaction is initiated by pyridyl guided B(3) activation, which forms a cyclicpalladium(II) species. Pd(IV) intermediate is generated after oxidation of Pd(II) by $AgNO_3$ and it triggers the subsequent B(3′)—H or B(6′)—H activation (Fig. 19).

In fact, the regioselectivity in pyridyl guided B—H functionalization is still elusive, and it is influenced by several factors like catalyst, steric hinderance, ligand, as well as structure of substances. In 2021 the Yan lab managed to prepare a series of B(4)-acylmethylated pyridylcarboranes enabled by Rh(III) catalysis (Fig. 20) *(66)*. Pyridyl carborane **188** were converted to product **190** smoothly in presence of 6 mol% $[Cp^*RhCl_2]_2$ or $Cp^*Rh(OAc)_2$ in HFIP under 100 °C. α-Carbonyl sulfoxoniums (**189**) have emerged as reliable source of acylmethyl groups. The reaction completed just in 30 min for alkyl α-carbonyl sulfoxonium starting materials and the reaction time should be elongated to 18 h when aryl α-carbonyl sulfoxoniums were employed. Various functionalities are compatible in this approach. Based on the detection of key intermediates (**Int 43, Int 45**) by

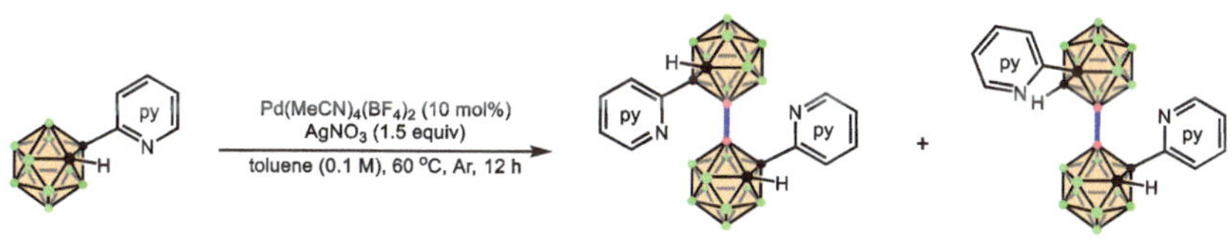

185 186 187

Fig. 19 Pd-catalyzed B(3)–H activation/oxidative dehydrogenative coupling for synthesis of bis(*o*-carborane)s.

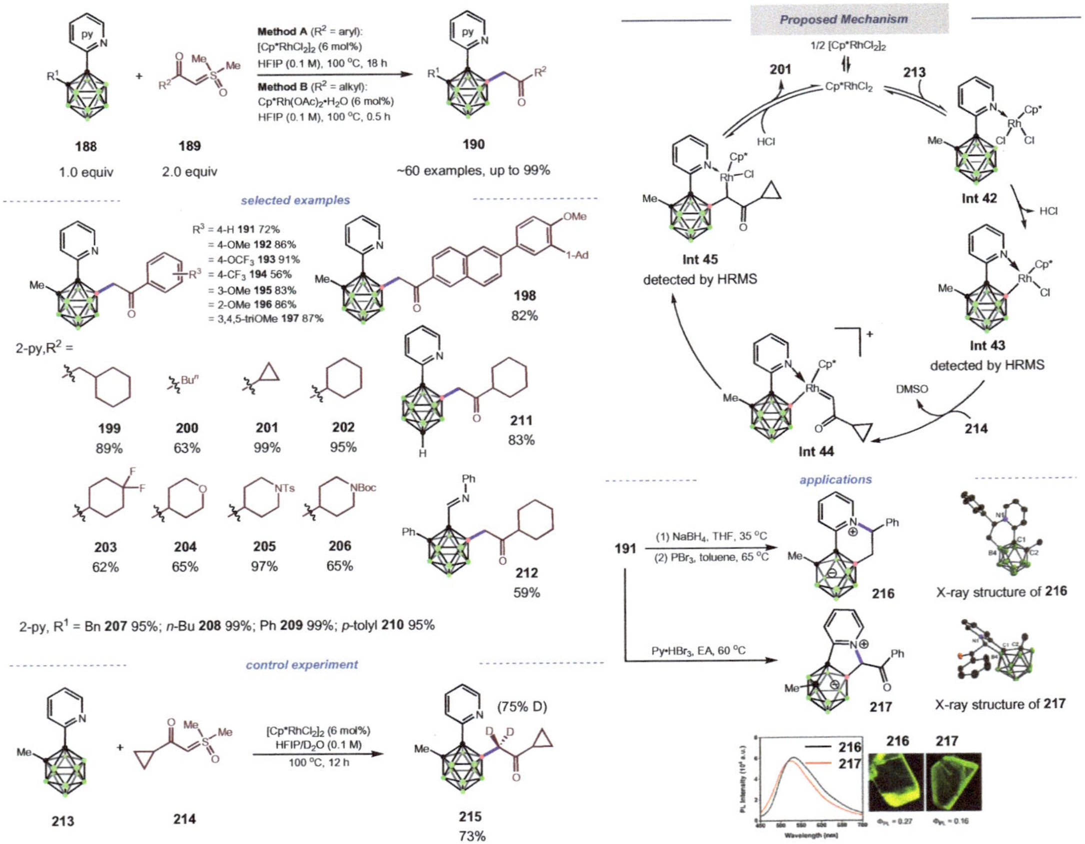

Fig. 20 Rh-catalyzed B(4)–H acylmethylation of pyridylcarboranes.

HRMS, a plausible mechanism was proposed. The reaction of **213** and **214** under standard conditions in a mixed solution of D_2O and HFIP gave rise to the incorporation with a deuterium level of 75%. However, deuterium incorporation at the enolizable position was not observed when **201** was stirred in D_2O and HFIP in absence of any other reagents. The control experiments demonstrate a reversible enolization and coordination process of the product with the rhodium catalyst. The equilibrium between **Int 45** and **201** was considered as the reason that prohibit the iterative B—H activation pathway, and thus renders regioselective mono-acylmethylation of *o*-carboranes. A drawback of pyridyl directing group is that it cannot be removed. While the investigators convert the product **191** to carborane-fused quinolizinium (**216**) and indolizinium (**217**) ingeniously. These two zwitterions have shown unexpected efficient emissions in their crystalline states. Similar works were also reported by the Lee and the Xie lab in 2022 *(67)*.

To expend the methodology of cross-coupling between *o*-carborane and *N*-acyl-glutarimides, pyridyl directed B—H arylation was investigated. In sharp contrast with imine directed B(3,6)-arylation, the use of 1-methyl-2-pyridyl-*o*-$C_2B_{10}H_{10}$ resulted in a series of B(3,5)-arylated products (**218**) *(60)*. Since B(3)-mono-arylated and B(4)-mono-arylated byproducts can be monitored individually in some cases, the reason of B(3,5)-selectivity presumably results from steric hinderance. B(4)-alkylated products with rearrangement of alkyl group was achieved when the methyl-containing tertiary *N*-acyl-glutarimides was tested. A rational exposition for this unique rearrangement is that the consecutive β-elimination and 1,2-migration can release the repulsion between third order alkyl and *o*-carborane in **Int 46**. According to Yan's article, no reaction occurred when methyl was replaced by larger alkyls. A plausible reason is that the sharply increased repulsion caused by the larger alkyl compromised the stability of **Int 46** (Fig. 21).

The first pyridyl assisted B—P construction emerged in 2022 (Fig. 22) *(68)*. Lu and co-workers harnessed 1.2 equiv. of phosphines **222** and 1.0 equiv. of *nido*-carborane **221** as coupling partners to construct B—P bond by Pd-catalysis. EWGs, EDGs, furyl, thienyl, as well as pyridyl were tolerated (**224–231**). The declined yield of **228** resulted from steric hinderance of 2-methyloxyl group. A tandem sequence of ligand exchange, electrophilic substitution, and recovery of Pd catalyst was proposed for the formation of **223**. Another *nido*-carborane–PPh_3 zwitterion was produced via palladium catalysis by the Yan group in 2014 *(69)*, and we will discuss this reaction in Section 2.2.1.

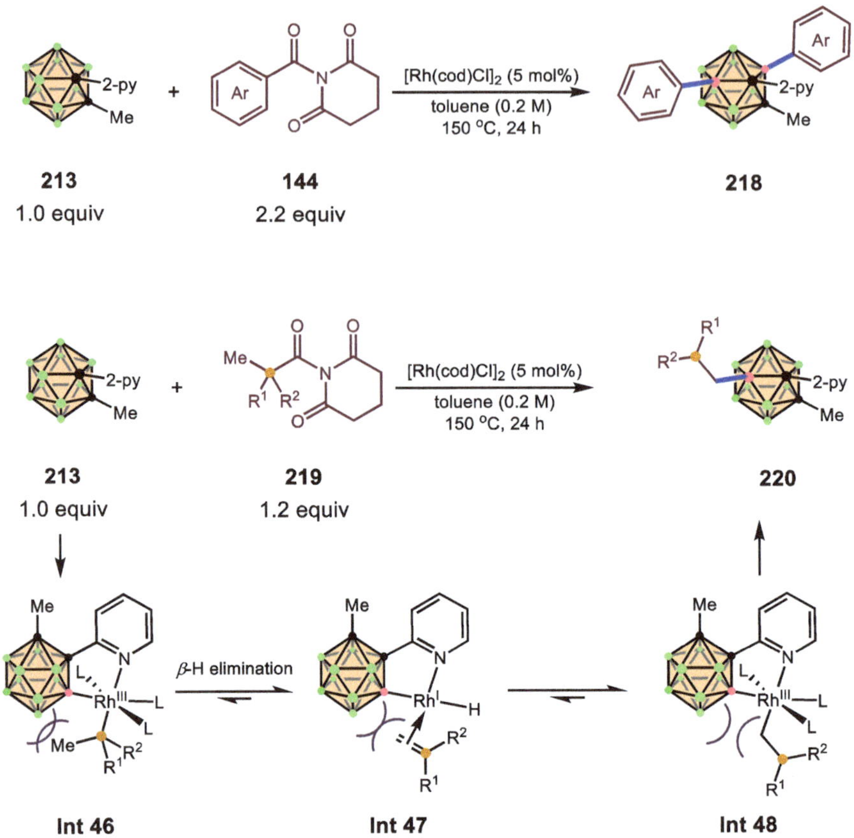

Fig. 21 Pyridyl directed B(3,5)-arylation and B(4)-alkylation with a rearrangement.

2.1.4 Amide directed B—H functionalization

Amide is a kind of important feedstock in organic chemistry, and either C(O) or N position in amide groups can be flexibly installed on carbon or boron vertices by conventional condensation reaction, Buchwald-Hartwig type B—N coupling, or even by recent developed carboxy group guided B(4) amination. With plentiful carborane-combined amides in hand, novel and copious transition-metal-catalyzed B—H functionalization have been unveiled in these years.

The year 2005 saw the pioneering use of 8-aminoquinoline moiety (8-AQ) as a directing group in Pd-catalyzed arylations of C(sp^2) and C(sp^3)—H by Daugulis et al., which has been cited more than 1000 times *(70)*. Enabled by chelation effect from the bidentate 8-AQ auxiliaries, cheap and abundant base-metals (iron, nickel, copper etc.) can replace noble metals

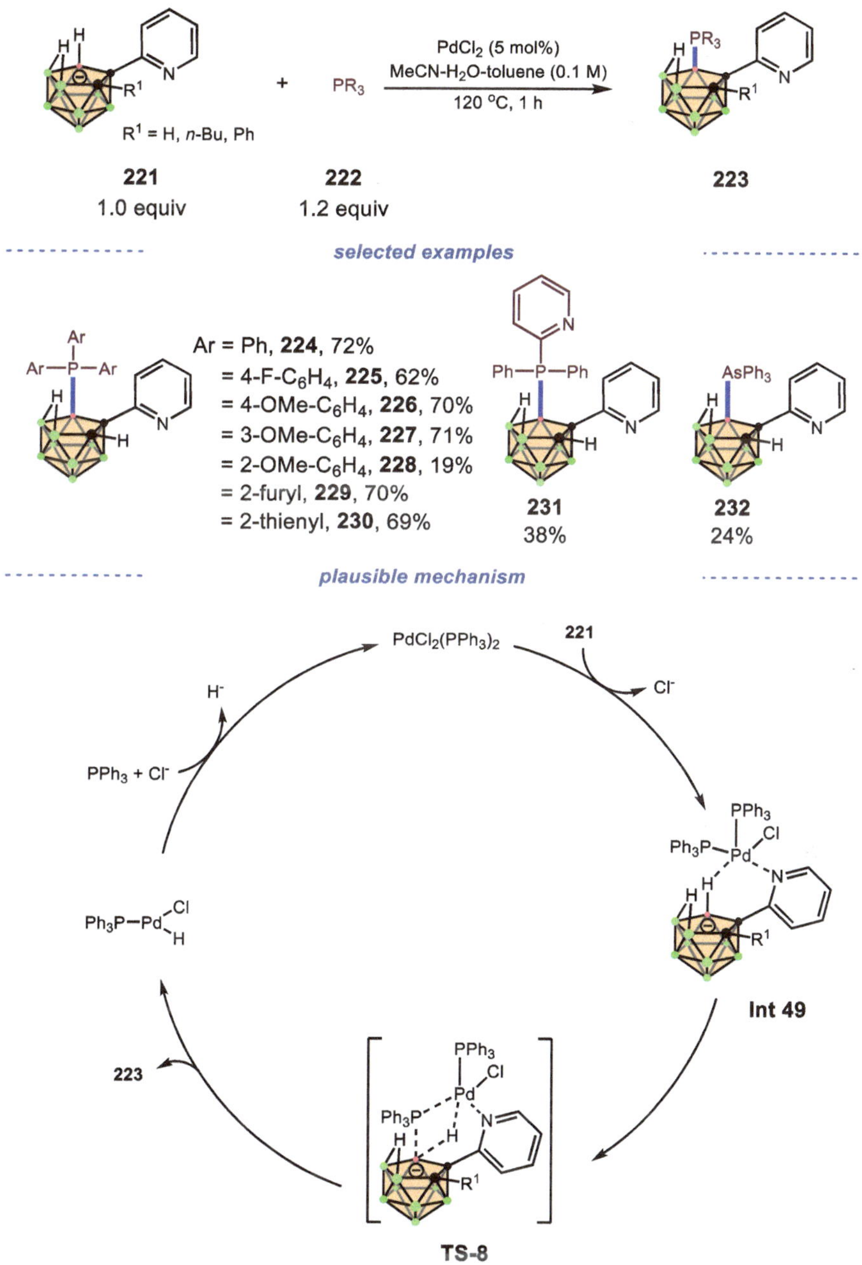

Fig. 22 Pyridyl directed Pd-catalyzed B—P formation of *nido*-carborane.

and catalyze a series of reactions via C—H activation. Triggered by superiority of metals in first transition series, the Xie group have uncovered a copper-catalyzed B(4) cyclization of 8-AQ embedded carborane **233** and diaryl acetylenes **45** (Fig. 23A) *(71)*. The Cu(III) in **Int 50** can be stabilized by bidentate DG and rendered the subsequent insertion of **45** to B(4)–Cu bond followed by reductive elimination. While a series B(4)-alkynylated products **236** were acquired when terminal aklynes and equimolar amount of copper triflate were used. **Int 51** is considered as the key intermediate in this transformation.

Copper mediated mono- or di-sulfenylations of phenyl moiety were proven effective previously and had also shed some light on B—H sulfenylation. By mixing *o*-carboranyl amides **233** and disulfides **237** with stoichiometric amount of cooper triflate, 2-phenyl-pyridine, as well as lithium *tert*-butoxide, the B(4,5)-disulfenylated product **238** were acquired in moderate to good yields *(72)*. The amide directing group can be removed in this reaction condition.

As a proof-of-concept investigation, Xie and co-workers realized a Cu-catalyzed electrochemical regioselective B(4,5)-dialkoxylation of *o*-carborane (Fig. 22C) *(73)*. The advantages of this protocol includes: (1) chemical oxidants are not essential for the reaction, (2) employment of cheap copper catalyst, (3) precise B(4)–OR and B(4,5)–OAr selectivity under room temperature, (4) broad substrate scope. Based on control experiments, the plausible mechanism was proposed. Generation of Cu(II) (**Int 53**) and Cu(III) (**Int 54**) were carried out by anodic oxidation. The authors attributed mono −/di-selectivity to steric bulkiness of the substituents.

Very recently, the first B(3,4)–H diarylation of *o*-carborane protocol was reported by the Ackermann group. The mechanism is similar to Pd-catalyzed B(4)-alkynylation (Fig. 7) *(74)*. Decreasing the amount of aryl iodides resulted in B(3)-arylation in moderate yield. According to DFT calculation, the sequence for B—H bond activation for this reaction should be: B(3)–H > B(4)–H > B(6)–H.

Molecules are chiral when they cannot be superposed on their mirror images, and this principle is also applicable to carborane derivatives. Notwithstanding the great progresses on transition-metal mediated B—H functionalization in last 10 years, enantioselective B—H functionalization is still anticipated. Based on previously reported carboxyl directed Ir-catalyzed B(4)-alkenylation (Fig. 4A), the Xie group performed enantioselective B(5)-alkenylation by employing a chiral ligand (*S*)-**L*** (Fig. 24) *(75)*. Up to 99% yield as well as 99% ee value were acquired when

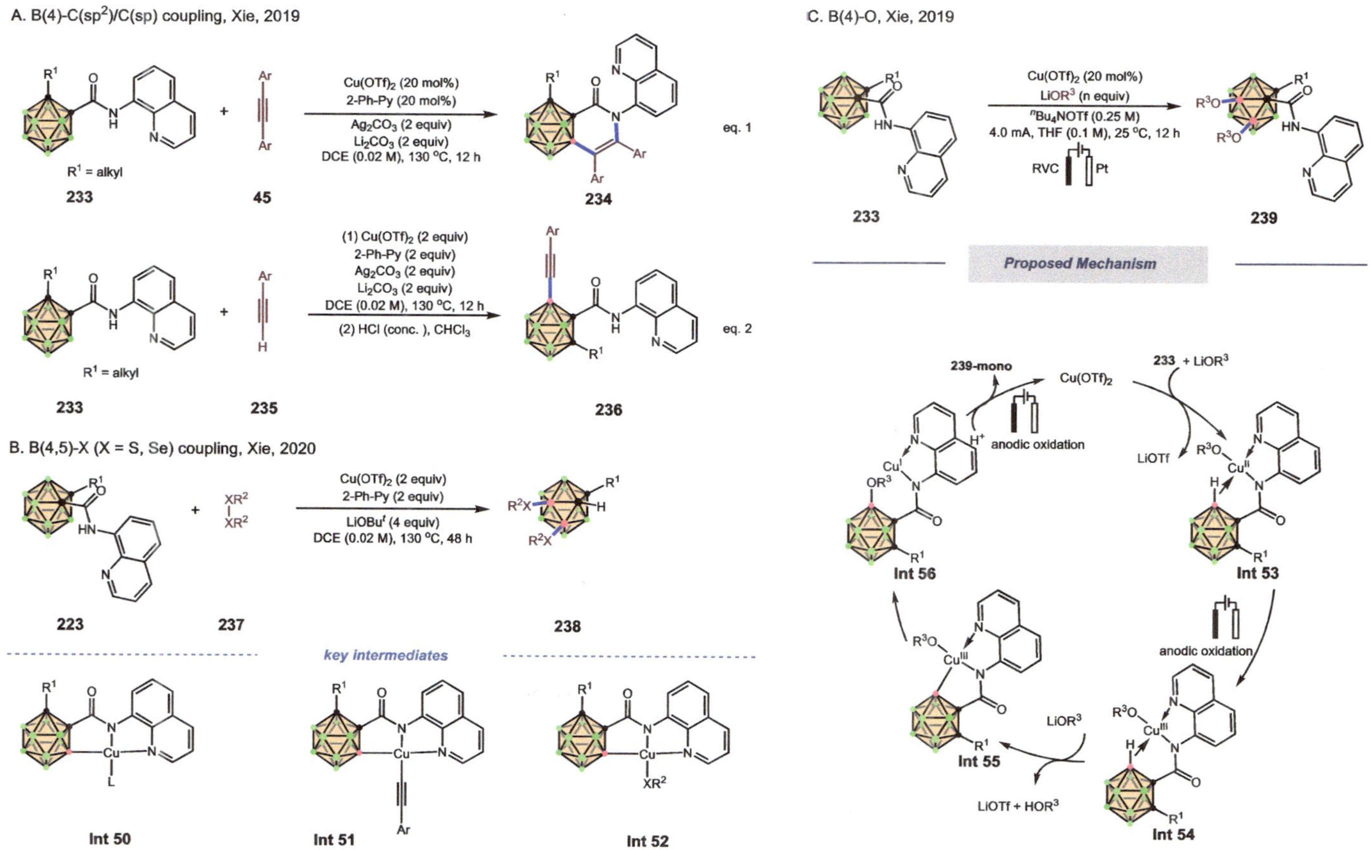

Fig. 23 B—H functionalization enabled by bidentate 8-aminoquinoline directing group.

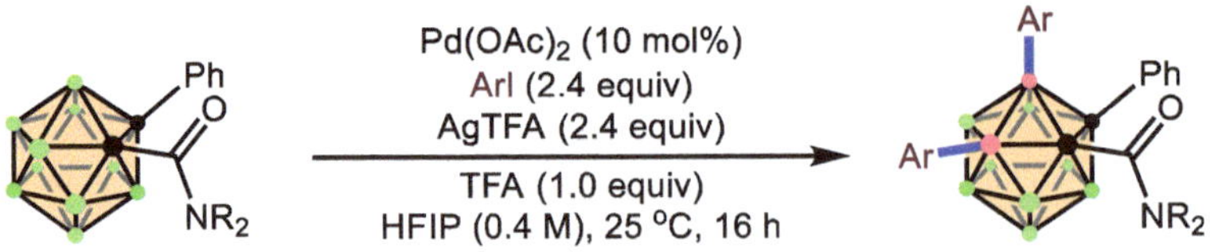

240 **241**

Fig. 24 Pd-catalyzed amide directed B(3,4) arylation.

1.0 equiv. of **242** and 1.2 equiv. of **45** were mixed in presence of [Ir(*cod*)Cl]$_2$ (2.5 mol%), (*S*)-**L*** (5.0 mol%), AgNTf$_2$ (12.5 mol%), benzenesulfonic acid (1.5 equiv.) in 0.05 M toluene at 50 °C. The symmetric 6e Cp* ligand was superseded by 4e cod ligand and a 2e chiral (*S*)-**L*** ligand, which precipitated B(5) selectivity. Better yields and ee values were attained when the carbon vertex is blocked (**244–257**). According to DFT calculation on the basis of concerted metalation-deprotonation (CMD) mechanism, the B(5)-activated transition state **TS 9-*S*** was more stable than its B(4)-activated enantiomer **TS 9-*R***. In addition, the unique C–H···π interaction in **TS 9-*S*** is able to stabilize the transition state further (Fig. 25).

In aforementioned cases, the amide directing groups are installed on carbon vertex that enabled functionalization of B(3), B(4), B(5) or B(6). Researchers deemed other boron vertices may be functionalized when DG was installed on boron vertex. For example, a B(9)-acyl amino group assisted Pd-catalyzed B(4)-arylation of *o*-carboranes with aryl boronic acids was uncovered by the Cao group *(76)*. A tandem sequence of CMD, ligand exchange, reductive elimination and recycling of catalyst was postulated for the formation of **259**. According to DFT calculation, Cao and co-workers ascribed the excellent B(4) selectivity to high polarization of B(4)—H bond. In 2019, a straightforward tactic for preparing *nido*-7,8-carborane fused oxazole **261** was unveiled by the Cao group *(77)* (Fig. 26). Moderate to good yields were attained when two cage carbons were not substituted. Installing steric bulky alkyls on cage carbons compromised the yields of products and resulted in the formation of B(11)—O or B(9)—O coupled byproducts. B(9)-, B(8)-, or B(4)-activated intermediates (**Int 57** – **Int 59**) can be formulated via a concerted metalation-deprotonation process, which may impact electron density on B(3) vertex. Based on experimental data, **Int 58** turned out to be the most favorable intermediate, which triggered an ensuing deboronation. Similar to Pd-catalyzed B(3)—H

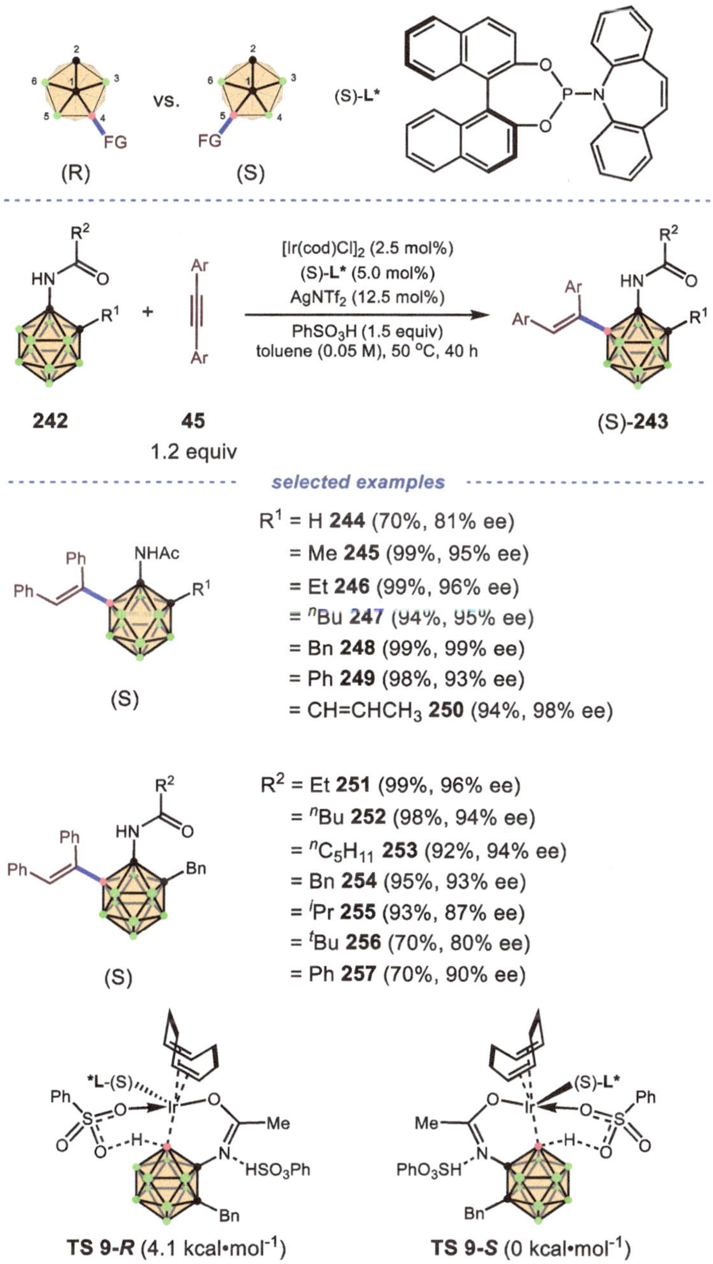

Fig. 25 Amide-assisted Ir-catalyzed enantioselective B(5)—H alkenylation of *o*-carboranes.

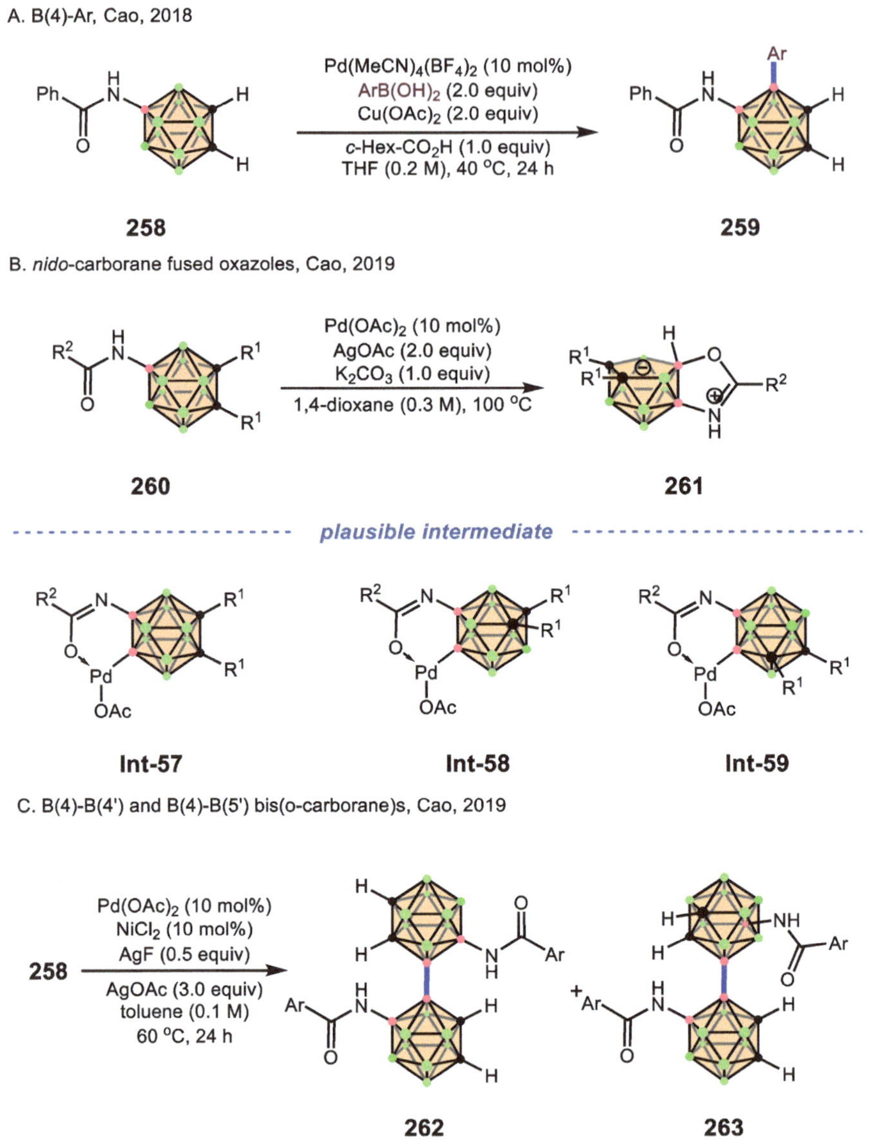

Fig. 26 B—H activation enabled by B(9) located amide directing group.

activation/oxidative dehydrogenative coupling for synthesis of bis(*o*-carborane)s, the daunting task for constructing B(4)–B(4′) or B(4)–B(5′) bis(*o*-carborane)s was accomplished well by the Cao group *(78)*. An efficient and succinct transformation under a mild condition was enabled by palladium-nickel dual catalysis. Functionalities such as F, Cl, CF_3, naphthalenyl, *c*-hexanyl etc. are intact after reaction (Fig. 27).

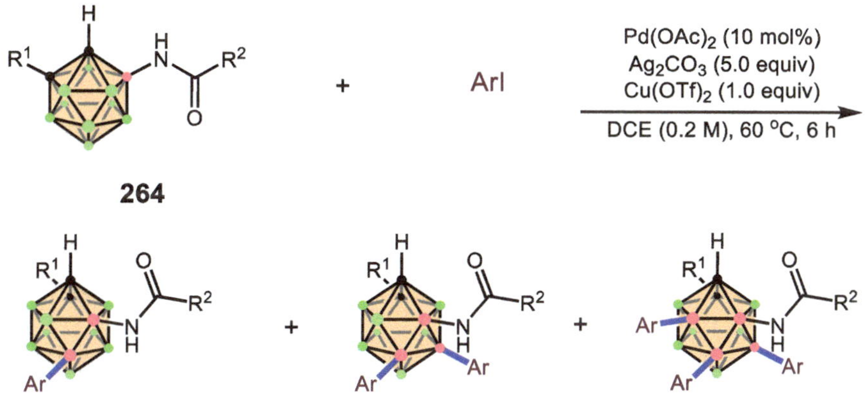

Fig. 27 Selective B(5,8,9)-tri-arylation of *o*-carboranes.

When amide directing group was installed on B(4) position, a Pd-catalyzed stepwise B(5,8,9)-tri-arylation was excavated by the Lee group in 2021 *(79)*. The B(5,8,9)-tri-arylated product **267** was achieved in a good yield when 5.0 equiv. aryl iodide and 5.0 equiv. of Ag_2CO_3 were employed. This reaction has a good tolerance of functional groups, 4-Me, 4-OMe, 4-F, 4-Cl, 4-Br, 4-COOEt, 4-Ph on aryl ring are tolerated. Moreover, electronic variation of the alkyl and aryl substituents of the B(4)-acylamino group and C(2)-substituents of **264** did not manifestly affect the efficiency. However, the B(9)-mono-arylated **265** and B(8,9)-di-arylated **266** were isolated when the loading of Ag_2CO_3 decreased to 3.0 equiv. or the temperature declined to room temperature. Although nature bond orbital (NBO) analyses showed that the vertex charge follows the order B(8) > B(9) > B(5) > B(3) *(80)*, the isolation of **265** to **267** clued that boron vertices on **264** were arylated in the sequence from B(9) to B(8), and B(5) was arylated in the last step.

An elaborately regioselective Pd-catalyzed B(8)-arylation was disclosed by the Xie group when the directing amide group was installed on B(3) vertex (Fig. 28) *(81)*. Furthermore, B(4,7,8)-tri-arylated product **275** or B(4,7,8)-tri-fluorinated product **276** were obtained when increasing load of catalyst and arylation/fluorination reagents were used. The isolation of B(4)-Pd complex **271** and the deuterium experiment attested that B(4)—H activation was prior to B(8)—H activation. Hence a justifiable "cage walking" mechanism (from **Int 60** to **Int 61**) was proposed.

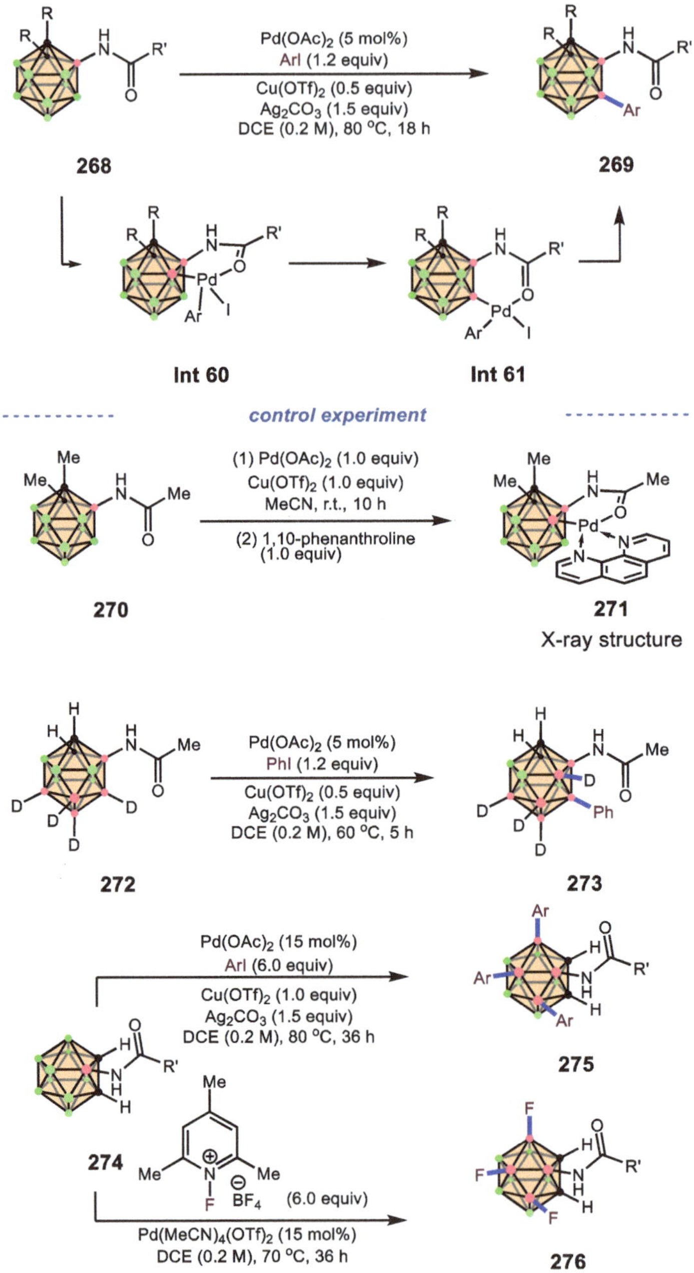

Fig. 28 Acyl amino group directed Pd-catalyzed B(8) arylation.

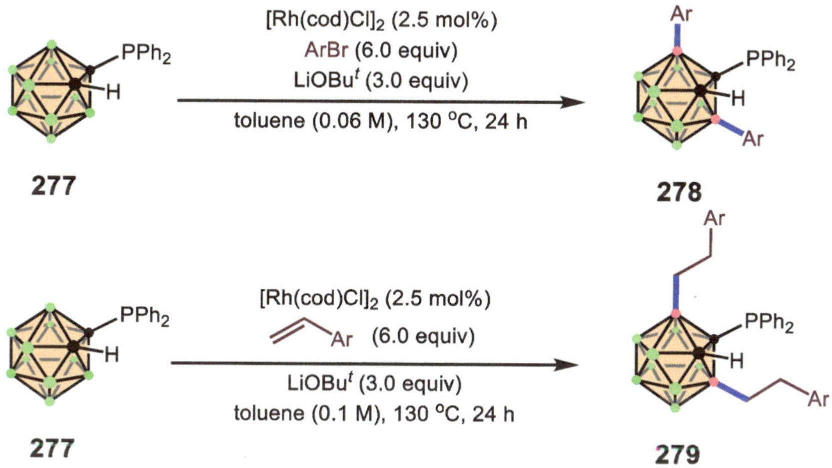

Fig. 29 Phosphine guided Rh-catalyzed B(3,6) di-functionalizations.

2.1.5 Phosphine directed B—H functionalization

Up till now, only one example of phosphine directed B—H activation has been reported. B(3,6)-di-arylated **278** and B(3,6)-di-alkenylated **279** were attained in moderate to good yields via rhodium catalysis (Fig. 29) *(82)*. Some potential luminescent groups are tolerated, though in 19% to 48% yield. Heteroaromatic rings such as thiophene and pyridine did not perform well in this reaction. The proposed mechanism is similar to carboxyl or imine directed Pd-catalyzed B(4,5)-di-arylation.

2.1.6 Hydroxyl directed B—H functionalization

In 1999, the process of B—H activation initiated by the insertion of methyl propiolate was discovered by Herberhold and Yan *(34a)*. The rigid structure of compound **280** rendered rhodium hard to approach to adjacent B—H bonds. However, when methyl propiolate inserted into Rh—S bond, the flexible structure of **Int 62** enabled the approach of Rh to B(3)—H and resulted in B(3)—Rh formation. Leveraging the similar principle, the Xie group developed hydroxyl guided coupling reactions of 1-hydroxy-*o*-carborane and alkynes or even carbon monoxide *(83)*. When cage carbon was substituted, products **282** with B(4) selectivity were obtained. When unblocked **283** was employed, products **284** with B(3) selectivity were acquired (Fig. 30).

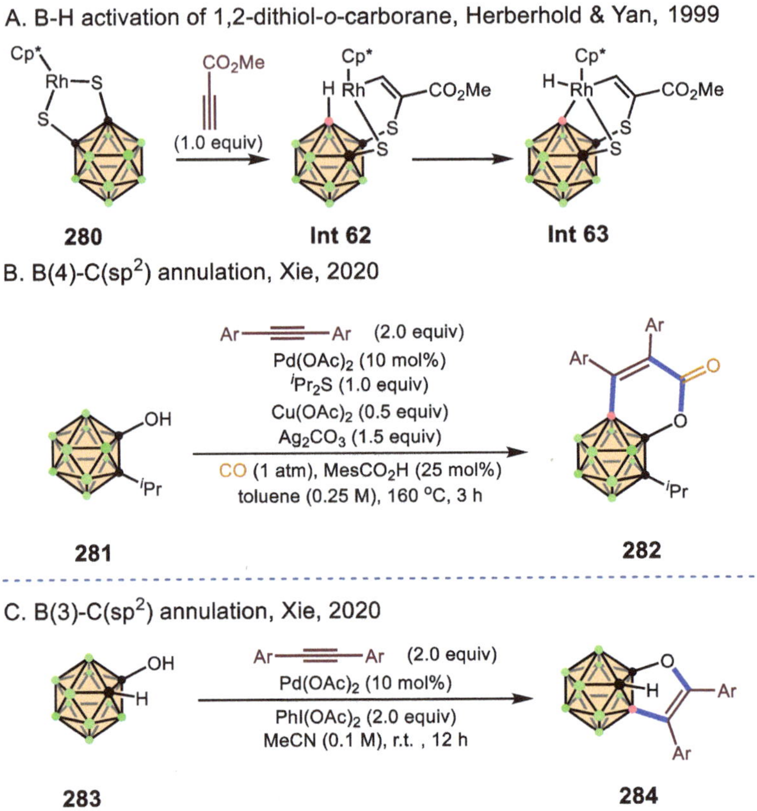

Fig. 30 Hydroxyl guided Pd-catalyzed B(4)- or B(3)-annulation.

2.1.7 Intramolecular B—H functionalization

Directing group assisted B—H functionalization has been well summarized from Sections 2.2.1 to 2.1.6. However, in some cases, the prefunctionalized groups such as C—Br bond or hydroxyl group was tethered at C_{cage} and incurred an intramolecular cross-coupling in presence of transition-metal catalysts. In this chapter, we categorized these kinds of works into DG guided B—H functionalization.

In 2015, an intramolecular coupling of *o*-carborane with aromatics via Pd-catalyzed B—H activation was unveiled by the Xie group *(84)*. The regioselectivity was influenced by steric hinderance at C(2). When C(2) was not substituted, mixtures consisted by B(3) functionalized **286** (major product) and B(4) functionalized **287** (minor product) were achieved. B(4)-arylated **289** was the only product of this Pd-catalyzed transformation when C(2) was blocked. On the basis of this intramolecular coupling, the first example of Pd-catalyzed enantioselective B—H functionalization was reported by the same group in 2018 when the chiral ligand **L** was opted *(85)*. The employment of chiral ligand **L** decreased the reaction temperature to 0–25 °C, a series of B(5)-arylated (*S*)-**291** were synthesized with a good yield and ee value. The congestion between the *t*-butyl group of ligand **L** and phenyl group in starting material **290** in transition state **TS 9-R** was observed by DFT calculation, which resulted in the B(5) selectivity of this reaction. In addition, an array of B(3)—O coupled **293** and B(4)—O coupled **294** were prepared by a oxidative Pd-catalyzed intramolecular cross-coupling of compound **292** *(86)* (Fig. 31).

Very recently, an Fe-catalyzed intramolecular selective B(4)—H/C—H dehydrogenative coupling of **233** was discovered by the Xie group *(87)*. The 8-aminoquinoline motif not only works as a bidentate directing group for B(4)—H activation, but also serves as an aryl source via a simultaneous ligand exchange and C—H activation. A broad substrate group was tolerated and the gave access to target molecule **295** in moderate to good yields (Fig. 32).

2.2 B—H functionalization without directing group

2.2.1 Direct functionalization of electron-rich B(8,9,10,11) position(s)

On the basis of nature orbital bond analysis calculated by the Xie group (Fig. 36A) *(80),* the electron densities order of *o*-carborane (**1**) is B(8,10) > B(9,12) > B(4,5,7,11) > B(3,6). Though electrophilic halogenation or methylation of electron-rich boron vertices via Lewis acid catalysis has been documented several decades ago, other functionalization strategies are still in urgent need. A protocol of Pd-catalyzed B(8,9,10,12)-tetra-fluorination was developed by the Xie group (Fig. 33) *(88)*. In the presence of 0.5 equiv. of $Pd(MeCN)_4(BF_4)_2$, the regioselective B(8,9,10,12)-tetra-fluorination was achieved when 10 equiv. 1-fluoro-2,4,6-trimethylpyridinium tetrafluoroborate (**296**) was employed. Mixtures of mono-, di-, tri-fluorinated by-products emerged when the dosage of catalyst, the dosage of **296**, or the temperature decreased. A Pd(II)-Pd(IV)-Pd(II) catalytic cycle was proposed to explain the regioselective B(8,9,10,12)-tetra-fluorination. Here, the key reactivity aspect is that the electrophilic B—H palladation at electron-rich B(8,9,10,12) vertices. Because the electron density of other boron positions decreased promptly, subsequent electrophilic B—H palladation did not occur. After the establish of electrophilic palladium catalysis mode, the Cao group discovered B(8,9,10,12)-tetra-acetoxylation, B(8)/B(9)-mono-acetoxylation, B(8)/B(9)-mono-arylation, as well as B(8)/B(9)-mono-alkenylation *(89)*.

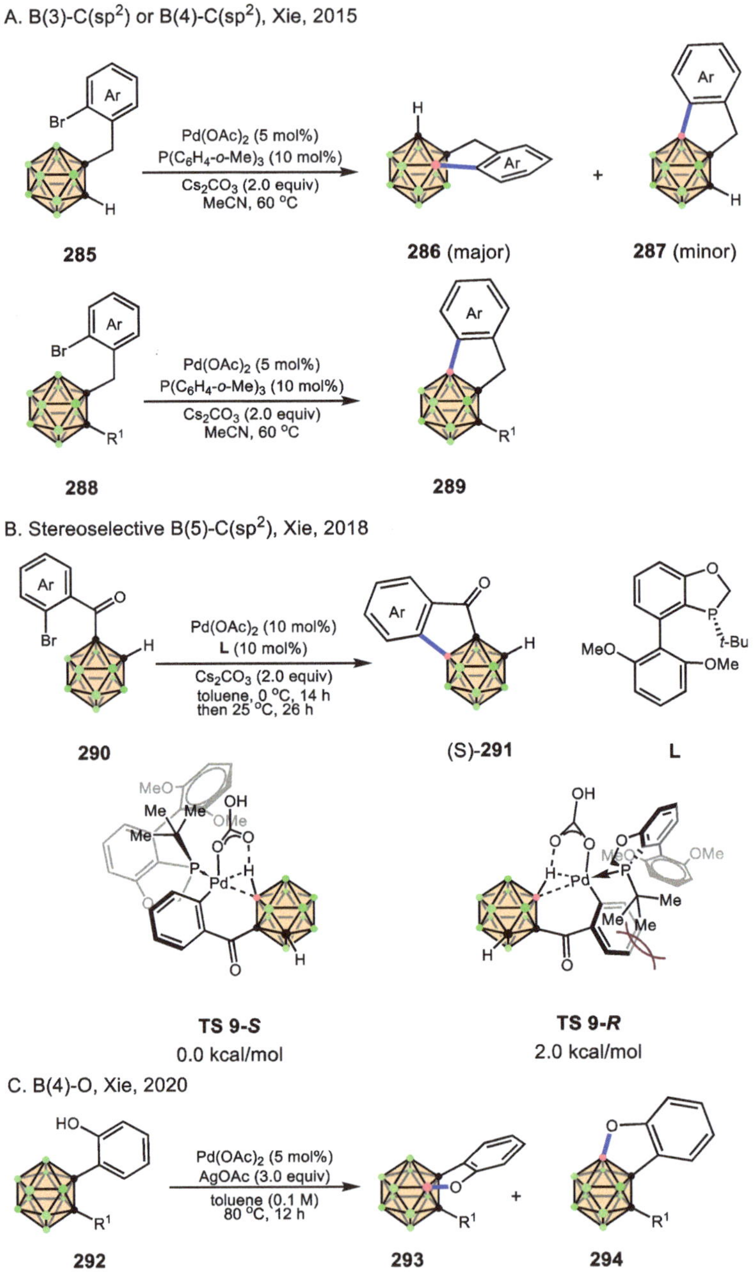

Fig. 31 Intramolecular B—H functionalization of prefunctionalized *o*-carboranes.

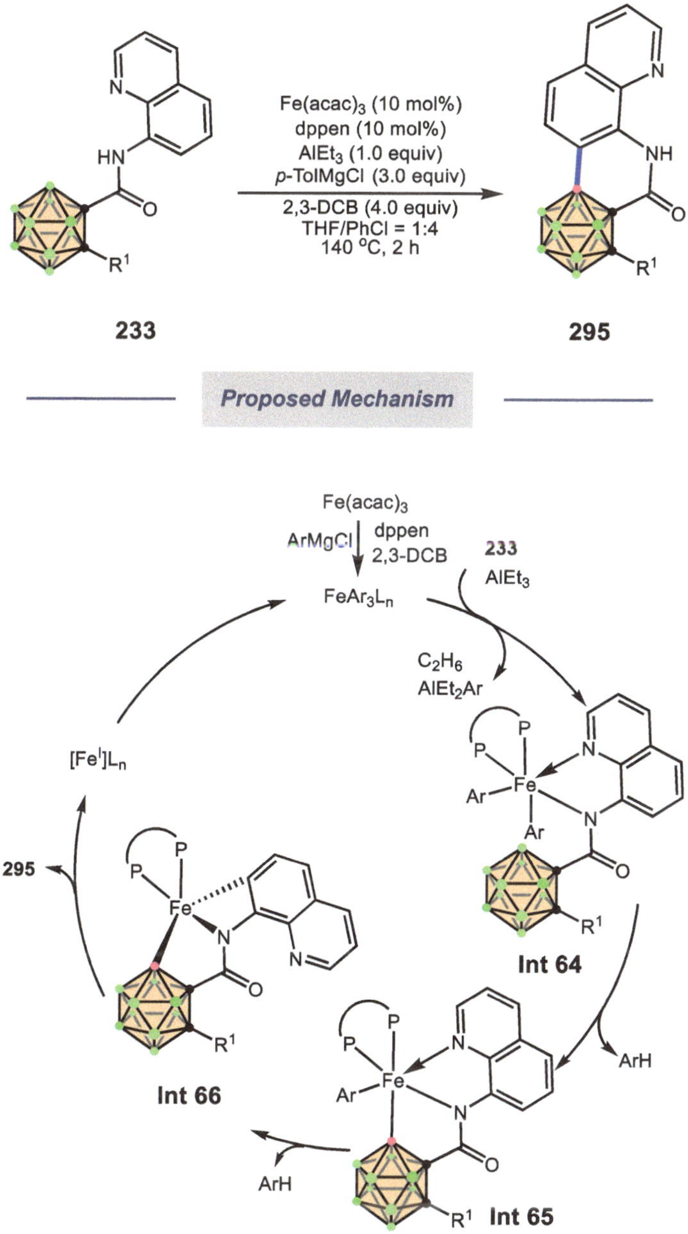

Fig. 32 Fe-catalyzed intramolecular B(4)-arylation.

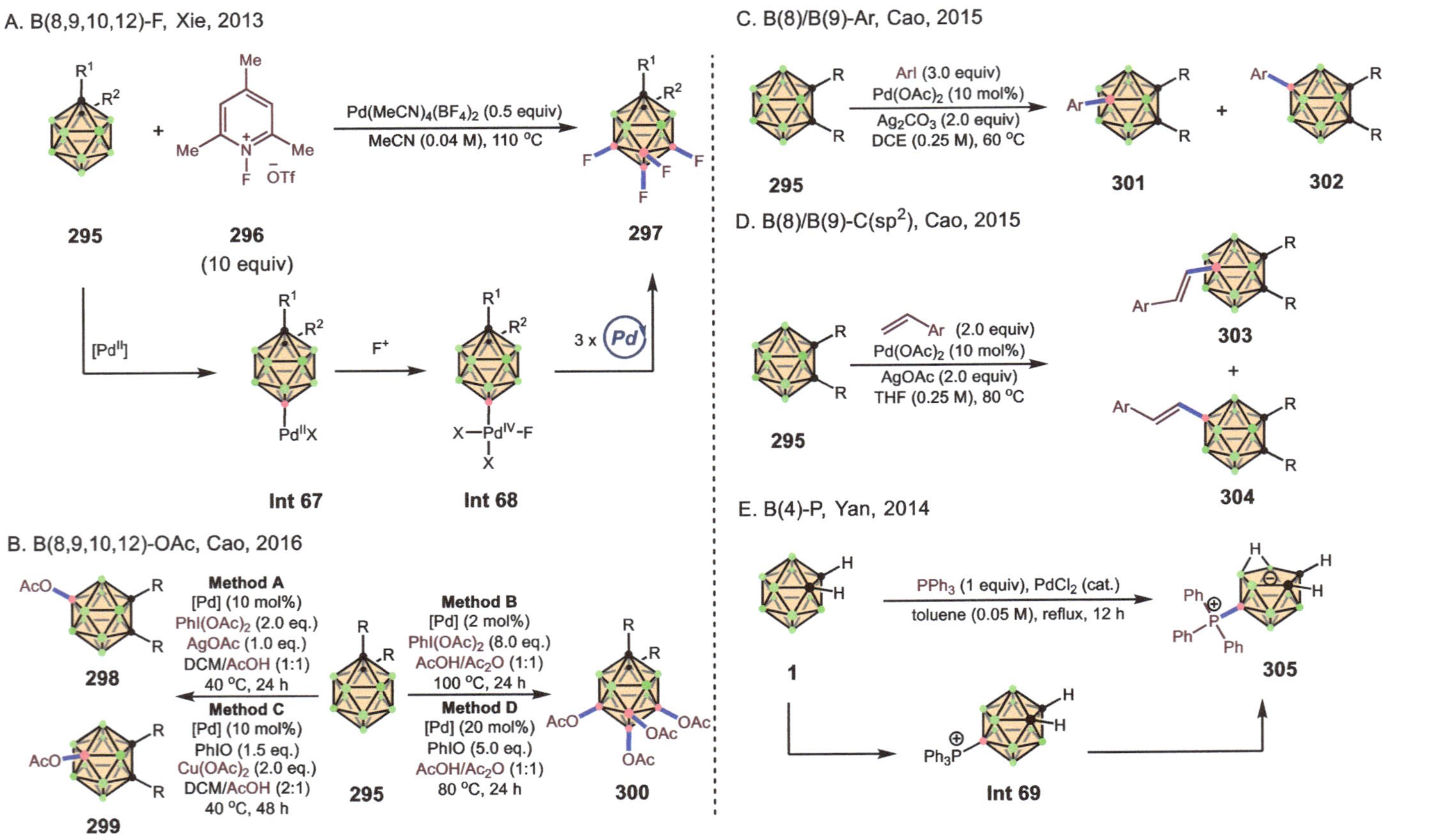

Fig. 33 Functionalization of B(8,9,10,12) via electrophilic B—H activation.

(Diacetoxyiodo)benzene or iodosobenzene was chosen as the oxidant, and tetra- or mono-selectivity was regulated by dosage of catalyst and oxidant, temperature, solvent, and auxiliary additives. Despite the electrophilic Pd-catalyzed B—H functionalization has shown a good application prospect, the nuance in electron density between B(8) and B(9) position results in B(8)/B(9) mixtures with a ratio laying between 1:1 and 1:2.7. In 2014, a B—P coupled zwitterion **305** was prepared by the Yan group (Fig. 33E) *(69)*. It is assumed that **Int 69** was generated by electrophilic B(9)—H palladation and an ensuing reductive elimination. The formation of phosphonium at B(9) position may further increase the electron density at B(3) position and caused the subsequent deboronation at B(3). The structure of compound **305** was confirmed by X-ray structure analysis.

We also note important contributions by Chen and co-workers in developing a palladium-catalysis for B(9)-mono-amination of *o*-carboranes and *m*-carboranes (Fig. 36) *(90)*. This B(9)—N protocol is carried out under air with diverse nitrogen sources, including *O*-benzoyl hydroxyl alkylamines, sulfamides, anilines, as well as imines. When free amines (R_1R_2NH) were employed, partial AgOAc served as oxidant to recover Pd(II) from Pd(0). Activated amine (R_1R_2NOBz) was not only a nitrogen precursor but was also deemed as an internal oxidant. The results of control experiments illustrated that the AgOAc works as a Lewis acid in both cases. HFIP turns out to play an important role in the transformation. A $PdAg(OAc)_3$ intermediate (**Int 70**) was presumed by the authors. The substituent effects at B(12) was well investigated (**307**–**312**). The yield is affected by both electronic and steric effects (Fig. 34).

B-halogen based carboranes are important building blocks for a good many of metal-catalyzed couplings. Very recently, the Chen group reported a catalyst-free B(9)—H halogenation of *o*-carboranes and *m*-carboranes *(91)*. The unique property of HFIP enabled the excellent selectivity as well as good yield of this transformation. This reaction can be set up under air condition with only trichloroisocyanuric (TCCA), or tribromoisocyanuric (TBCA), or *N*-iodosuccinimide (NIS) as the halogenation reagent. Pure products can be acquired by simple extraction with *n*-hexane, which shows the ease of operation (Fig. 35).

2.2.2 Transition-metal catalyzed functionalization of electron-deficient B(3,6) position(s)

In contrast, B(3,6) are the most electron-deficient positions and thus can be activated by oxidative addition of electron-rich transition-metals.

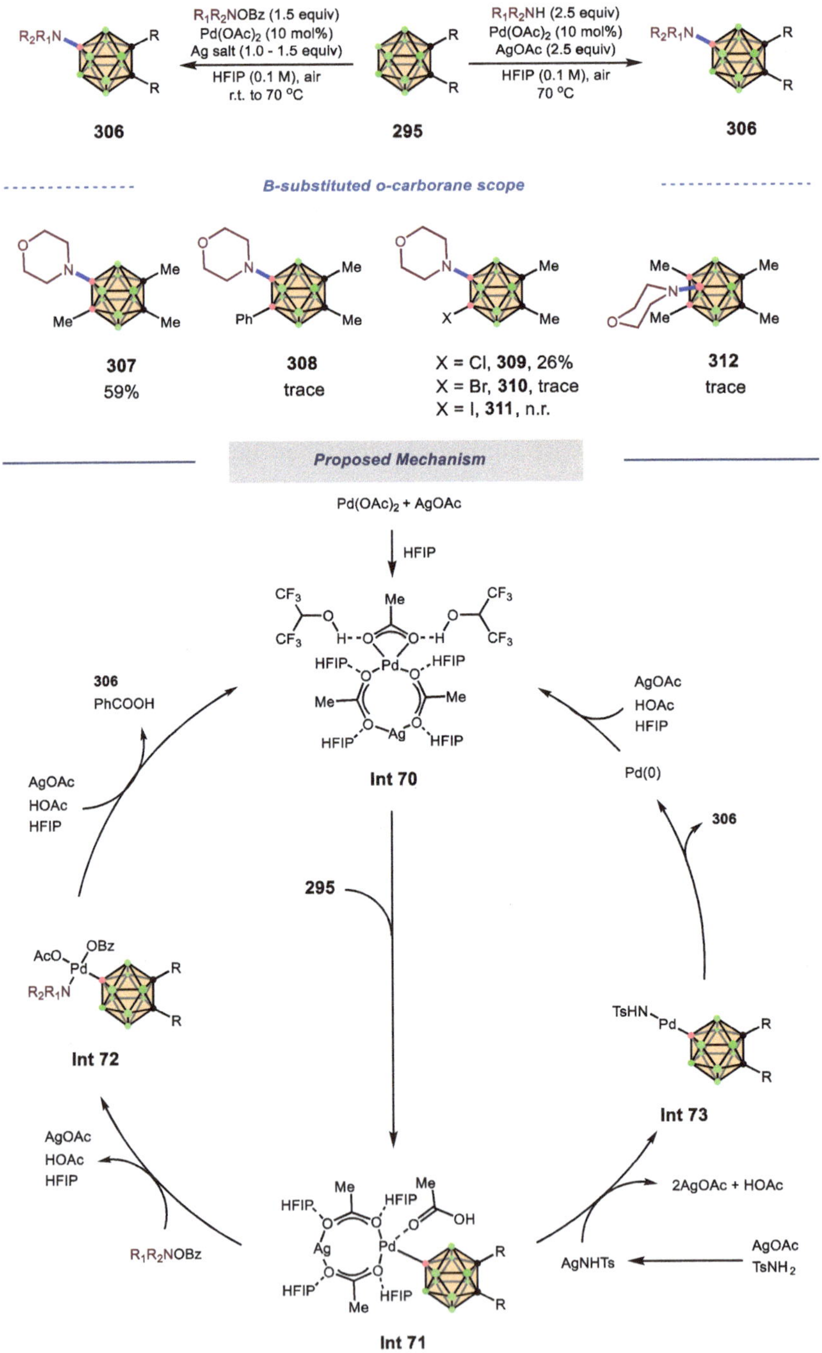

Fig. 34 Palladium-catalyzed B(9)-amination of *o*-carboranes and *m*-carboranes.

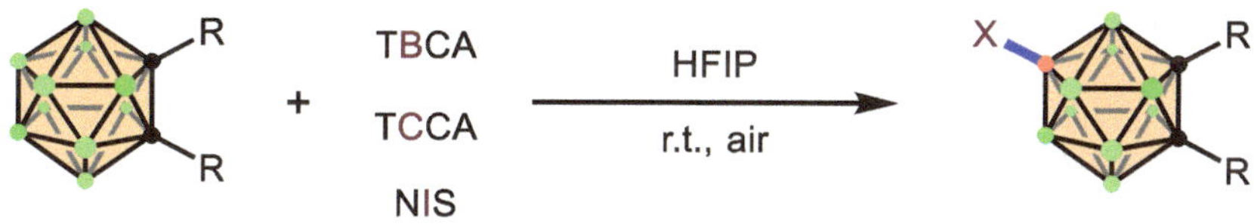

295

X = Cl, **313**
X = Br, **314**
X = I, **315**

Fig. 35 Catalyst-free B(9) halogenation.

The B(3)–Ir coupled complex **316**, was prepared from *o*-carborane **1** and low valent iridium complex $[Ir(cod)Cl]_2$ by Hawthorne and co-workers *(32)*. Leveraging Ir-enabled B(3)—H activation, the Snedden group realized the first transition-metal catalyzed B—H functionalization *(92)*.

When *o*-carborane **1** and excess amount of propyne were mixed in the presence of 5 mol% $[Cp^*IrCl_2]_2$ and 20 mol% proton sponge, 3-propenyl-*o*-carborane (**317**) was afforded, though merely in 40% GC yield. In spite of limited applications, the previous works shed some light on electron-rich transition-metal-catalyzed B(3,6) functionalization. The Xie group disclosed an efficient B(3)-mono-borylation or B(3,6)-di-borylation when *o*-carboranes are treated with 4 equiv. of bis(pinacolato)diboron (B_2pin_2) in THF at 110 °C for 5 h in presence of 3.5 mol% $[Ir(cod)Cl]_2$ and 21 mol% 2-methyl-pyridine *(93)*. According to experimental data and literatures, a plausible mechanism was proposed in Fig. 36B. Monomeric **Int 74** was generated by dissociation of $[Ir(cod)Cl]_2$ in the presence of 2-methyl-pyridine ligand. After oxidative addition with B_2pin_2, which attended by reductive elimination, **Int 76** was generated and entered catalytic cycle. The six-coordinated octahedral intermediate **Int 77** formed after oxidative addition of **Int 76** by B_2pin_2. The author assumed that key intermediate **Int 79** was produced by direct B(3) electrophilic substitution (path a) or oxidative addition on B(3)–H (yielded Ir(V) **Int-78**) followed by reductive elimination (path b). When one carbon is substituted by bulky *tert*-butyldimethylsilyl (TBDMS) group, B(4)-mono-borylated compound **322** was produced. A catalytic B(3,6)-*trans*-di-alkenylation (alkyl-substituted alkynes as substrates) and B(3)-*cis*-mono-alkenylation (aryl-substituted alkynes as substrates) was described by the Xie group (Fig. 36C) *(94)*. The EE:EZ ratio of B(3,6)-dialkenylated products increased to >99:1 when bulky *tert*-butyl and *c*-hexyl tethered on one side of terminal alkynes. The carborane-containing Ir(V) species was proposed as

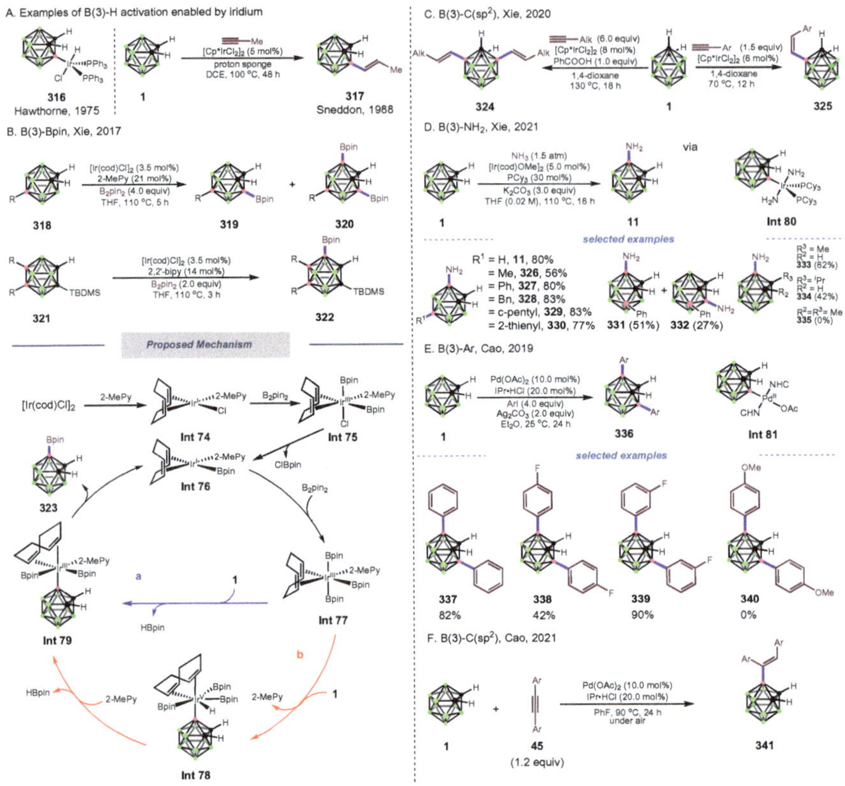

Fig. 36 Functionalization of B(3,6) via oxidative addition of B—H bond.

a crucial intermediate. Another excellent work by Xie et al. demonstrated that $[Ir(cod)(OMe)]_2$ was capable to mediate a regioselective B(3) amination of *o*-carborane **1** with NH_3 (gas) via an oxidant-free dehydrogenative B–H/N–H cross-coupling assisted by PCy_3 and K_2CO_3, offering an efficient and safer method to synthesis 3-NH_2-*o*-carborane (**11**) and its derivatives (Fig. 36D). The addition of K_2CO_3 drastically lifted the reaction efficiency, owning to interactions between K^+ and compound **11** inhibited the poisoning effect from NH_2 group. The conscientious mechanism studies indicated the possible involvement of Ir(V) species **Int 80**. Several alkyl or aromatic rings were tolerated in this transformation (**326**–**332**). Installation of bulky functionalities on one cage carbon declined the reaction yield (**333, 334**). No reaction occurred when two cage carbons were substituted (**335**).

Even though the B(3,6)-arylated *o*-carborane can be readily prepared by imine directed Rh-catalyzed B(3,6)-di-arylation or by Pd-catalyzed Suzuki-Miyaura coupling of aryl iodide with B(3,6)-borylated *o*-carboranes, some disadvantages such as: (1) the multi-step chemical sequences to install and remove DGs, (2) high reaction temperatures (above the boiling point of the solvent under 1 atm) in the sealed tube, (3) employment of expensive rhodium or iridium catalysts are involved in these strategies. Leveraging the strong σ-donating capacity of *N*-heterocyclic carbene (NHC) ligand, the Cao group developed a Pd/NHC catalyzed direct B(3,6)-di-arylation of *o*-carborane (Fig. 36E) *(95)*. In the presence of 10 mol% $Pd(OAc)_2$ and 20 mol% 1,3-bis-(2,6-diisopropylphenyl)-imidazolinium chloride (IPr·HCl), treatment of *o*-carborane **1** with various iodobenzenes at room temperature afforded a series of B(3,6)-di-arylated *o*-carborane derivatives in moderate to good yield. The efficiency of this transformation was influenced by electronic effect and steric hinderance of functionalities on phenyl rings. The B(3) activated **Int 81** was assumed as the key intermediate in this reaction. The same group later found the aforementioned Pd/NHC catalytic system also worked well for knitting diaryl acetylenes onto B(3) position (Fig. 36F) *(96)*. The reaction was set up in fluorobenzene under 90 °C and oxygen from air served as the oxidant.

Even if approaches for preparing versatile carborane derivatives have improved a lot in recent 10 years, most excellent cases shown in this chapter were based on prefunctionalized carborane, including: (1) installation of a directing group; (2) construction of B—I, B—Br, B—N_2BF_4, B—BF_3K or even B—Pt bonds as carboranyl donors; (3) installation of tunable bulky or electron-rich/deficient functional group on cage carbons. In 2021, the Cao group unveiled a novel B(3)—H/C—H oxidative dehydrogenative coupling (Fig. 37) *(97)*. This reaction, which employed unfunctionalized *o*-carborane **1** as the carboranyl donor, should not be deemed as B(3) arylation but rather a *N*-directed *ortho*-carboranylation@B(3) via Ir-catalyzed C—H activation. Functionalities including Me, ^{t}Bu, OMe, F, Cl, Br, CF_3, NO_2 etc. are well tolerated in this transformation. The catalysis is initiated by oxidative addition of electron-deficient B(3)—H on Ir(III) center, followed by reductive elimination of Ir(V) **Int 82** and release acid HX. Then Ir(III) intermediate **Int 83** was oxidized by Cu(II) and the attendant N-directed ligand exchange/electrophilic C—H fracture occurred in sequence. Target molecule **342** was released by reductive elimination.

2.2.3 B—H functionalization via nucleophilic substitution

According to the distribution of electronic density on *o*-carborane cage (Fig. 38A), the most electron-deficient B(3,6) positions are vulnerable. Strong Lewis bases such as alkoxides, amines, and fluorides were found to

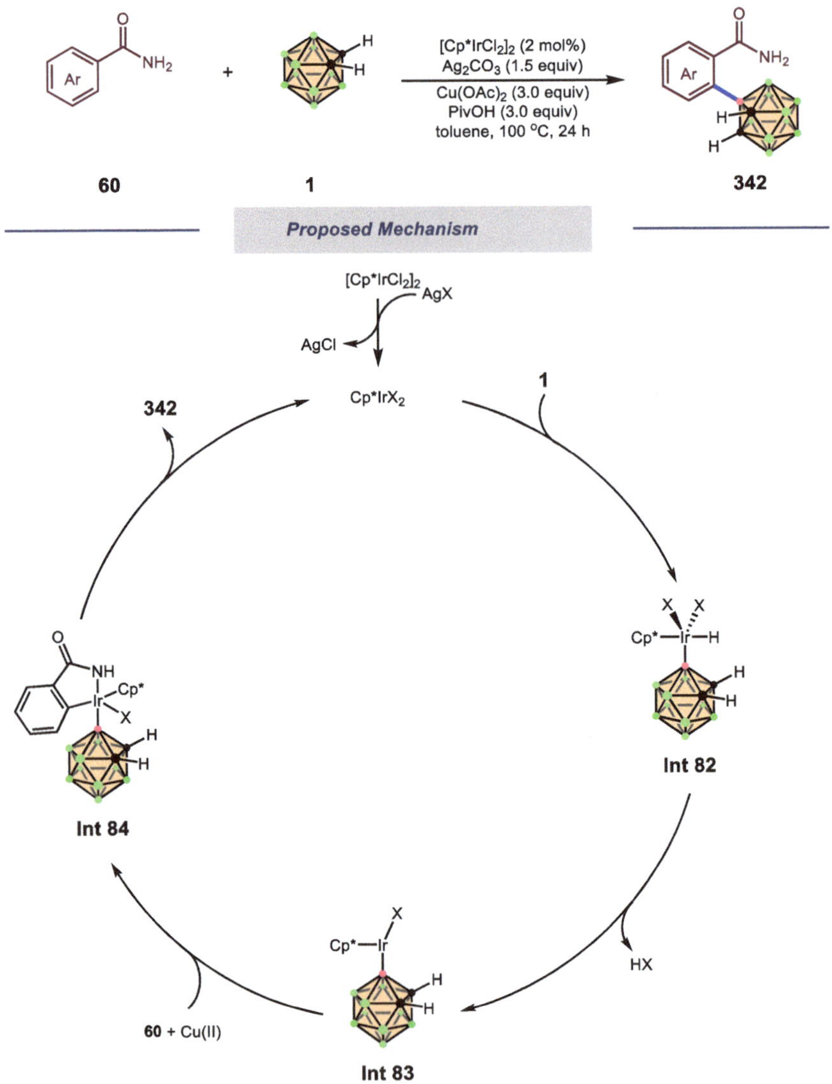

Fig. 37 *N*-directed *ortho*-carboranylation@B(3) via Ir-catalyzed C—H activation.

deboronate B(3) vertex of *o*-carboranes forming their *nido*-7,8-$C_2B_9H_{12}$ counterparts *(1c,98)*. The stepwise deboronation mechanism is proposed in Fig. 36B. It is proposed that the hard nucleophiles first attacks the most electron-deficient B(3) vertex that opens the boron cage *(99)*. The same boron atom is attacked by the second nucleophile hereafter, which leads

A. Nature bond orbital (NBO) analysis, Xie, 2018

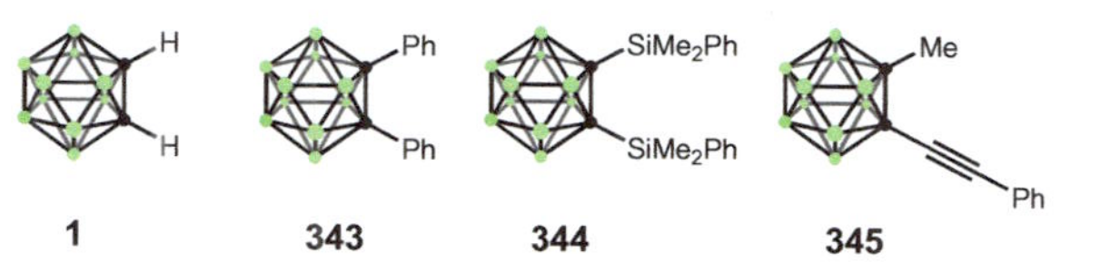

	1	343	344	345
CR	-0.195	-0.234	-0.185	-0.331/-0.412
B(3)H	0.214	0.217	0.203	0.241
B(4)H/B(5)H	0.068	0.092/0.079	0.060/0.063	0.082/0.080
B(8)H	-0.088	-0.092	-0.088	-0.089
B(9)H	-0.065	-0.061	-0.062	-0.063

B. Deboronation reaction

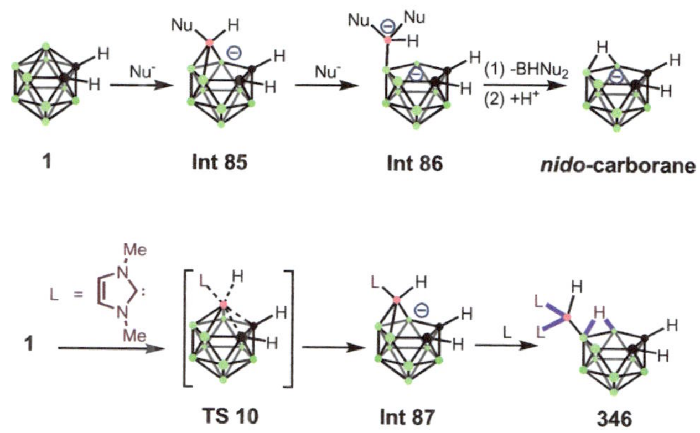

C. B(4)-alkyl, Xie, 2017

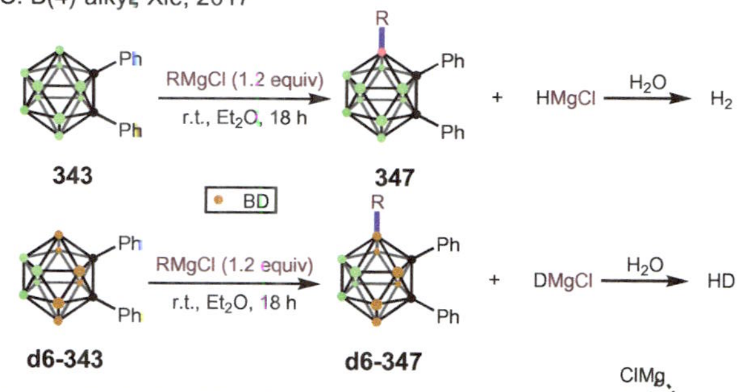

D. B(3)/B(9)-alkyl, Xie, 2018

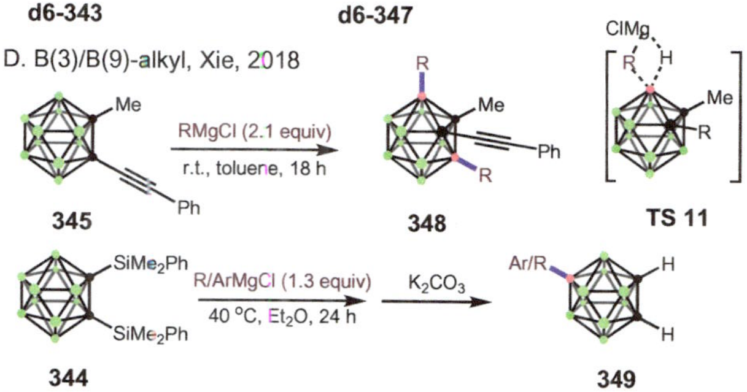

E. Intramolecular annulation, Xie, 2020

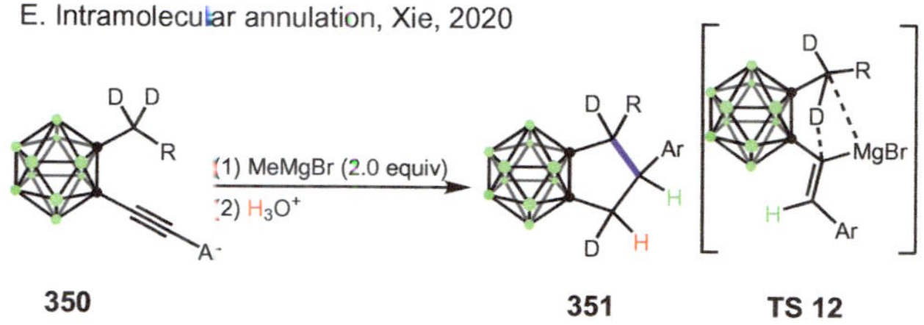

Fig. 38 B—H functionalization via nucleophilic substitution.

to elimination of $BHNu_2$ and formation of *nido*-carboranes. When **1** was treated with soft nucleophile NHC, the deboronation process halted at the second stage and compound **346** was afforded *(100)*. DFT calculation indicates that when nucleophile approaches to B(3) position, two B(3)—C bonds are weakened (**TS 10**) which result in cage opening of carborane (**Int 87**) *(100a)*.

The Xie group first uncovered nucleophilic B—H substitution when Grignard reagents were alternated as nucleophile *(101)*. This nucleophilic substitution features departure of H^- other than cage opening when Grignard reagent approaches to B—H bond. The Grignard nucleophile tends to attack the most electro-deficient boron atom, while the regio-selectivity is regulated by steric hinderance in the same breath. Although B(3) vertex is the most electron-deficient, the nucleophilic substitution of **343** and **344** occurred at B(4) and B(9), respectively. The steric hinderance of 1,2-diphenylated **343** prevented nucleophiles from approaching B(3) vertex, thus accordingly B(4) functionalized **347** was derived (Fig. 38C) *(101)*. The generation of HMgCl was observed by NMR, and H_2 and HD were monitored by GC when the reaction mixture was treated with acid. This evidence demonstrated the generation of hydride via nucleophilic substitution. The huge hinderance from $SiMe_2Ph$ group blocks carbon anion from attacking B(3,6) and B(4,5,7,11) positions, therefore more electron-deficient B(9) was functionalized eventually *(80)*. Alkynyl turns out to be a good EWG with a smaller steric hinderance thus led to an efficient nucleophilic B(3,6)-di-substitution of 1-arylethynyl-2-methyl-*o*-carborane **345** (Fig. 38D) *(80)*.

However, when 1-arylethynyl-2-alkyl-*o*-carborane **350** ($R \neq H$) was selected as a substrate, the anticipated B(3,6)-di-alkylation products was not obtained. After scrutinizing the composition of products and optimizing the reaction condition, the Xie group found when **350** was treated with 2 equiv. of MeMgBr in THF/toluene (1,10) at 80 °C and quenched by water after 36 h, an unexpected annulated product **351** was attained (Fig. 38E) *(102)*. Based on the results acquired from control experiments, the mechanism was proposed in Fig. 39. Five to eight-fold nucleophilic substitution of B—H bonds in partial **350** with MeMgBr result in the formation of multi-methylated ($n = 5$–8) byproduct **352** and five to eight times stoichiometric amount of HMgBr (based on **352**). The produced HMgBr trigged the subsequent cascade cyclization process, which was finally quenched by water.

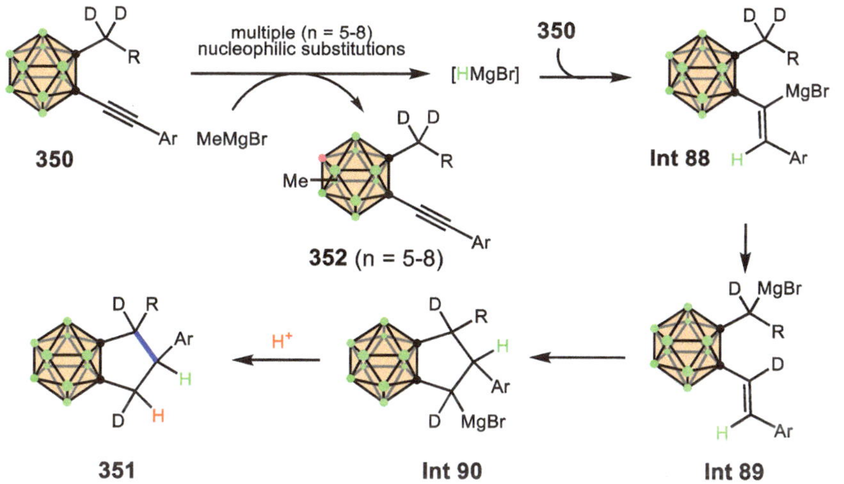

Fig. 39 Proposed mechanism for cascade cyclization of 1-arylethynyl-2-alkyl-*o*-carboranes.

2.2.4 *B—H functionalization of* nido-*7,8-$C_2B_9H_{12}$ via single electron transfer process*

In contrast to great progress achieved in B—H functionalization of *o*-carborane, approaches for B—H functionalization of *nido*-carborane are rare. A pioneer $FeCl_3$ enabled oxidative substitution was reported by Hawthorne and Kang in the 1990s, an electron-neutral B—N coupled **355** was prepared (Fig. 40A) *(103)*. This reaction, however, suffers from high temperature as well as limited substrate scope. The Bregadze group explored a series of displacement reaction between potassium dodecahydro-7, 8-dicarba-*nido*-undecaborate **346** and nucleophiles (I^-, Br^-, SCN^-) enabled by electrocatalysis under either divided or undivided cell. The B(9)-mono-substituted product **357** exists as an anion and is thus soluble in water (Fig. 40B) *(104)*.

Enlightened by previous works, the Yan group poured their heart into seeding a straightforward and metal-free protocol for the oxidative functionalization of cage B—H. This kind of protocol would be highly desirable in regard of the functional group tolerance, environmental sustainability as well as the avoidance of metal-removal process in medicinal industry.

The optimized reaction conditions were characterized by 2 equiv. of 2,3-dichloro-5,6-dicyano-1,4-benzoquinone (DDQ) as the only oxidant

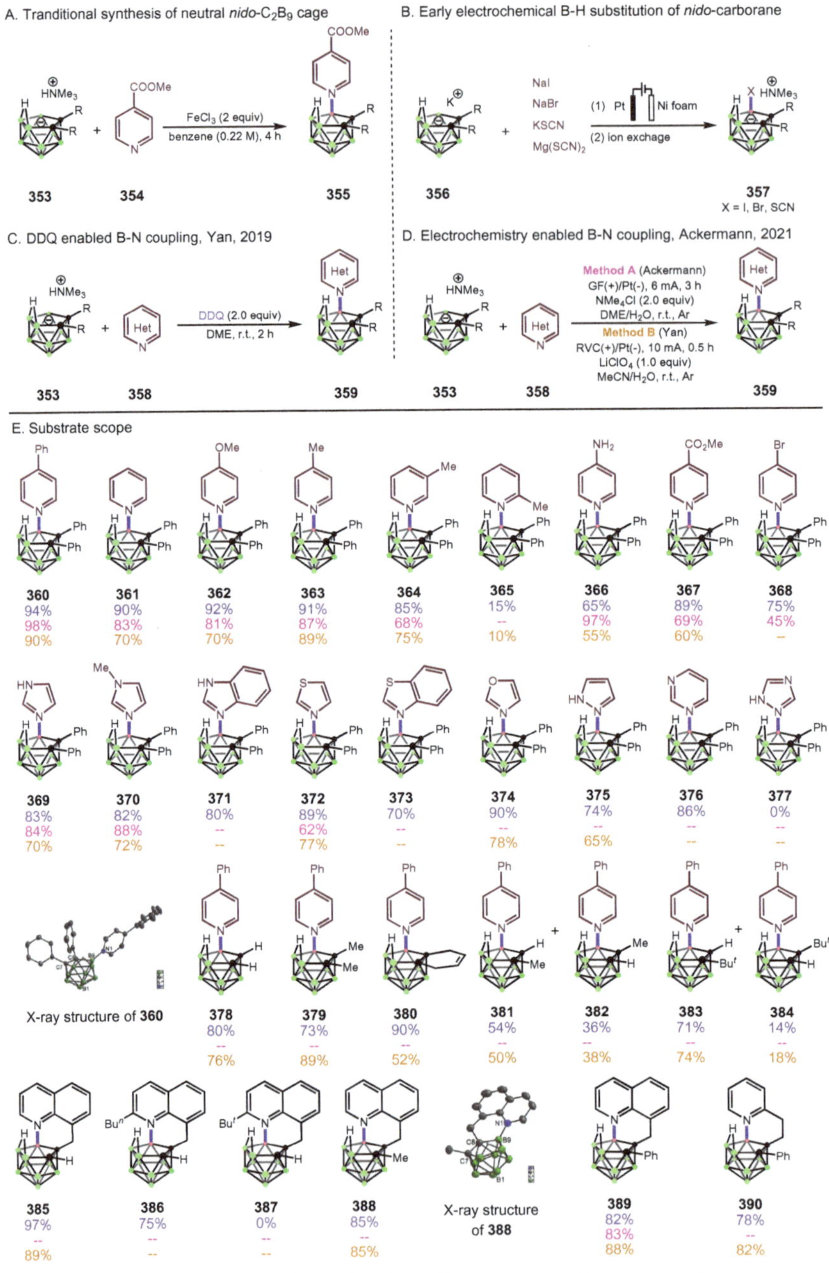

Fig. 40 Metal-free cross coupling of *nido*-carboranes and *N*-heterocycles.

and dimethoxyethane as the best solvent, delivering the desired neutral B—N oxidative coupling product **359** in good to high yield (up to 97%) in 2 h under room temperature. It noteworthy that the neutral coupling product **359** is hydrophobic, even though the opening boron cluster fragment exists in the product. A vast array of pyridines, quinolines as well as other *N*-heterocycles with diverse functionalities could react smoothly with *nido*-carboranes to provide efficient access to B—N coupling products in good yields (noted in purple, Fig. 40). Electron-donating substituents give rise to higher yields than electron-withdrawing groups on the pyridine ring. This transformation is sensitive to steric hinderance, only 15% **365** was acquired when 2-methyl-pyridine was employed. Starting materials with free N—H bonds are well tolerated (**366**, **369**, **371**, **375**). The coupling of *nido*-carborane proceeds well with imidazole, oxazole, thiazole, pyrimidine, and azaindole, which allows good to excellent yield (**369**–**376**). However, no reaction was observed when trizole was employed (**377**), which presumably due to its lower nucleophilicity (for pyridine, Nu value = 11.5 in MeCN; for triazole, Nu value = 7.7; from Mayr's database). Different substituents on cage carbons are also well tolerated (**378**–**384**). When unsymmetric carbon substituents were tested, inseparable B(9)/B(11) mixtures such as methyl-involved **381** and **382** (ratio 3:2), *t*-butyl-involved **383** and **384** (ratio 5:1) were afforded. Basically, the substitution favors the less sterically hindered boron atom. At the same time, this transformation is validated fit for intramolecular B—N couplings as well (**385**–**390**).

It is confirmed by deuterium-labeled experiment (Fig. 41A) that (1) hydrogen atom on B(9/11) of reactant **391** transferred to bridging position (B–H–B) of the product **378**; (2) the bridging atom (B–D–B) of reactant **391** was grabbed by DDQ in this transformation. The plausible mechanism was proposed with the help of theoretical calculation. The *nido*-carborane **393** was oxidized by DDQ via an endergonic single electron transfer (SET) process to generate **Int 91** ($\Delta G_{sol} = +0.3$ kal/mol) at first. In next step, **Int 91** is oxidized by the second DDQ coupling with a deprotonation process to form an open-cage intermediate **Int 92**. Owing to the electron-deficient nature, the vulnerable B(9/11) site of **Int 92** was attacked by pyridine with the departure of H^-. The released hydride could be captured by C_2B_3 plane and forged a new B–H–B bridge. Fig. 41D displays the intrinsic reaction coordinate (IRC) plots of key B—N coupling transition state.

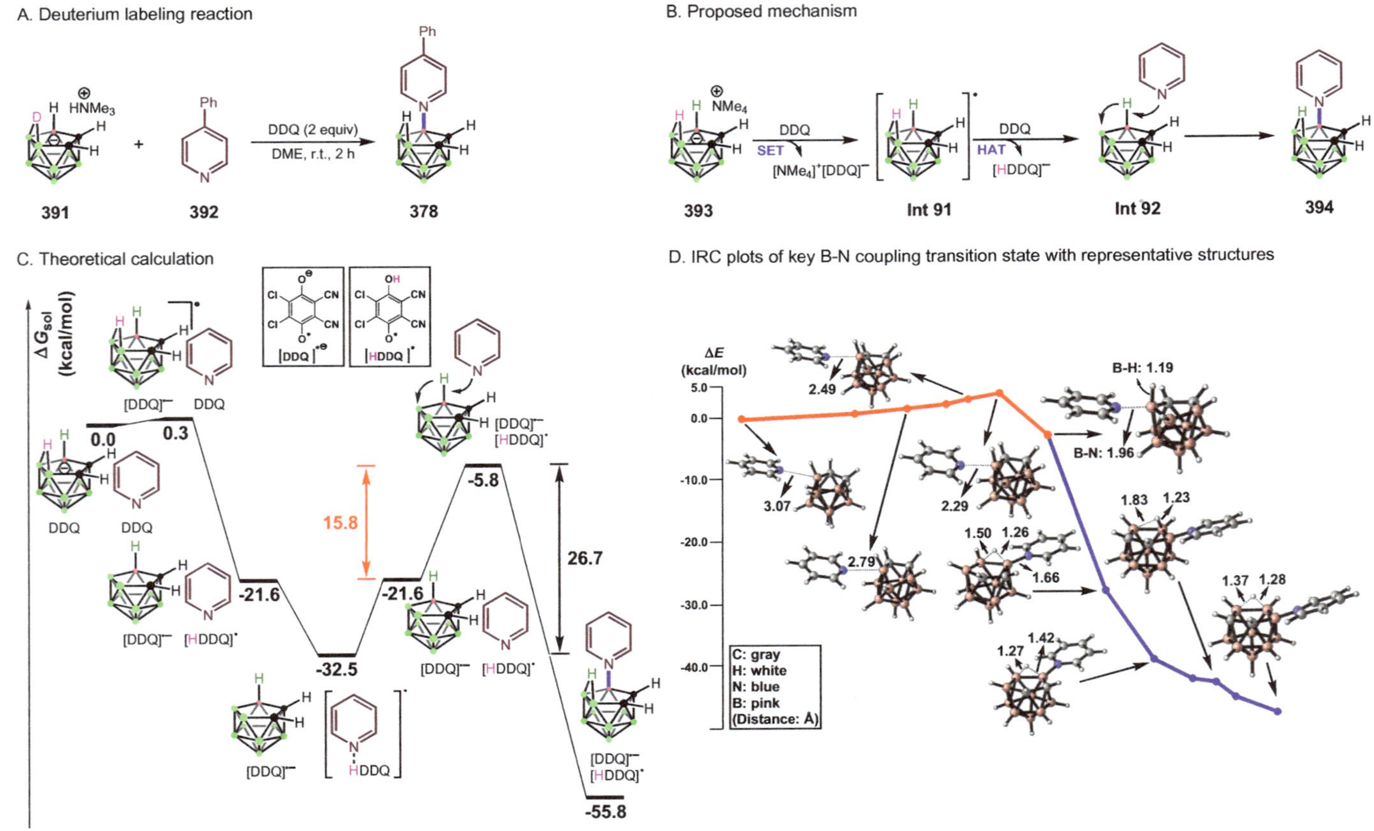

Fig. 41 Mechanistic studies of DDQ oxidized B—N coupling.

Recognizing single electron transfer process is a key to resolve various B—Nu bond constructions, the Yan group realized the same B—N formation by electrochemical synthesis (Fig. 38D, Method B, unpublished). By using 1.0 equiv. of $LiClO_4$ as the electrolyte and $MeCN/H_2O$ as co-solvents, 10 mA constant current electrolysis with reticulated vitreous carbon (RVC) anode and platinum plate cathode in an undivided cell for 0.5 h, regioselective B—N coupling was achieved in moderate to good yields. At almost the same time, the Arkermann group reported a breakthrough electrocatalyzed B—N construction. The optimized condition featured NMe_4Cl as the electrolyte and dimethoxyethane-water as the best solvent system in an undivided cell (Fig. 38D, Method A) *(105)*. The yields in these two electrooxidative B—N couplings are in consist with the DDQ enabled B–N formation.

Based on mechanistic investigations and literature reports, the plausible mechanism for electrooxidative B–Nu formation was depicted in Fig. 42. Initially, the *nido*-carborane anion was oxidized successively and generated cationic **Int 93**. After deprotonation, the nucleophile such as pyridine attacked the vulnerable B(9/11) and yielded the target molecule **394**. Generation of hydrogen on cathode was detected by headspace gas chromatographic analysis.

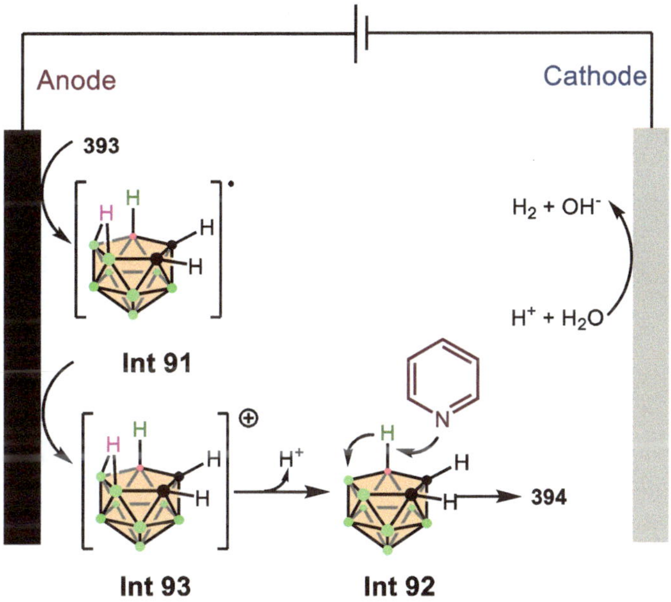

Fig. 42 Proposed mechanism for electrooxidative B—H functionalization.

The robustness of these oxidative B—N couplings were validated by late-stage carboranylations (Fig. 43). The carborane cage could be grafted to *N*-heterocycle-containing clotrimazole (**395**) and miconazole (**396**) in good yields. In addition, several luminent groups just like boron-dipyrrin (BODIPY) and tetraphenylethylene could also be installed on carborane unit. On the other hand, neither *nido*-carborane nor 8-aminoisoquinoline fragment is a chromophore. But when they combined with the help of oxidative B—N cross-coupling, the product **399** exhibited a positive solvatochromism phenomenon. The emission peak displayed a red shift from 492 nm (*n*-hexane) to 547 nm (DMSO), and the quantum yield launched to a maximum value ($\Phi = 80\%$) in THF (Fig. 44).

When the optimized condition of electrooxidative B—N coupling was modified a little, Yan and Chen developed a new approach for coupling *nido*-carborane with chalcogens (S, Se, Te) and pnictogens (P, As, Sb) *(106)*. The nucleophilicity of heteroatom and the steric hinderance are still two important factors that influence the yield. For example, although dialkyl or cyclic thioethers led to moderate to good yield (**405**, **406**, **408**–**410**), no product was obtained when diphenyl thioether was employed (**407**). One rational reason is that the conjugation of phenyl rings with the lone pair of sulfur atom declined its nucleophilicity. When thioethers with one alkyl and one aryl were tested, good yields were acquired (**411** and **412**). Despite no reactivity of diphenyl sulfide, selenides and tellurides were amenable substrates in the electrooxidative B—H functionalization (**415**–**417**), with descending efficiency from Te (**417**) to Se (**415**) to S (**407**), which is in consistence with nucleophilicity of chalcogens. Enantiomers of compound **417** were observed in its single crystal structure. Pnictogens also performed well in this transformation. Not only triarylphosphines (PAr_3), but also trialkylphosphines (PR_3) and triaryl phosphite ($P(OR)_3$) afforded moderate to good yields (**418**–**421**). Notably, the previously unreported B—As (**422**) and B—Sb (**423**) coupled products were generated in moderate yields. However, when triphenyl bismuth was examined, no reaction occurred. The gram-scale reaction reflected the synthetic utility of this protocol. In addition, drug fragments (ibuprofen, indomethacin, ciprofibrate etc.) were tolerated. These synthesized products can be viewed as mimics of thioester or sulfoxide fragments in drug-like molecules.

Cyclic voltammetry experiments (Fig. 45) displayed four oxidation peaks of *nido*-carborane **400** (0.68 V, 0.87 V, 1.02 V, 1.57 V). The result of

A. Late-stage carboranylation

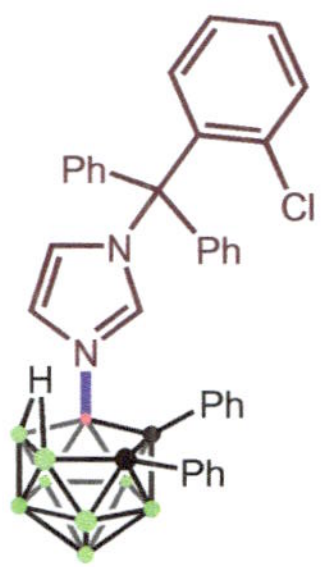

395

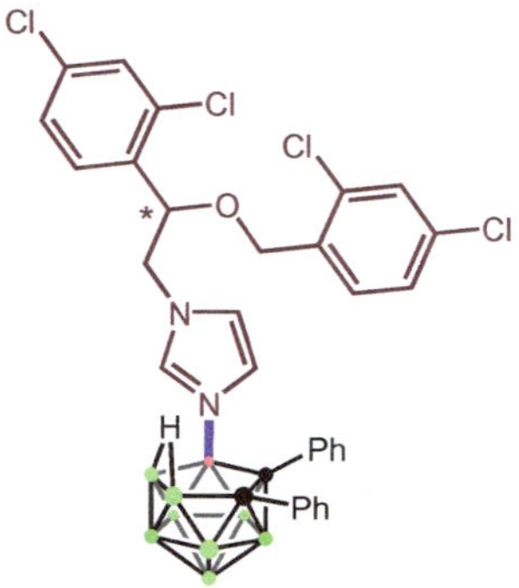

396

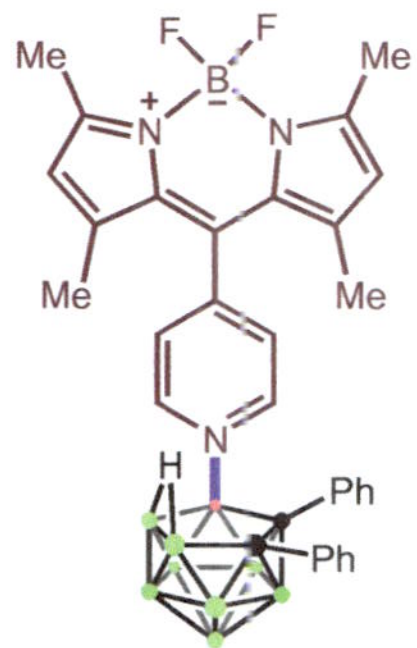

397

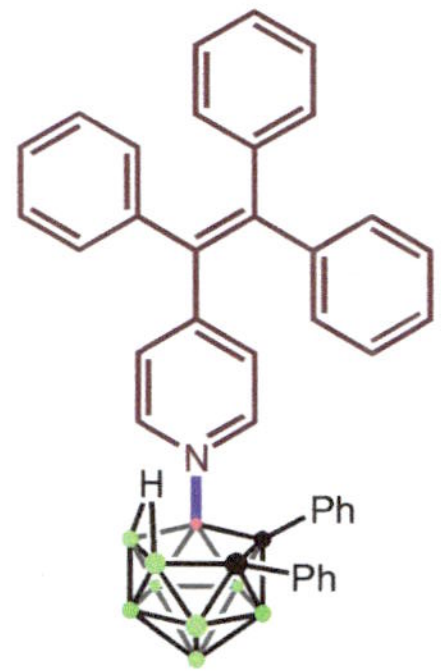

398

B. Preparation of novel luminescent materials

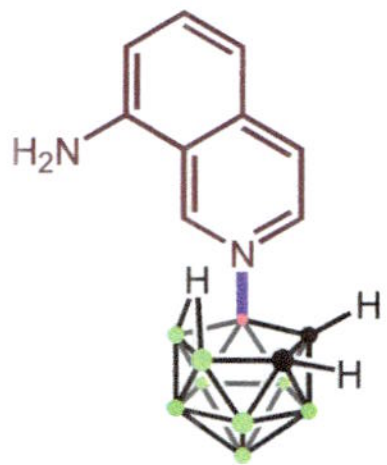

399

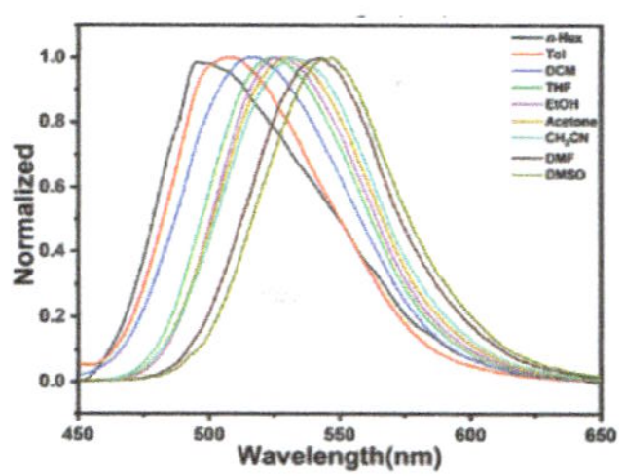

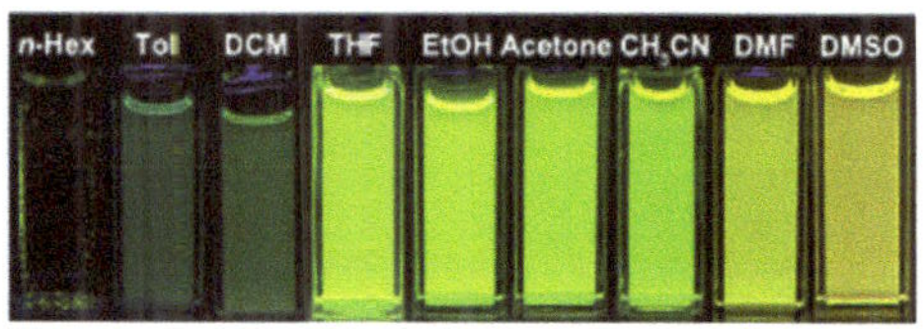

Fig. 43 Showcases of carboranylation in late stage via single electron transfer.

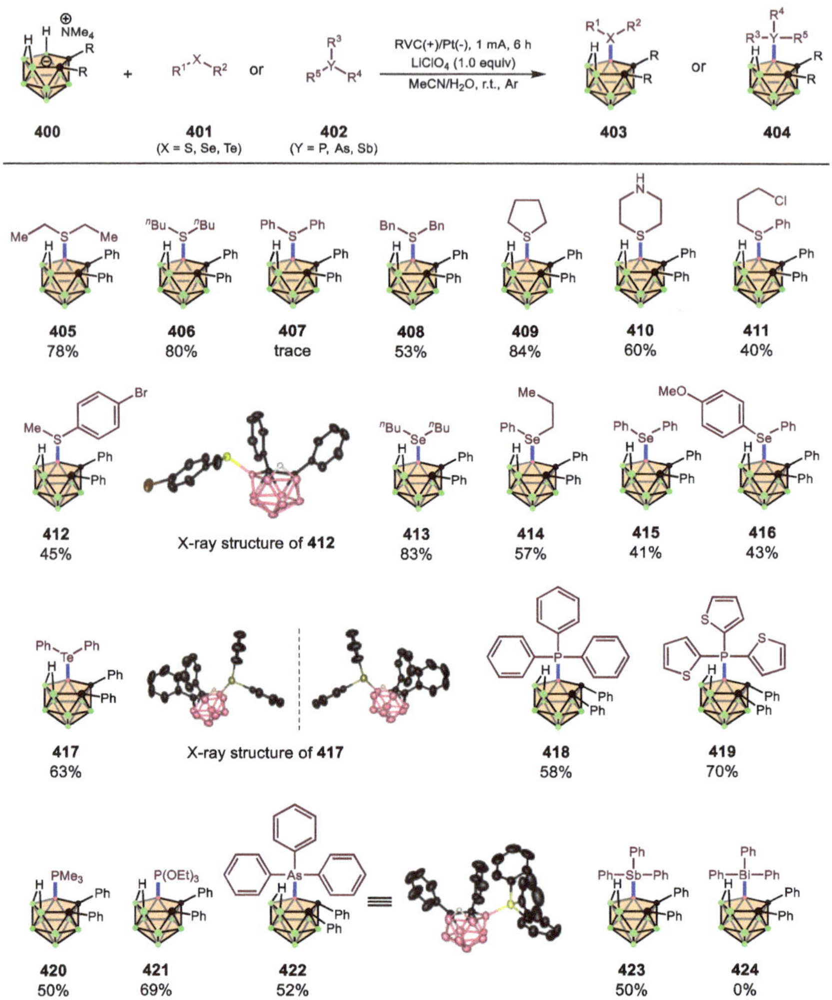

Fig. 44 Electrooxidative B—H functionalization with chalcogens and pnictogens.

deuterium labeling experiment is similar as DDQ-mediated B—N formation. It turns out that *nido*-carborane should be oxidized before diethyl sulfide on the anode to generate a cationic intermediate. The mechanism is similar as electrooxidative B—N coupling depicted in Fig. 42.

In the aforementioned couplings of heteroatom and *nido*-carborane, the amino group were well tolerated (**366**, **399**, **410**). This may be attributed to

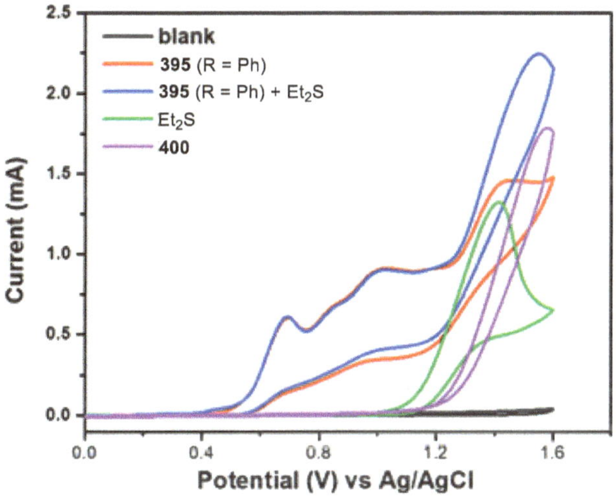

Fig. 45 Cyclic voltammetry at 100 mV/s, *n*-Bu_4NPF_6 (0.1 M in MeCN), concentration of substrates 10 mM.

the stronger nucleophilicity of S and N(sp^2) in heterocycles than N(sp^3). Very recently, a breakthrough in visible-light-induced photocatalysis for B—N(sp^3) formation was established by the Yan group (Fig. 46) *(107)*. This open-to-air protocol employed 5 mol% of organic 9-mesityl-10-methylacridinium perchlorate as the photocatalyst (**PC**), and the products were obtained in 2 h at room temperature under irradiation of 3.0 W blue LED light. Secondary cyclic amines were examined and afforded corresponding products in moderate to good yields (**427**–**431**). Apart from secondary cyclic amines, linear secondary amines (**432**), aliphatic primary amines (**433**, **434**), benzylamine (**435**) are also optional substrates for this transformation. Anilines with electron-rich (**436**) or electron-deficient (**437**) aromatic heterocycles also performed well. In addition, this method is also applicable to sulfonamides (**438**). A plausible mechanism for this photocatalysis is proposed according to experimental data and literature reports. Initially, the excited **PC*** ($E_{1/2}$ red*: 2.06 V vs. SCE in MeCN) oxidizes the amine to its radical cation, which interact with *nido*-carborane anion to form **Int 94**. Then the highly reactive oxygen grabbed one electron from **Int 94** thus formed **Int 95** and superoxide radical anion. The superoxide radical anion catched a H atom in **Int 95** immediately and afforded the product and a peroxyl radical, which can oxidize **PC$^•$** and accomplished the catalytic cycle.

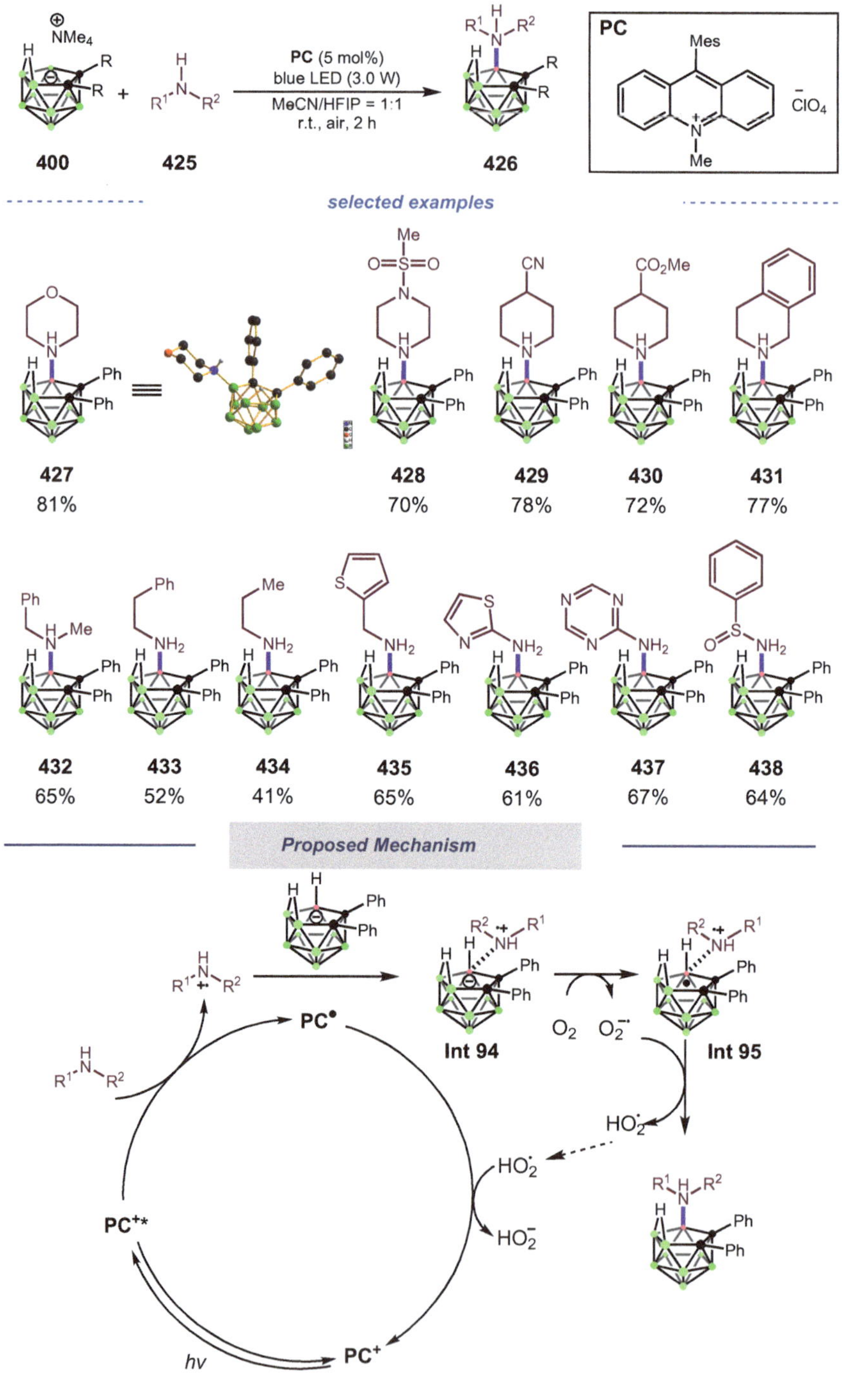

Fig. 46 Photo-catalytic B–N(sp^3) coupling.

3. Carboranylation from prefunctionalized carboranes

3.1 Synthesis of carborane derivatives via B-halogen bond activation

Arguably, Pd-catalyzed functionalization via B—I transformation was the most significant method for derivatization of carboranes before the surge of direct B—H functionalization emerged *(1c, 108)*. Not surprisingly, approaches to forge B–halogen bond, which have been achieved several decades ago, have been ascribed unambiguously in Fig. 3. Nevertheless, only B-iodo-carboranes have been used in Pd-catalyzed carboranylation to date. The B-bromo-carboranes are always pointless for synthesis until the Spokoyny group disclosed the first Pd-catalyzed B—N, B—O, and B—CN bond coupling via B—Br activation *(109)*. It is substantiated that $Pd_2(dba)_3$ is suitable for preparing compound **441** from 9-I-*m*-carborane rather than 9-Br-*m*-carborane (**139**) *(110)*. In fact, stable Pd(0) sources such as $Pd_2(dba)_3$ comprises dibenzylideneacetone (dba) ligands, which in some cases can impede the catalytic cycle. In addition, Pd(II) species like $Pd(OAc)_2$ and $PdCl_2$ need to be reduced to Pd(0) in-situ before participating Pd(0)-Pd(II) cycle. The key to resolve this problem is alternating precatalyst devised by Buchwald and co-workers *(111)* (Fig. 47).

With the robust catalyst in hand, various carborane derivatives have been produced. DavePhos turned out to be the best ligand for incorporation of 2,2,2-trifluoroacetamide (**441**), and XPhos and tBuBrettPhos are the best choice for installing cyano group (**443**) and cyanate group (**444**) on B(9) position, respectively *(109, 112)*. SPhos is proper for a broad substrate scope such as, primary and secondary alcohols and amines, water, ammonia, aniline, sulfonamide as well as phosphoramide *(109, 112)*. It is noteworthy that compound **444** was achieved in 84% yield, albeit its phenyl congener is unstable. The 9-OAr-10-CH_2CH_3–1,7-dicarba-*closo*-dedocaborane **448** were prepared with the help of sequentially Pd-catalyzed B—I and B—Br functionalization. When the precatalyst was treated with base, the highly reactive LPd(0) species was generated in situ, thus activated B—Br bond via oxidative addition. After successive transmetalation and reductive elimination, product **440** was obtained and reactive LPd(0) was released synchronously.

After thoroughly investigation, the Spokoyny group isolated four isomers of **450** when the DavePhos or SPhos was replaced by bulkier XPhos congener in the presence of alcohol (Fig. 48) *(113)*. The structures

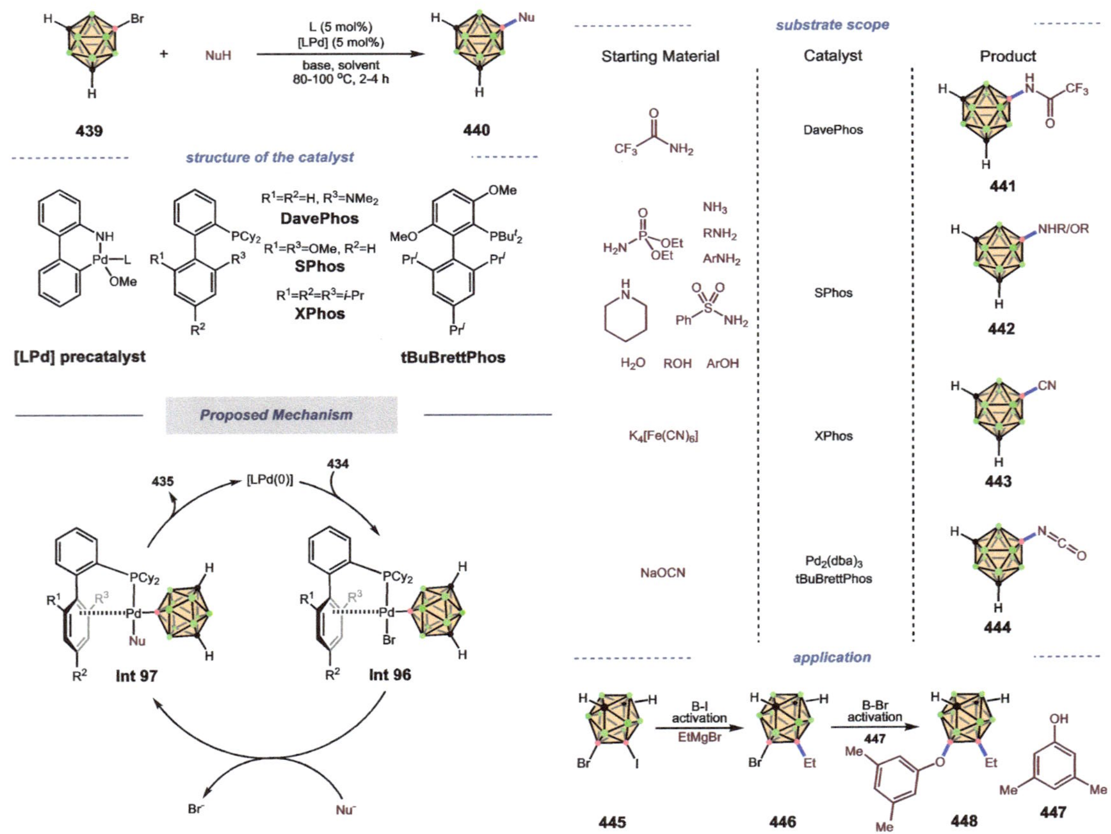

Fig. 47 Pd-catalyzed B—Br activation.

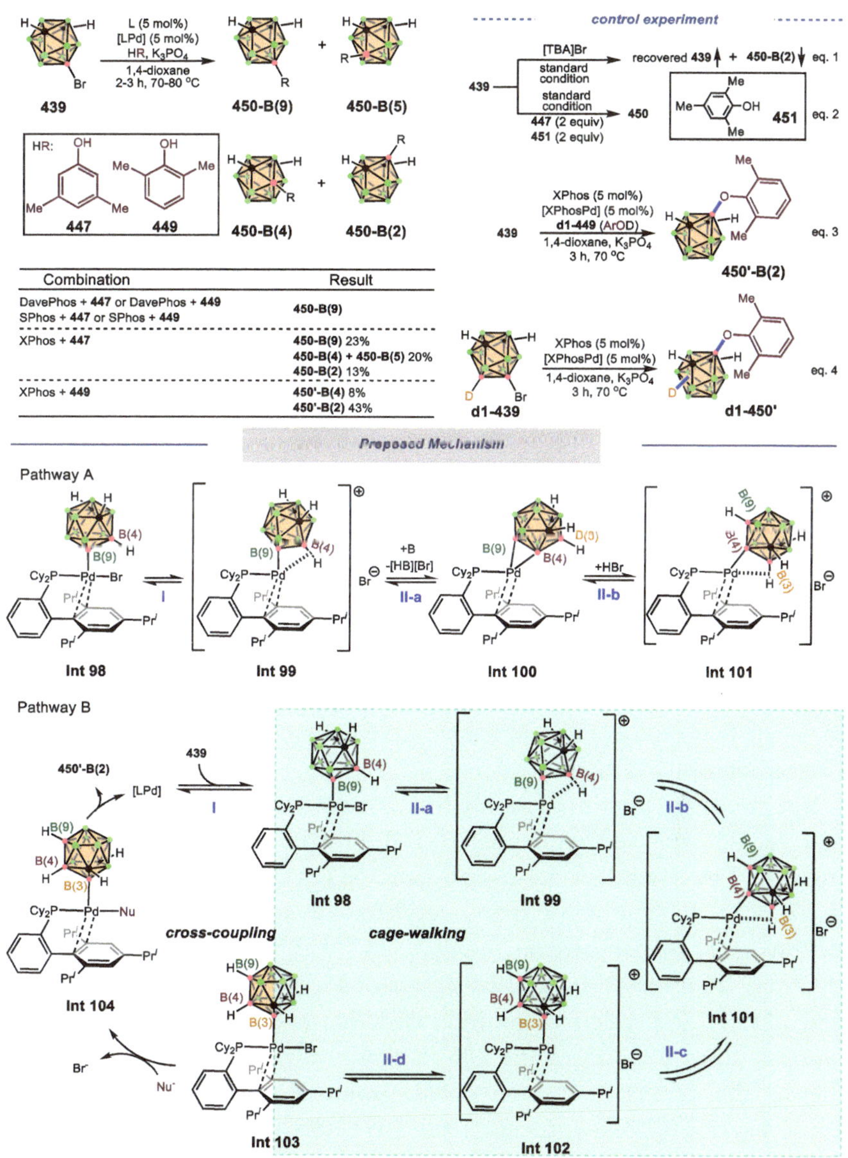

Fig. 48 XPhos-promoted isomerization and proposed "cage-walking" mechanism.

of **450-B(9)**, **450-B(5)**, **450-B(4)**, **450-B(2)** were confirmed by X-ray single-crystal analysis. **450-B(9)** was obtained.

in 23% yield, the inseparable **450-B(5)** and **450-B(4)** mixtures were obtained in 20% yield, and the **450-B(2)** isomer was attained in 13% yield.

When the bulkier substrate 2,6-xylenol (**449**) was examined, the yield of B(2) substituted **450'-B(2)** elevated to 43%, and **450'-B(4)** was just obtained in 8% yield, which implies the steric hinderance of substrate promotes the isomerization process. According to control experiment, **439–B(9)** forms its bromo-*meta*-carborane isomers in presence of [XPhosPd] precatalyst and triethylamine, which insinuates that isomerization process might operate separately from the main cross-coupling cycle. Addition of tetrabutylammonium bromide ([TBA]Br) inhibits isomerization. This validates the dissociation of Pd—Br bond, which benefits for generation of a palladium intermediate with a vacant coordination site (**Int 99**), is vital (Fig. 48, Eq. 1). The competitive experiment indicates increasing the steric hinderance can inhibit transmetalation and leads to isomerization (Eq. 2). One possible pathway (A) for "cage-walking," which assumes "deprotonation- B(4)–H activation-reprotonation" (from **Int 99** to **Int 100**, then to **Int 101**) is ruled out since no deuterated **450′** was monitored in the presence of 2,6-$Me_2C_6H_4OD$ (Eq. 3). Another prospective pathway (B) features a stepwise Pd-induced B—H activation that results in an intramolecular β-hydride shift. The feasibility of pathway B was confirmed by transformation that employed 9-Br-10-D-*meta*-$C_2B_{10}H_{10}$. Hence a complete mechanism, including cross-coupling and cage-walking, can be proposed. This protocol is an efficient implement to functionalize B(2) vertex of *m*-carborane.

Very recently, a series of highly functionalized *o*-carboranes was synthesized via Pd-catalyzed sequential B–I/B–H activation by the Xie group (Fig. 49). The 3-alkenyl-4-I-*o*-carborane **451** was prepared when **14** was treated with 5.0 equiv. of diarylacetylene **45** and 10 mol% amount of $Pd(PPh_3)_4$ *(114)*. However, when **14** was mixed with 10 mol% $Pd(OAc)_2$, 20 mol% PPh_3, and 2.0 equiv. of **45**, compound **452** was synthesized *(115)*. DFT calculations indicated that the migration of the [Pd] fragment (from **Int 105** to **Int 106**) is thermodynamically favorable *(114)*, whereas the presence of base could promote a concerted metalation-deprotonation (CMD) process (**Int 107**) that afford compound **452**. Likewise, a palladium-catalyzed highly selective functionalization of 4,7-I_2-o-carborane (**18**) and 3,6-I_2-o-carborane (**454**) have been developed later, providing 4,7-dialkenyl-5,11-I_2-*o*-carboranes (**453**) and 3,6-dialkenyl-4,11-I_2-*o*-carborane (**455**) *(116)*. However, 4,5-I_2-*o*-carbrane (**456**) is not reactive in the same condition. The newly produced B—I bonds can be transferred to other functionalities thus further expands the product scope.

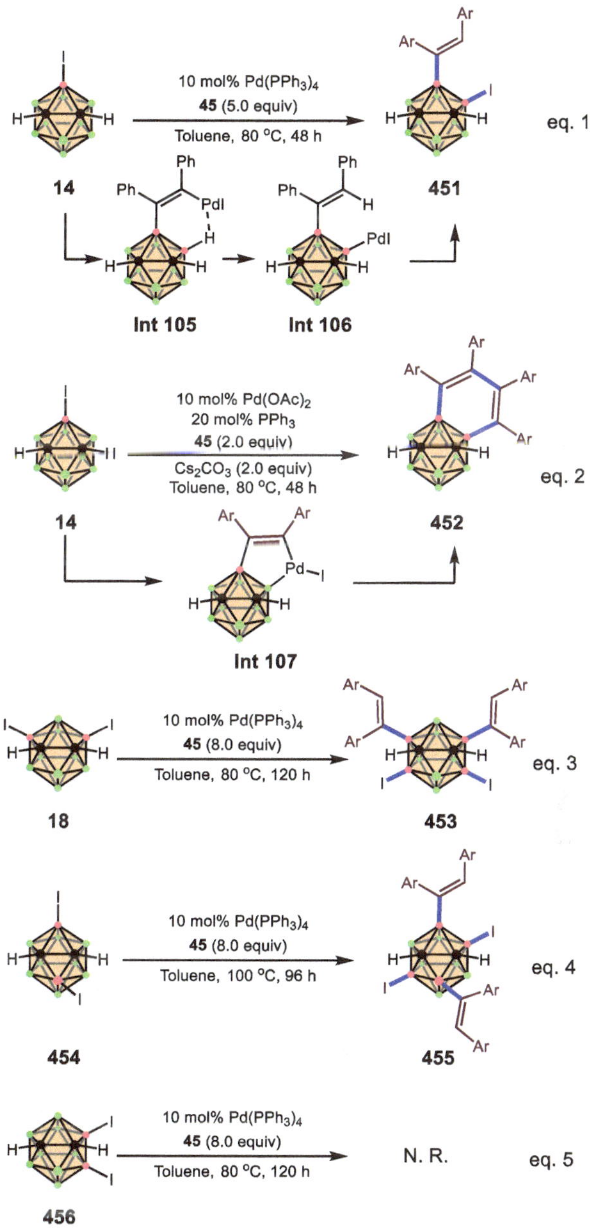

Fig. 49 Pd-catalyzed cascade B–I/B–H activation.

3.2 Carboranylation enabled by organometallic reagent

Bioconjugation, mostly involves chemical reactions to produce covalent attachments between biomolecule with another molecular entity, has been facilitated to create modified biomolecules with enhancing functions than what nature intended. As far as peptide and protein are concerned, cysteine (Cys) residues represent conspicuous positions for chemical modification due to their soft nucleophilicity as well as low natural abundance which enable them to undergo site-selective modification *(117)*. Despite chemical selective modification of cysteine have been well achieved by metal free and swiftly developed metal-mediated route, the 3D organomimetic (Fig. 50).

m-carborane cage has not been incorporated onto unprotected peptides until Spokoyny et al. realized Cys-carboranyl cross-coupling by utilizing bench-stable Pt(II)-based organometallic reagent **457** in 2021 *(21)*. The reaction was set up in Tris-HCl buffer/DMF system under room temperature, a lot of functional groups in unprotected peptide or protein have been well tolerated. The affinity between soft nucleophilic Cys sulfur residue hastened the ligand exchange process thus combined Cys with Pt(II) center. The Cys-carboranylated peptides were afforded after reductive elimination. The boron cage can be installed not only on linear peptides but also on DARPin (designed ankyrin repeat protein), which is a sophisticated substrate. Compare with its unsubstituted and Cys-arylated counterparts, the Cys-carboranylated peptide exhibited better stability in acid or alkaline environment, which clues that the Cys-carboranylated peptides is capable to be biorthogonal even towards harsh biological milieu. Another Cys-carboranylated has shown excellent inhibitory to Proteinase K, while its Cys-H and Cys-arylated congeners were degraded. The interaction between carborane cage and protein via unique B—H ··· H—N dihydrogen bond was found by molecular dynamic simulations, which can account for the reason of the stability of Cys-carboranylated peptide.

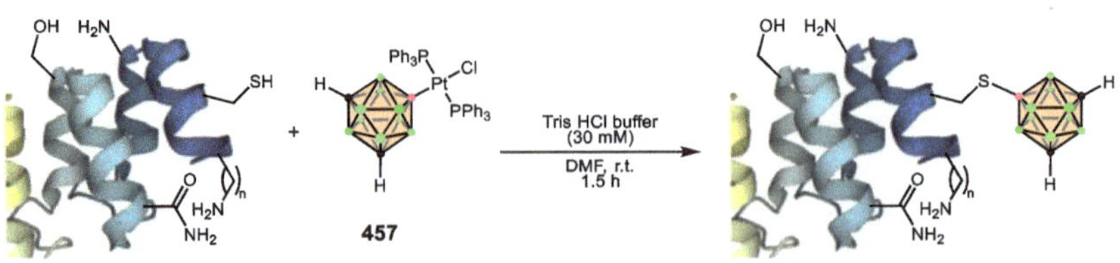

Fig. 50 Cysteine-selective carboranylation of peptides and proteins.

3.3 Carboranylation enabled by electrophile or radical precursors

Except for B–halogen and Pt(II)-based organometallic reagent, the carboranylation could be achieved by employing electrophile or radical precursors. In 2016, the Xie group developed prepared a series of B(3)-arylated *o*-carboranes via visible-light-induced photocatalysis *(118)*. Cheap 2-(2,4,5,7-tetrabromo-6-oxido-3-oxo-3*H*-xanthen-9-yl)benzoate (eosin Y) was used as photocatalyst, and [3-N_2-*o*-$C_2B_{10}H_{11}$]BF_4 acted as a B(3) radical precursor. B(3) centered radical was captured by (2,2,6,6-Tetramethylpiperidin-1-yl) oxyl (TEMPO), and the possibility of electrophilic substitution was ruled out. The electrophilicity of [3-N_2-*o*-$C_2B_{10}H_{11}$]BF_4 (**458**) was utilized well by the same group. Various 3-X-*o*-$C_2B_{10}H_{11}$ (X=F, Cl, Br, I, OAc, ONO_2, NO_2, N_3, NH_2, $OP(O)(OH)_2$, OSO_2H, CF_3, CN, OEt, SCN, aryl, phenyl, NR_2, OR, OC(O)R, NHC(O)R, $OSMe_2OH$, $P(O)Ar_2$) were obtained just in 5 min under room temperature when **458** was treated with a serious of nucleophiles in acetonitrile *(119)* (Figs. 51 and 52).

In 2020, the Spokoyny group utilized boron-substituted carborane (e.g., —BF_3K, $B(OH)_2$) as starting materials and realized carboranylation via boron-centered radical generated by Mn(III) oxidation. The radicals were seized by TEMPO *(120)*.

Very recently, the Xie group disclosed a B—Ar coupling enabled by visible-light-promoted nickel-catalysis (Fig. 53) *(121)*. The generation of

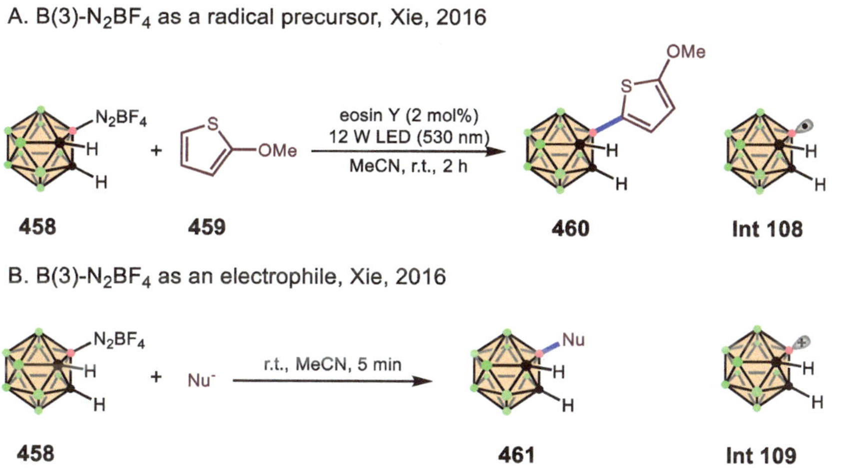

Fig. 51 Reactivity of [3-N_2-*o*-$C_2B_{10}H_{11}$]BF_4.

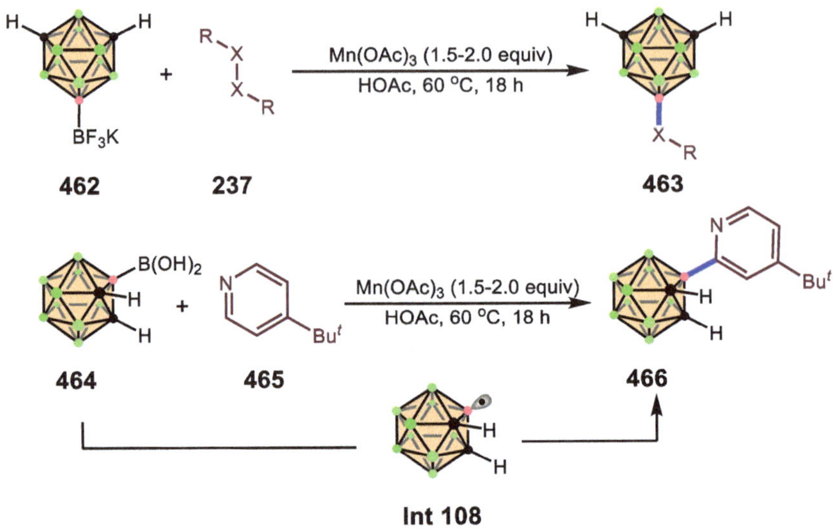

Fig. 52 Carboranylation via a boron-centered carboranyl radical.

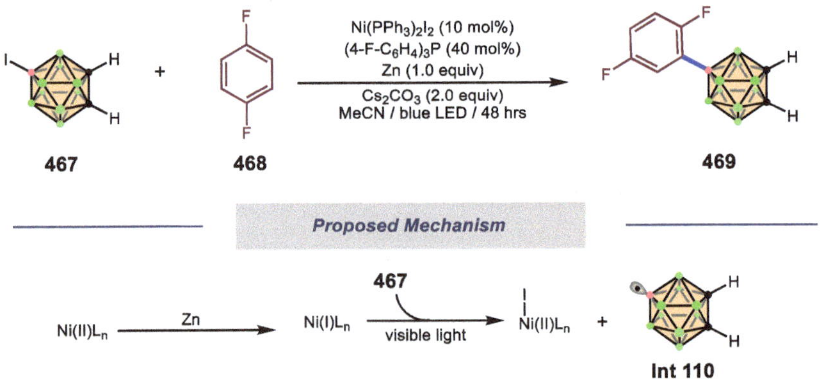

Fig. 53 Visible-light-promoted Ni-catalyzed B—Ar coupling via B centered carboranyl radicals.

boron centered radical **Int 110** is proposed to be the abstraction of iodine from iodo-*o*-carboranes by low valent and excited Ni*(0)L_n or Ni(I)*L_n species. Either visible light or nickel catalyst is essential for this transformation. The carboranyl radical **Int 110** can be captured by TEMPO.

Carboxylic acids are feedstock chemicals with broad structural diversity and generally high chemical stability. In the recent decade,

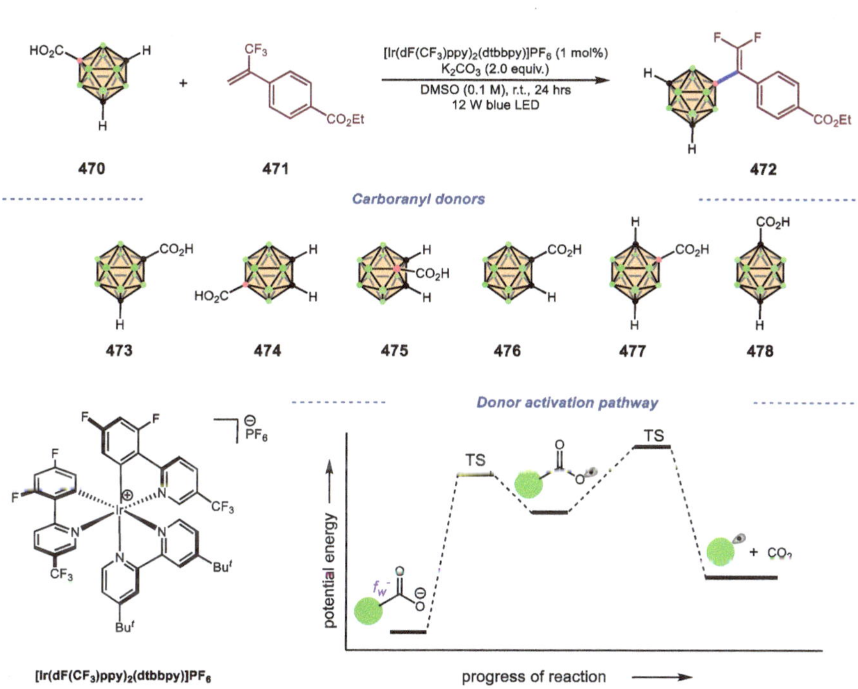

Fig. 54 Photocatalytic decarboxylative B—C coupling via a carboranyl cage radical.

MacMillan *(122)*, Ritter *(123)*, Yoon *(124)*, and other groups have disclosed that carboxylic acid is an ideal donor of alkyl fragment in photoredox-enabled cross couplings. Very recently, the Yan group developed a visible-light-induced decarboxylative carboranylation via a carboranyl cage radical (Fig. 54). This work is in submission. A series of carboranylated geminal difluoromethyl products, (hetero)arenes, alkenes as well as alkanes were produced in moderate to good yield. Unlike corresponding works of the Xie and the Spokoyny lab *(120,121)*, the reactivity is greatly affected by electronic density of vertex that connected with the carboxyl. To evaluate the electronic density of the highest occupied molecular orbital (HOMO) of different vertices, Fukui indices (f_w^-) was induced. A larger f_w^- value means a larger electron density on corresponding vertices. According to computational and experimental data, the reaction energy barrier of SET oxidation is negatively correlated with the f_w^-, while the experimental yield shows a positively correlated with the f_w^-. Therefore, the 9-COOH-*m*-carborane (**470**) provided a higher yield than other carboranyl donors (**473**–**478**) (Table 1).

Table 1 NPA charge, Fukui incides (f_w^-) of C_{cage}/B_{cage} connected with carboxyl; energy barrier of decarboxylation; and yield.

Compound	NPA charge	f_w^-	Energy barrier (kcal mol^{-1})	Yield (%)
470	−0.19151	0.09358	115.2	85
473	−0.61207	0.03057	139.2	0
474	−0.16274	0.09577	122.8	23
475	0.09567	0.04365	125.4	13
476	−0.48611	0.02413	149.9	0
477	−0.05003	0.06833	115.4	30
478	−0.59252	0.03378	136.2	0

4. Cross-coupling on cage carbon

Despite a large number of groundbreaking works on functionalization of boron vertex in the past decade, investigations on functionalization of cage carbon have also progressed. In contrast to transition-metal catalyzed carboxylic acid directed B—H functionalization, Yan and Li found that when 1-COOH-2-R-*o*-$C_2B_{10}H_{10}$ (**29**) was treated with 2.5 equiv. of diazo salts and 2.0 equiv. of potassium acetate in acetonitrile under room temperature, *trans*-azocarboranes (**480**) were afforded in moderate to high yields (Fig. 55) *(125)*. It is plausible that carbanion **Int 111** was generated by base-promoted decarboxylation and then incurred a C—N construction.

In 2015, the Xie group reported a $NiCl_2$-catalyzed cross-coupling of carborane-based Grignard reagent and aryl iodide. The *mono-* or *di-*arylation was controlled by stoichiometry (Fig. 56) *(126)*. In 2016, Jia and Xie reported a facile phosphine-catalyzed alkenylation of cage carbon, where they used electron-deficient alkynes as coupling partner and tris(4-methoxyphenyl)phosphine as catalyst *(127)*. This strategy is compatible with various aromatic rings present in alkynes. The plausible mechanism is depicted in Fig. 57, the allene **Int 113** is the key intermediate in this transformation.

In recent years, the Xie group have come up with C_{cage}—C(sp^2) and C_{cage}—C(sp^3) constructions via homolytic cleavage of C_{cage}—I bond under

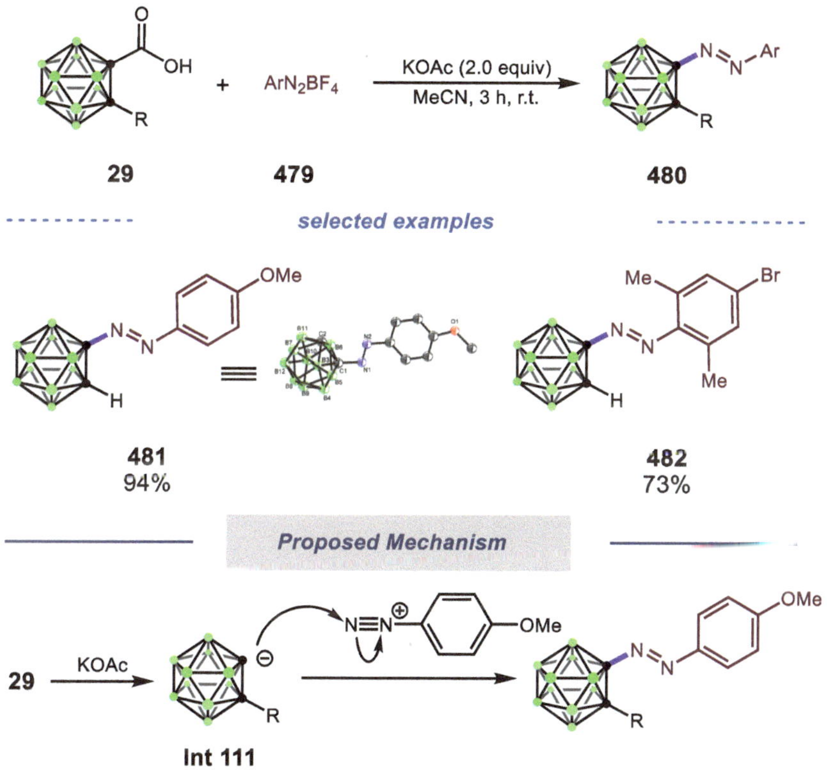

Fig. 55 Base-promoted decarboxylative azo-coupling.

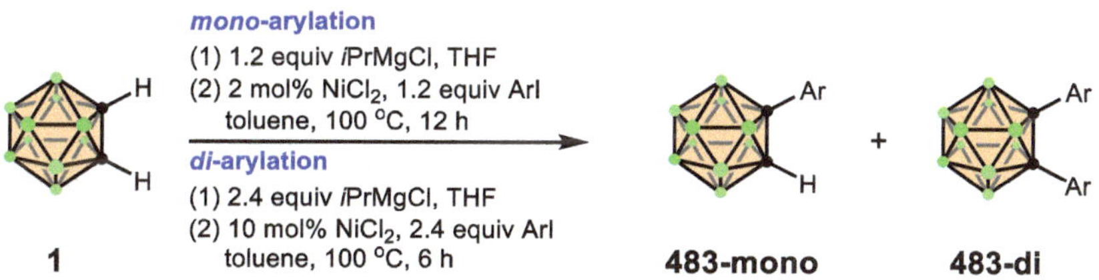

Fig. 56 Arylation of cage carbon.

ultra-violet radiation (Fig. 58) *(128)*. The carborane cage turned out to be stable under UV irradiation, and the tested functionalities are intact as well.

The elegant advantages of electrocatalysis have been well embedded in cage carbon functionalization by Ackermann and co-workers,

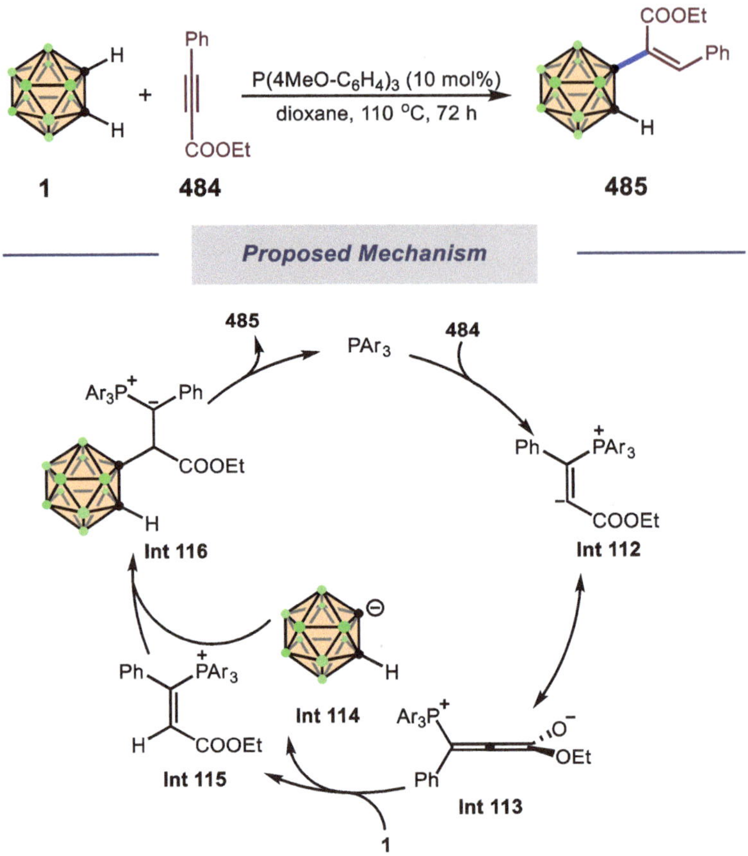

Fig. 57 Synthesis of C-alkenylcarboranes via B—H activation.

which provided a fine example of C_{cage}—S, C_{cage}—Se formation enabled by electrocatalysis (Fig. 59) *(129)*. Thiols/selenols bearing EWG and EDG were hereby well tolerated. This transformation cannot be quenched by TEMPO or 1,1-diphenylethylene. In addition, a signal suspected to a thiol radical was found by electron paramagnetic resonance (condition: 4-methoxybenzenethiol, LiOBu^t, in THF with electricity under air). Detailed with the data of control experiments, a plausible mechanism was hence proposed. Thiol radical was generated by anodic oxidation in alkaline condition and then captured by Cu(I) that form **Int 119**. A base-mediated ligand exchange resulted in formation of **Int 120**, which was then oxidized by iodine. In the last step, the target molecule **495** was produced by reductive elimination.

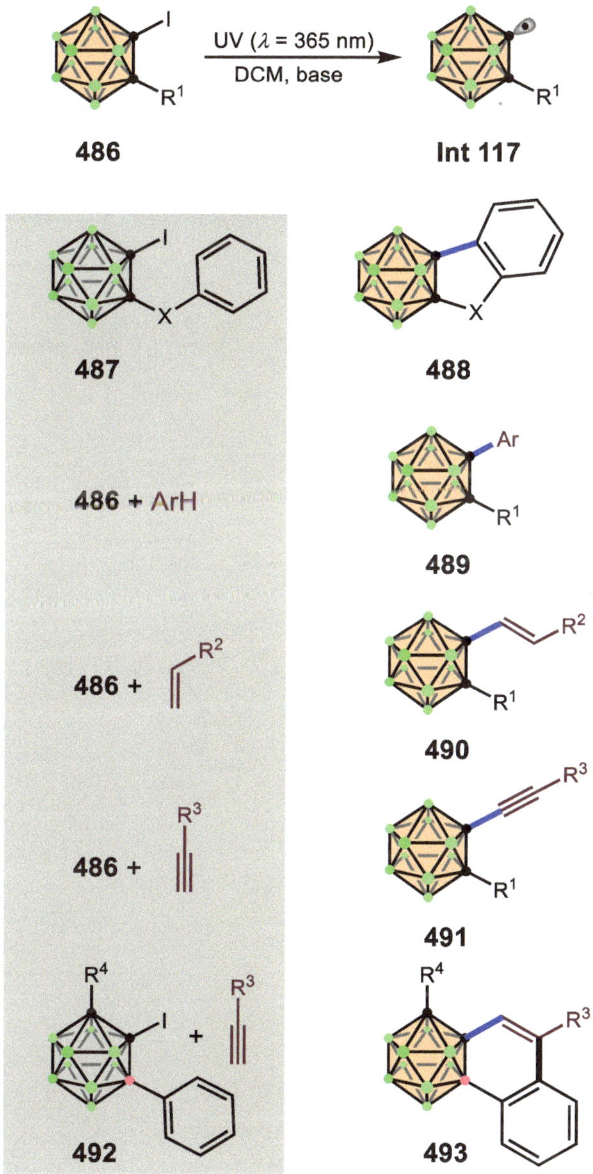

Fig. 58 Construction of C_{cage}—C bond via B—I activation.

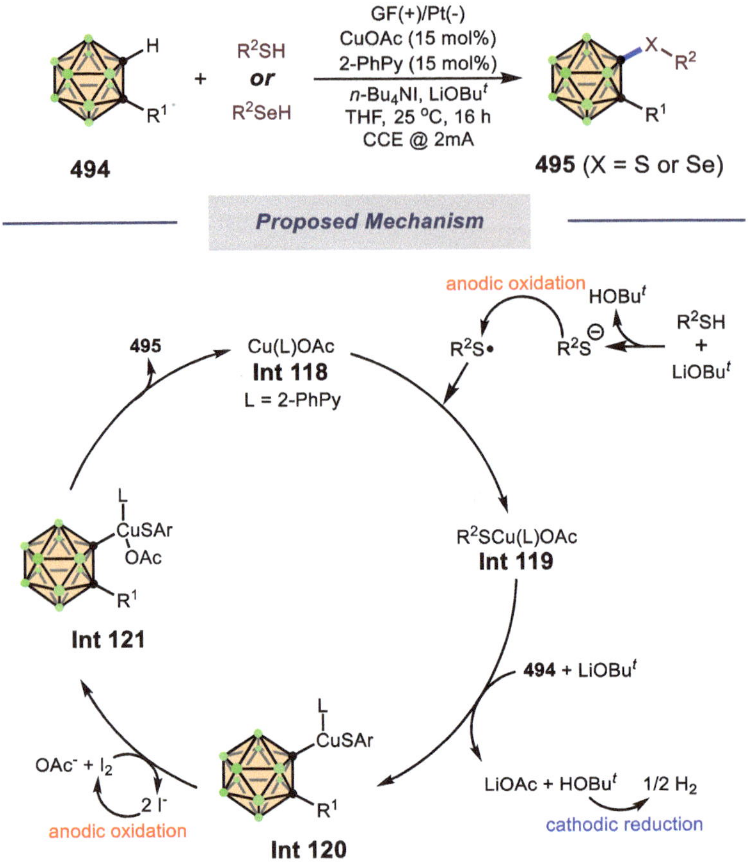

Fig. 59 Eletrochemistry enabled C_{cage}—S, C_{cage}—Se formation.

5. Summary and outlooks

Traditionally, functionalization of carborane has been dominated by (1) reaction between $B_{10}H_{14}$ or $B_{10}H_{12}L_2$ (L = SMe_2 or MeCN) with alkynes; (2) Lewis acid-mediated electrophilic B(9)–H functionalization; (3) deboronation and reboronation of B(3) vertex; (4) Pd-catalyzed B—I transformation; and (5) base-promoted C_{cage}—H cleavage and further nucleophilic reaction for decades. However, these classical approaches plagued with very limited product variety, as the majority of B—H bonds cannot be efficiently functionalized.

In this chapter, we summarized the innovative approaches for functionalization of carborane (excluding transformations via carboryne intermediates) that have emerged over the past decade. Innovative synthetic methods, from transition metal catalysis, metal-free catalysis, electrocatalysis, visible-light-induced photocatalysis, mechanochemical ball-milling reaction, to asymmetric catalysis, afforded a lot of valuable highly functionalized carborane derivatives. According to the driving force of functionalization, these methods can be categorized into: (1) directing group (including carboxy, imine, pyridyl, amide, hydroxyl, phosphine) assisted *ortho*-B–H activation; (2) direct regioselective B—H functionalization without a directing group (including transition-metal-induced electron-rich B(8,9,10,12) or electron-deficient B(3,6) functionalization and nucleophilic B—H substitution); (3) functionalization of *nido*-7,8-carborane via single electron transfer process; (4) transformation of pre-functionalized carborane on boron site; (5) functionalization of cage carbon. With these tools in hand, a large class of highly functionalized carborane derivatives are now more accessible than ever.

Although the synthesis of carborane has come a long way in past 10 years, the field is still in its infancy. Nobel metals such as palladium, rhodium, and iridium are always employed for transformation. The examples of cheap metal catalysis or even metal-free catalysis are still rare. The regioselective B(8,10) functionalization should be further investigated. In some instances, the reaction temperature exceeds the boiling point of the solvent, which may impede its application in large-scale synthesis or industry. Furthermore, a more complex substrate scope, especially for $C(sp^3)$ fragments, and realization of catalyzed biorthogonal reactions are expected. By reviewing the progresses in functionalization of carboranes, we are looking forward to applying these methods to the research of medicine, luminescent materials, and more.

Reference

1. (a) Miessler, G. L. F.; Fischer, P. J.; Tarr, D. A. *Inorganic Chemistry*; Pearson, 2014; (b) Grimes, R. N. *Carboranes*; Elsevier: Amsterdam, 2016; (c) Scholz, M.; Hey-Hawkins, E. *Chem. Rev.* **2011**, *111*, 7035–7062.
2. Poater, J.; Sola, M.; Vinas, C.; Teixidor, F. *Angew. Chem. Int. Ed.* **2014**, *53*, 12191–12195.
3. Zhang, X.; Dai, H.; Yan, H.; Zou, W.; Cremer, D. *J. Am. Chem. Soc.* **2016**, *138*, 4334–4337.
4. (a) Shi, C.; Sun, H.; Tang, X.; Lv, W.; Yan, H.; Zhao, Q.; Wang, J.; Huang, W. *Angew. Chem. Int. Ed.* **2013**, *52*, 13434–13438; (b) Zhu, L.; Tang, X.; Yu, Q.; Lv, W.; Yan, H.; Zhao, Q.; Huang, W. *Chem. Eur. J.* **2015**, *21*, 4721–4730; (c) Axtell, J. C.; Kirlikovali, K. O.; Djurovich, P. I.; Jung, D.; Nguyen, V. T.; Munekiyo, B.; Royappa, A. T.; Rheingold, A. L.; Spokoyny, A. M. *J. Am. Chem. Soc.* **2016**, *138*, 15758–15765; (d) Shi, C.; Tu, D.; Yu, Q.; Liang, H.; Liu, Y.; Li, Z.; Yan, H.; Zhao, Q.; Huang, W. *Chem. Eur. J.* **2014**, *20*, 16550–16557; (e) Lee, Y. H.; Park, J.; Lee, J.; Lee, S. U.; Lee, M. H. *J. Am. Chem. Soc.* **2015**, *137*, 8018–8021; (f) Tu, D.; Leong, P.; Li, Z.; Hu, R.; Shi, C.; Zhang, K. Y.; Yan, H.; Zhao, Q. *Chem. Commun.* **2016**, *52*, 12494–12497; (g) Tu, D.; Cai, S.;

Fernandez, C.; Ma, H.; Wang, X.; Wang, H.; Ma, C.; Yan, H.; Lu, C.; An, Z. *Angew. Chem. Int. Ed.* **2019**, *58*, 9129–9133; (h) Wei, X.; Zhu, M.-J.; Cheng, Z.; Lee, M.; Yan, H.; Lu, C.; Xu, J.-J. *Angew. Chem. Int. Ed.* **2019**, *58*, 3162–3166; (i) Prokhorov, A. M.; Hofbeck, T.; Czerwieniec, R.; Suleymanova, A. F.; Kozhevnikov, D. N.; Yersin, H. *J. Am. Chem. Soc.* **2014**, *136*, 9637–9642; (j) Kirlikovali, K. O.; Axtell, J. C.; Gonzalez, A.; Phung, A. C.; Khan, S. I.; Spokoyny, A. M. *Chem. Sci.* **2016**, 7, 5132–5138; (k) Furue, R.; Nishimoto, T.; Park, I. S.; Lee, J.; Yasuda, T. *Angew. Chem. Int. Ed.* **2016**, *55*, 7171–7175; (l) Nguyen Van, N.; Jana, S.; Sujith, S.; Ryu, J. Y.; Lee, J.; Lee, S. U.; Lee, M. H. *Angew. Chem. Int. Ed.* **2018**, *57*, 12483–12488; (m) Abdelgawwad, A. M. A.; Xavier, J. A. M.; Roca-Sanjuan, D.; Vinas, C.; Teixidor, F.; Frances-Monerris, A. *Angew. Chem. Int. Ed.* **2021**, *60*, 25753–25757.

5. (a) Tan, F.; Lopez-Periago, A.; Light, M. E.; Cirera, J.; Ruiz, E.; Borras, A.; Teixidor, F.; Vinas, C.; Domingo, C.; Planas, J. G. *Adv. Mater.* **2018**, *30*, e1800726; (b) Rodriguez-Hermida, S.; Tsang, M. Y.; Vignatti, C.; Stylianou, K. C.; Guillerm, V.; Perez-Carvajal, J.; Teixidor, F.; Vinas, C.; Choquesillo-Lazarte, D.; Verdugo-Escamilla, C.; Peral, I.; Juanhuix, J.; Verdaguer, A.; Imaz, I.; Maspoch, D.; Giner Planas, J. *Angew. Chem. Int. Ed.* **2016**, *55*, 16049–16053.
6. Keener, M.; Hunt, C.; Carroll, T. G.; Kampel, V.; Dobrovetsky, R.; Hayton, T. W.; Menard, G. *Nature* **2020**, *577*, 652–655.
7. Buades, A. B.; Sanchez Arderiu, V.; Olid-Britos, D.; Vinas, C.; Sillanpaa, R.; Haukka, M.; Fontrodona, X.; Paradinas, M.; Ocal, C.; Teixidor, F. *J. Am. Chem. Soc.* **2018**, *140*, 2957–2970.
8. Liu, G.; Zhang, J.; Wu, B.; Wang, J. *Org. Lett.* **2007**, *9*, 4263–4266.
9. (a) Suzuki, M. *Int. J. Clin. Oncol.* **2020**, *25*, 43–50; (b) Hawthorne, M. F. *Angew. Chem. Int. Ed. Eng.* **1993**, *32*, 950–984.
10. (a) Lesnikowski, Z. J. *J. Med. Chem.* **2016**, *59*, 7738–7758; (b) Wang, Y.; Spokoyny, A. M. *ACS Cent. Sci.* **2022**. https://doi.org/10.1021/acscentsci.1022c00187; (c) Varkhedkar, R.; Yang, F.; Dontha, R.; Zhang, J.; Liu, J.; Spingler, B.; van der Veen, S.; Duttwyler, S. *ACS Cent. Sci.* **2022**. https://doi.org/10.1021/acscentsci.1021c01132; (d) Wilkinson, S. M.; Gunosewoyo, H.; Barron, M. L.; Boucher, A.; McDonnell, M.; Turner, P.; Morrison, D. E.; Bennett, M. R.; McGregor, I. S.; Rendina, L. M.; Kassiou, M. *ACS Chem. Neurosci.* **2014**, *5*, 335–339.
11. (a) Fujii, S.; Masuno, H.; Taoda, Y.; Kano, A.; Wongmayura, A.; Nakabayashi, M.; Ito, N.; Shimizu, M.; Kawachi, E.; Hirano, T.; Endo, Y.; Tanatani, A.; Kagechika, H. *J. Am. Chem. Soc.* **2011**, *133*, 20933–20941; (b) Brynda, J.; Mader, P.; Sicha, V.; Fabry, M.; Poncova, K.; Bakardiev, M.; Gruener, B.; Cigler, P.; Rezacova, P. *Angew. Chem. Int. Ed.* **2013**, *52*, 13760–13763; (c) Otero, R.; Seoane, S.; Sigueiro, R.; Belorusova, A. Y.; Maestro, M. A.; Perez-Fernandez, R.; Rochel, N.; Mourino, A. *Chem. Sci.* **2016**, 7, 1033–1037.
12. (a) Lovering, F.; Bikker, J.; Humblet, C. *J. Med. Chem.* **2009**, *52*, 6752–6756; (b) Subbaiah, M. A. M.; Meanwell, N. A. *J. Med. Chem.* **2021**, *64*, 14046–14128.
13. Measom, N. D.; Down, K. D.; Hirst, D. J.; Jamieson, C.; Manas, E. S.; Patel, V. K.; Somers, D. O. *ACS Med. Chem. Lett.* **2017**, *8*, 43–48.
14. Reekie, T. A.; Williams, C. M.; Rendina, L. M.; Kassiou, M. *J. Med. Chem.* **2019**, *62*, 1078–1095.
15. Wang, L.; Doherty, G. A.; Judd, A. S.; Tao, Z. F.; Hansen, T. M.; Frey, R. R.; Song, X.; Bruncko, M.; Kunzer, A. R.; Wang, X.; Wendt, M. D.; Flygare, J. A.; Catron, N. D.; Judge, R. A.; Park, C. H.; Shekhar, S.; Phillips, D. C.; Nimmer, P.; Smith, M. L.; Tahir, S. K.; Xiao, Y.; Xue, J.; Zhang, H.; Le, P. N.; Mitten, M. J.; Boghaert, E. R.; Gao, W.; Kovar, P.; Choo, E. F.; Diaz, D.; Fairbrother, W. J.; Elmore, S. W.; Sampath, D.; Leverson, J. D.; Souers, A. J. *ACS Med. Chem. Lett.* **2020**, *11*, 1829–1836.

16. Aguilar, A.; Lu, J.; Liu, L.; Du, D.; Bernard, D.; McEachern, D.; Przybranowski, S.; Li, X.; Luo, R.; Wen, B.; Sun, D.; Wang, H.; Wen, J.; Wang, G.; Zhai, Y.; Guo, M.; Yang, D.; Wang, S. *J. Med. Chem.* **2017**, *60*, 2819–2839.
17. Caldwell, R. D.; Qiu, H.; Askew, B. C.; Bender, A. T.; Brugger, N.; Camps, M.; Dhanabal, M.; Dutt, V.; Eichhorn, T.; Gardberg, A. S.; Goutopoulos, A.; Grenningloh, R.; Head, J.; Healey, B.; Hodous, B. L.; Huck, B. R.; Johnson, T. L.; Jones, C.; Jones, R. C.; Mochalkin, I.; Morandi, F.; Nguyen, N.; Meyring, M.; Potnick, J. R.; Santos, D. C.; Schmidt, R.; Sherer, B.; Shutes, A.; Urbahns, K.; Follis, A. V.; Wegener, A. A.; Zimmerli, S. C.; Liu-Bujalski, L. *J. Med. Chem.* **2019**, *62*, 7643–7655.
18. Tse, E. G.; Houston, S. D.; Williams, C. M.; Savage, G. P.; Rendina, L. M.; Hallyburton, I.; Anderson, M.; Sharma, R.; Walker, G. S.; Obach, R. S.; Todd, M. H. *J. Med. Chem.* **2020**, *63*, 11585–11601.
19. Yin, Y.; Ochi, N.; Craven, T. W.; Baker, D.; Takigawa, N.; Suga, H. *J. Am. Chem. Soc.* **2019**, *141*, 19193–19197.
20. Kracke, G. R.; VanGordon, M. R.; Sevryugina, Y. V.; Kueffer, P. J.; Kabytaev, K., Jalisatgi, S. S.; Hawthorne, M. F. *ChemMedChem* **2015**, *10*, 62–67.
21. Waddington, M. A.; Zheng, X.; Stauber, J. M.; Hakim Moully, E.; Montgomery, H. R.; Saleh, L. M. A.; Kral, P.; Spokoyny, A. M. *J. Am. Chem. Soc.* **2021**, *143*, 8661–8668.
22. Teixidor, F.; Barbera, G., Vaca, A.; Kivekas, R.; Sillanpaa, R.; Oliva, J.; Vinas, C. *J. Am. Chem. Soc.* **2005**, *127*, 10158–10159.
23. (a) Herzog, A.; Maderna, A.; Harakas, G. N.; Knobler, C. B.; Hawthorne, M. F. *Chem. Eur. J.* **1999**, *5*, 1212–1217; (b) Teixidor, F.; Barbera, G.; Vinas, C.; Sillanpaa, R.; Kivekas, R. *Inorg. Chem.* **2006**, *45*, 3496–3498.
24. (a) Grafstein, D.; Dvorak, J. *Inorg. Chem.* **1963**, *2*, 1128–1133; (b) Wiesboeck, R. A.; Hawthorne, M. F. *J. Am. Chem. Soc.* **1964**, *86*, 1642; (c) Tebbe, F.; Philip, M. G.; Hawthorne, M. F. *J. Am. Chem. Soc.* **1964**, *89*, 4222–4223.
25. Hawthorne, M. F.; Wegener, P. A. *J. Am. Chem. Soc.* **1968**, *90*, 896–901.
26. (a) Safronov, A. V.; Hawthorne, M. F. *Inorg. Chim. Acta* **2011**, *375*, 308–310; (b) Aizawa, K.; Ohta, K.; Endo, Y. *Heterocycles* **2010**, *80*, 369–377.
27. Ramachandran, B. M.; Knobler, C. B.; Hawthorne, M. F. *Inorg. Chem.* **2006**, *45*, 336–340.
28. Zakharkin, L. I.; Kalinin, V. N.; Gedymin, V. V. *J. Organomet. Chem.* **1969**, *16*, 371–379.
29. Bondarev, O.; Sevryugina, Y. V.; Jalisatgi, S. S.; Hawthorne, M. F. *Inorg. Chem.* **2012**, *51*, 9935–9942.
30. Safronov, A. V.; Sevryugina, Y. V.; Jalisatgi, S. S.; Kennedy, R. D.; Barnes, C. L.; Hawthorne, M. F. *Inorg. Chem.* **2012**, *51*, 2629–2637.
31. (a) Zhang, R.; Zhu, L.; Liu, G.; Dai, H.; Lu, Z.; Zhao, J.; Yan, H. *J. Am. Chem. Soc.* **2012**, *134*, 10341–10344; (b) Wang, Z.; Ye, H.; Li, Y.; Li, Y.; Yan, H. *J. Am. Chem. Soc.* **2013**, *135*, 11289–11298.
32. Hoel, E. L.; Hawthorne, M. F. *J. Am. Chem. Soc.* **1975**, *97*, 6388–6395.
33. Yao, Z.-J.; Yu, W.-B.; Lin, Y.-J.; Huang, S.-L.; Li, Z.-H.; Jin, G.-X. *J. Am. Chem. Soc.* **2014**, *136*, 2825–2832.
34. (a) Herberhold, M.; Yan, H.; Milius, W.; Wrackmeyer, B. *Angew. Chem., Int. Ed.* **1999**, *38*, 3689–3691; (b) Fey, N.; Haddow, M. F.; Mistry, R.; Norman, N. C.; Orpen, A. G.; Reynolds, T. J.; Pringle, P. G. *Organometallics* **2012**, *31*, 2907–2913.
35. Liu, D.; Dang, L.; Sun, Y.; Chan, H. S.; Lin, Z.; Xie, Z. *J. Am. Chem. Soc.* **2008**, *130*, 16103–16110.
36. Quan, Y.; Xie, Z. *J. Am. Chem. Soc.* **2014**, *136*, 15513–15516.
37. Cheng, B.; Chen, Y.; Xie, Z. *J. Org. Chem.* **2021**, *86*, 12412–12418.
38. Baek, Y.; Cheong, K.; Ko, G. H.; Han, G. U.; Han, S. H.; Kim, D.; Lee, K.; Lee, P. H. *J. Am. Chem. Soc.* **2020**, *142*, 9890–9895.
39. Quan, Y.; Tang, C.; Xie, Z. *Chem. Sci.* **2016**, 7, 5838–5845.

40. Quan, Y.; Xie, Z. *Angew. Chem. Int. Ed.* **2016**, *55*, 1295–1298.
41. Quan, Y.; Lyu, H.; Xie, Z. *Chem. Commun.* **2017**, *53*, 4818–4821.
42. Lyu, H.; Quan, Y.; Xie, Z. *Angew. Chem. Int. Ed.* **2015**, *54*, 10623–10626.
43. Wang, Q.; Tian, S.; Zhang, C.; Li, J.; Wang, Z.; Du, Y.; Zhou, L.; Lu, J. *Org. Lett.* **2019**, *21*, 8018–8021.
44. Au, Y. K.; Lyu, H.; Quan, Y.; Xie, Z. *J. Am. Chem. Soc.* **2019**, *141*, 12855–12862.
45. (a) Chen, Y.; Quan, Y.; Xie, Z. *Chem. Commun.* **2020**, *56*, 7001–7004; (b) Au, Y. K.; Lyu, H.; Quan, Y.; Xie, Z. *Chin. J. Chem.* **2020**, *38*, 383–388.
46. Lyu, H.; Quan, Y.; Xie, Z. *Angew. Chem. Int. Ed.* **2016**, *55*, 11840–11844.
47. Ham, H.; Shin, S.; Ko, G. H.; Han, S. H.; Han, G. U.; Maeng, C.; Kim, T. H.; Noh, H. C.; Lee, K.; Kim, H.; Yang, H.; Lee, P. H. *J. Org. Chem.* **2021**, *86*, 15153–15163.
48. Lyu, H.; Quan, Y.; Xie, Z. *J. Am. Chem. Soc.* **2016**, *138*, 12727–12730.
49. Li, H.; Bai, F.; Yan, H.; Lu, C.; Bregadze, V. I. *Eur. J. Org. Chem.* **2017**, 1343–1352.
50. (a) Beak, Y.; Kim, S.; Son, J.-Y.; Lee, K.; Kim, D.; Lee, P. H. *ACS Catal.* **2019**, 9; (b) Han, G. U.; Baek, Y.; Lee, K.; Shin, S.; Chan Noh, H.; Lee, P. H. *Org. Lett.* **2021**, *23*, 416–420.
51. Han, G. U.; Shin, S.; Baek, Y.; Kim, D.; Lee, K.; Kim, J. G.; Lee, P. H. *Org. Lett.* **2021**, *23*, 8622–8627.
52. Lyu, H.; Quan, Y.; Xie, Z. *Chem. Eur. J.* **2017**, *23*, 14866–14871.
53. Zhang, X.; Zheng, H.; Li, J.; Xu, F.; Zhao, J.; Yan, H. *J. Am. Chem. Soc.* **2017**, *139*, 14511–14517.
54. (a) Yang, K.; Li, Q.; Liu, Y.; Li, G.; Ge, H. *J. Am. Chem. Soc.* **2016**, *138*, 12775–12778; (b) Liu, X.-H.; Park, H.; Hu, J.-H.; Hu, Y.; Zhang, Q.-L.; Wang, B.-L.; Sun, B.; Yeung, K.-S.; Fang-Lin, Z.; Yu, J.-Q. *J. Am. Chem. Soc.* **2017**, *139*, 888.
55. Cao, H.-J.; Chen, M.; Sun, F.; Zhao, Y.; Lu, C.; Zhang, X.; Shi, Z.; Yan, H. *ACS Catal.* **2021**, *11*, 14047–14057.
56. Liu, Y.; Ge, H. *Nat. Chem.* **2017**, *9*, 26–32.
57. Zhang, X.; Yan, H. *Chem. Sci.* **2018**, *9*, 3964–3969.
58. Lin, F.; Yu, J. L.; Shen, Y.; Zhang, S. Q.; Spingler, B.; Liu, J.; Hong, X.; Duttwyler, S. *J. Am. Chem. Soc.* **2018**, *140*, 13798–13807.
59. Lyu, H.; Quan, Y.; Xie, Z. *Chem. Sci.* **2018**, *9*, 6390–6394.
60. Li, C.-X.; Ning, Q.; Zhao, W.; Cao, H. J.; Wang, Y. P.; Yan, H.; Lu, C. S.; Liang, Y. *Chem. Eur. J.* **2021**, *27*, 2699–2706.
61. (a) Meng, G.; Szostak, M. *Angew. Chem. Int. Ed.* **2015**, *54*, 14518–14522; (b) Meng, G.; Szostak, M. *Org. Lett.* **2016**, *18*, 796–799.
62. Li, C.-X.; Zhang, H. Y.; Wong, T. Y.; Cao, H. J.; Yan, H.; Lu, C. S. *Org. Lett.* **2017**, *19*, 5178–5181.
63. Fu, Y.; Li, Y.; Luo, D.; Lu, Y.; Huang, J.; Yang, Z.; Lu, J.; Jiang, Y. Y.; Lu, J. Y. *Inorg. Chem.* **2022**, *61*, 911–922.
64. Yang, Z. Y.; Wu, Y. B.; Fu, Y. T.; Yang, J. X.; Lu, J.; Lu, J. Y. *Chem. Commun.* **2021**, *57*, 1655–1658.
65. Wu, J.; Cao, K.; Zhang, C. Y.; Xu, T. T.; Wen, X. Y.; Li, B.; Yang, J. *Inorg. Chem.* **2020**, *59*, 17340–17346.
66. Cao, H. J.; Wei, X.; Sun, F.; Zhang, X.; Lu, C.; Yan, H. *Chem. Sci.* **2021**, *12*, 15563–15571.
67. (a) Ko, G. H.; Lee, J. K.; Han, S. H.; Lee, P. H. *Org. Lett.* **2022**, *24*, 1507–1512; (b) Zhang, L. B.; Xie, Z. W. *Org. Lett.* **2022**, *24*, 1318–1322; (c) Ko, G. H.; Um, K.; Noh, H. C.; Kim, J. Y.; Jeong, H.; Maeng, C.; Han, S. H.; Han, G. U.; Lee, P. H. *Org. Lett.* **2022**, *24*, 1604–1609.
68. Yang, Z.; Sun, C.; Wei, X.; Lu, J.; Lu, J.-Y. *ChemCatChem* **2022**. https://doi.org/10.1002/cctc.202101571.

69. Zhu, L.; Jiang, Q.-B.; Yan, H. *Chin. J. Inorg. Chem.* **2014**, *30*, 2246–2252.
70. Zaitsev, V. G.; Shabashov, D.; Daugulis, O. *J. Am. Chem. Soc.* **2005**, *127*, 13154–13155.
71. Chen, Y.; Au, Y.-K.; Quan, Y.; Xie, Z. *Sci. Chin. Chem.* **2019**, *62*, 74–79.
72. Chen, Y.; Quan, Y. J.; Xie, Z. W. *Chem. Commun.* **2020**, *56*, 12997–13000.
73. Au, Y. K.; Lyu, H. R.; Quan, Y. J.; Xie, Z. W. *J. Am. Chem. Soc.* **2020**, *142*, 6940–6945.
74. Liang, Y.-F.; Yang, L.; Jei, B. B.; Kuniyil, R.; Ackermann, L. *Chem. Sci.* **2020**, *11*, 10764–10769.
75. Cheng, R. F.; Zhang, J.; Zhang, H. F.; Qiu, Z. Z.; Xie, Z. W. *Nat. Commun.* **2021**, 12.
76. Xu, T. T.; Cao, K.; Zhang, C. Y.; Wu, J.; Jiang, L. H.; Yang, J. X. *Chem. Commun.* **2018**, *54*, 13603–13606.
77. Zhang, C. Y.; Cao, K.; Xu, T. T.; Wu, J.; Jiang, L. H.; Yang, J. X. *Chem. Commun.* **2019**, *55*, 830–833.
78. Wu, J.; Cao, K.; Zhang, C. Y.; Xu, T. T.; Ding, L. F.; Li, B.; Yang, J. X. *Org. Lett.* **2019**, *21*, 5986–5989.
79. Baek, Y.; Cheong, K.; Kim, D.; Lee, P. H. *Org. Lett.* **2021**, *23*, 1188–1193.
80. Tang, C.; Zhang, J.; Zhang, J.; Xie, Z. *J. Am. Chem. Soc.* **2018**, *140*, 16423–16427.
81. Lyu, H.; Zhang, J.; Yang, J.; Quan, Y.; Xie, Z. *J. Am. Chem. Soc.* **2019**, *141*, 4219–4224.
82. Zhang, Z.-Y.; Zhang, X.; Yuan, J.; Yue, C.-D.; Meng, S.; Chen, J.; Yu, G.-A.; Che, C.-M. *Chem. Eur. J.* **2020**, *26*, 5037–5050.
83. (a) Au, Y. K.; Quan, Y. J.; Xie, Z. W. *Chem. AsianJ.* **2020**, *15*, 2170–2173; (b) Cheng, R. F.; Qiu, Z. Z.; Xie, Z. W. *Chin. J. Chem.* **2020**, *38*, 1575–1578.
84. Quan, Y. J.; Xie, Z. W. *J. Am. Chem. Soc.* **2015**, *137*, 3502–3505.
85. Cheng, R. F.; Li, B. W.; Wu, J.; Zhang, J.; Qiu, Z. Z.; Tang, W. J.; You, S. L.; Tang, Y.; Xie, Z. W. *J. Am. Chem. Soc.* **2018**, *140*, 4508–4511.
86. Cui, C. X.; Zhang, J.; Qiu, Z. Z.; Xie, Z. W. *Dalton Trans.* **2020**, *49*, 1380–1383.
87. Chen, Y.; Lyu, H. R.; Quan, Y. J.; Xie, Z. W. *Org. Lett.* **2021**, *23*, 4163–4167.
88. Qiu, Z.; Quan, Y.; Xie, Z. *J. Am. Chem. Soc.* **2013**, *135*, 12192–12195.
89. (a) Cao, K.; Xu, T.-T.; Wu, J.; Jiang, L.; Yang, J. *Chem. Commun.* **2016**, *52*, 11446–11449; (b) Xu, T. T.; Cao, K.; Wu, J.; Zhang, C. Y.; Yang, J. X. *Inorg. Chem.* **2018**, *57*, 2925–2932.
90. Ma, Y.-N.; Gao, Y.; Ma, Y.; Wang, Y.; Ren, H.; Chen, X. *J. Am. Chem. Soc.* **2022**. https://doi.org/10.1021/jacs.1022c03031.
91. Guo, W.; Guo, C.; Ma, Y. N.; Chen, X. *Inorg. Chem.* **2022**, *61*, 5326–5334.
92. Mirabelli, M. G. L.; Sneddon, L. G. *J. Am. Chem. Soc.* **1988**, *110*, 449–453.
93. Cheng, R.; Qiu, Z.; Xie, Z. *Nat. Commun.* **2017**, *8*.
94. Cheng, R.; Qiu, Z.; Xie, Z. *Chem. Eur. J.* **2020**, *26*, 7144.
95. Xu, T. T.; Cao, K.; Zhang, C. Y.; Wu, J.; Ding, L. F.; Yang, J. X. *Org. Lett.* **2019**, *21*, 9276–9279.
96. Cao, K.; Xu, T. T.; Wu, J.; Zhang, C. Y.; Wen, X. Y.; Yang, J. X. *Inorg. Chem.* **2021**, *60*, 1081–1086.
97. Cao, K.; Wu, J.; Zhang, C. Y.; Ding, L. F.; Yang, J. X. *Chem. Select* **2021**, *6*, 10178–10181.
98. Hawthorne, M. F.; Young, D. C.; Wegner, P. A. *J. Am. Chem. Soc.* **1965**, *87*, 1818–1819.
99. Taoda, Y.; Sawabe, T.; Endo, Y.; Yamaguchi, K.; Fujii, S.; Kagechika, H. *Chem. Commun.* **2008**, *2049-2051*.
100. (a) Willans, C. E.; Kilner, C. A.; Fox, M. A. *Chem. Eur. J.* **2010**, *16*, 10644–10648; (b) Zheng, F. R.; Xie, Z. W. *Dalton Trans.* **2012**, *41*, 12907–12914.
101. Tang, C.; Zhang, J.; Xie, Z. *Angew. Chem., Int. Ed.* **2017**, *56*, 8642–8646.

102. Zhang, J.; Tang, C.; Xie, Z. *Chem. Sci.* **2020**, *11*, 9925–9929.
103. Kang, H. C.; Lee, S. S.; Knobler, C. B.; Hawthorne, M. F. *Inorg. Chem.* **1991**, *30*, 2024–2031.
104. (a) Shirokii, V. L.; Rudakov, D. A.; Bazhanov, A. V.; Zalesskaya, E. G.; Potkin, V. I.; Maier, N. A.; Timofeev, S. V.; Petrovskii, P. V.; Bregadze, V. I. *Russ. J. Electrochem.* **2004**, *40*, 212–214; (b) Rudakov, D. A.; Shirokii, V. L.; Potkin, V. I.; Maier, N. A.; Bragin, V. I.; Petrovskii, P. V.; Sivaev, I. B.; Bregadze, V. I.; Kisin, A. V. *Russ. Chem. Bull.* **2005**, *54*, 1599–1602; (c) Rudakov, D. A.; Potkin, V. I.; Lantsova, I. V. *Russ. J. Electrochem.* **2009**, *45*, 813–817.
105. Yang, L.; Jei, B. B.; Scheremetjew, A.; Kuniyil, R.; Ackermann, L. *Angew. Chem. Int. Ed.* **2021**, *60*, 1482–1487.
106. Chen, M.; Zhao, D. S.; Xu, J. K.; Li, C. X.; Lu, C. S.; Yan, H. *Angew. Chem. Int. Ed.* **2021**, *60*, 7838–7844.
107. Huang, R.-H.; Zhao, W. J.; Xu, S. W.; Xu, J. K.; Li, C.-X.; Lu, C. S.; Yan, H. *Chem. Commun.* **2021**, *57*, 8580–8583.
108. Dziedzic, R. M.; Spokoyny, A. M. *Chem. Commun.* **2019**, *55*, 430–442.
109. Dziedzic, R. M.; Saleh, L. M. A.; Axtell, J. C.; Martin, J. L.; Stevens, S. L.; Royappa, A. T.; Rheingold, A. L.; Spokoyny, A. M. *J. Am. Chem. Soc.* **2016**, *138*, 9081–9084.
110. Sevryugina, Y.; Julius, R. L.; Hawthorne, M. F. *Inorg. Chem.* **2010**, *49*, 10627–10634.
111. Bruno, N. C.; Tudge, M. T.; Buchwald, S. L. *Chem. Sci.* **2013**, *4*, 916–920.
112. Mu, X.; Hopp, M.; Dziedzic, R. M.; Waddington, M. A.; Rheingold, A. L.; Sletten, E. M.; Axtell, J. C.; Spokoyny, A. M. *Organometallics* **2020**, *39*, 4380–4386.
113. Dziedzic, R. M.; Martin, J. L.; Axtell, J. C.; Saleh, L. M. A.; Ong, T. C.; Yang, Y. F.; Messina, M. S.; Rheingold, A. L.; Houk, K. N.; Spokoyny, A. M. *J. Am. Chem. Soc.* **2017**, *139*, 7729–7732.
114. Ge, Y. X.; Zhang, J.; Qiu, Z. Z.; Xie, Z. W. *Angew. Chem. Int. Ed.* **2020**, *59*, 4851–4855.
115. Ge, Y. X.; Zhang, J.; Qiu, Z. Z.; Xie, Z. W. *Dalton Trans.* **2021**, *50*, 1766–1773.
116. Ge, Y. X.; Qiu, Z. Z.; Xie, Z. W. *Chem. Commun.* **2021**, *57*, 8071–8074.
117. (a) Zhu, S.; Samala, G.; Sletten, E. T.; Stockdill, J. L.; Nguyen, H. M. *Chem. Sci.* **2019**, *10*, 10475–10480; (b) Wan, L. Q.; Zhang, X.; Zou, Y. K.; Shi, R.; Cao, J. G.; Xu, S. Y.; Deng, L. F.; Zhou, L.; Gong, Y. Q.; Shu, X. L.; Lee, G. Y.; Ren, H. Y.; Dai, L. Z.; Qi, S. Q.; Houk, K. N.; Niu, D. W. *J. Am. Chem. Soc.* **2021**, *143*, 11919–11926; (c) Messina, M. S.; Stauber, J. M.; Waddington, M. A.; Rheingold, A. L.; Maynard, H. D.; Spokoyny, A. M. *J. Am. Chem. Soc.* **2018**, *140*, 7065–7069; (d) Vara, B. A.; Li, X.; Berritt, S.; Walters, C. R.; Petersson, E. J.; Molander, G. A. *Chem. Sci.* **2018**, *9*, 336–344; (e) Vinogradova, E. V.; Zhang, C.; Spokoyny, A. M.; Pentelute, B. L.; Buchwald, S. L. *Nature* **2015**, *526*, 687–691.
118. Zhao, D.; Xie, Z. W. *Angew. Chem. Int. Ed.* **2016**, *55*, 3166–3170.
119. Zhao, D.; Xie, Z. W. *Chem. Sci.* **2016**, 7, 5635–5639.
120. Mills, H. A.; Martin, J. L.; Rheingold, A. L.; Spokoyny, A. M. *J. Am. Chem. Soc.* **2020**, *142*, 4586–4591.
121. Li, S.; Xie, Z. *J. Am. Chem. Soc.* **2022**. https://doi.org/10.1021/jacs.1022c02329.
122. Johnston, C. P.; Smith, R. T.; Allmendinger, S.; MacMillan, D. W. C. *Nature* **2016**, *536*, 322–325.
123. Sun, X.; Chen, J. T.; Ritter, T. *Nat. Chem.* **2018**, *10*, 1229–1233.
124. Li, Q. Y.; Gockel, S. N.; Lutovsky, G. A.; DeGlopper, K. S.; Baldwin, N. J.; Bundesmann, M. W.; Tucker, J. W.; Bagley, S. W.; Yoon, T. P. *Nat. Chem.* **2022**, *14*, 94–99.
125. Li, H. H.; Bai, F.; Yan, H.; Lu, C. S. *Org. Lett.* **2017**, *19*, 862–865.
126. Tang, C.; Xie, Z. W. *Angew. Chem. Int. Ed.* **2015**, *54*, 7662–7665.

127. Zheng, F. R.; Leung, T. F.; Chan, K. W.; Sung, H. H. Y.; Williams, I. D.; Xie, Z. W.; Jia, G. C. *Chem. Commun.* **2016**, *52*, 10767–10770.
128. (a) Ni, H. C.; Qiu, Z. Z.; Xie, Z. W. *Angew. Chem. Int. Ed.* **2017**, *56*, 712–716; (b) Ni, H. C.; Lu, Z. P.; Xie, Z. W. *J. Am. Chem. Soc.* **2020**, *142*, 18661–18667; (c) Ni, H. C.; Lu, Z. P.; Xie, Z. W. *Dalton Trans.* **2021**, *51*, 104–110.
129. Yang, L.; Jei, B. B.; Scheremetjew, A.; Yuan, B. B.; Stuckl, A. C.; Ackermann, L. *Chem. Sci.* **2021**, *12*, 12971–12976.

About the authors

Dr. Chun-Xiao Li (李春曉) obtained his B.S. degree (2012) in pharmacy at China Pharmaceutical University (China). He received his Ph.D. from Nanjing University, China (2019) under the supervision of Professor Dr. Hong Yan. He joined Nguyen lab at Wayne State University, USA in 2021, and now he focusses on preparation of sulfated glycans that can attenuate the combined risk of COVID-19 and stereoselective functionalization of monosaccharides enabled by photocatalysis.

Professor Dr. Hong Yan (燕紅) received her Ph.D. from Nanjing University, China (1993). After working at the State Key Laboratory of Coordination Chemistry as a lecturer, she was a postdoctoral fellow at the University of Neuchâtel, Switzerland (1996–1997), the University of Bayreuth, Germany (1997–2000), and the University of Notre Dame, USA (2000–2004). In 2005, she joined Nanjing University as a professor. Her current interests focus on synthetic organic chemistry and luminescent materials.

CHAPTER SIX

Derivatization of monocarborane and dodecaborate anions by controlled B—H activation

Simon Duttwyler*
Department of Chemistry, Zhejiang University, Hangzhou, China
*Corresponding author: e-mail address: duttwyler@zju.edu.cn

Contents

Abstract

Significant progress in the synthesis of substituted icosahedral boron clusters by transition metal catalysis has been made in the past years. This development has enabled the exploration of an increasing number of applications, along with a deepened understanding of the organometallic chemistry of polyhedral boranes. Direct conversion of B—H bonds to functionalized boron vertices is an attractive strategy to synthesize cage compounds with desired properties. In this chapter, the derivatization of the anionic boron clusters $[B_{12}H_{12}]^{2-}$ and $[CB_{11}H_{12}]^{-}$ by controlled B—H activation is presented.

1. Introduction

Cage-shaped boranes based on the 12-vertex clusters *closo*-dodecaborate $[B_{12}H_{12}]^{2-}$, often referred to as "dodecaborate," and monocarba-*closo*-dodecaborate $[CB_{11}H_{12}]^{-}$, often referred to as "monocarborane," are an

Advances in Catalysis, Volume 71
ISSN 0360-0564
https://doi.org/10.1016/bs.acat.2022.05.001

intriguing class of main group compounds that exhibit icosahedral symmetry and delocalized electron systems *(1–9)*. In this chapter, an account of recent progress in the controlled modification of such cages is provided. Emphasis is placed on direct and controlled functionalization of B—H vertices, primarily enabled by directing groups and transition metal catalysis. Following some background information, the strategy of B—H activation and selected examples of organometallic species containing boron–metal bonds are presented. The main section encompasses studies on the derivatization and applications of anionic boron clusters whose synthesis has become possible by B—H activation.

Dodecaborates and monocarboranes are named according to the number of vertices and the three-dimensional geometry of the cage; "dodeca" indicates 12 corners, and *closo* means that the compounds possess a closed-cage structure with triangular faces (Fig. 1). The numbering system for the core atoms follows the IUPAC recommendations for polyhedral molecules *(10)*. In the coming sections on the synthetic methodologies and applications, gray spheres either represent a B—H unit or a boron atom when bonded to a specified substituent.

The sphere-like electron delocalization of *closo* clusters has been inferred from X-ray diffraction and neutron diffraction data, along with computational studies. Often called "σ aromaticity," it is closely related to the two-dimensional π aromaticity of classical organic arenes such as benzene *(11–16)*. Theoretical studies on the dianion $[B_{12}H_{12}]^{2-}$ propose a highly delocalized bonding within the cluster core including 26 electrons that originate from valence electrons of the boron centers plus two electrons from the charge. In this electron count, electrons from *exo*-skeletal B—H bonds are disregarded. The 26 electrons occupy 13 bonding molecular orbitals, affording exceptional thermodynamic stability of the dodecaborate dianion. Similar conclusions have resulted from calculations on the monocarborane anion, although in this case a dipole moment is induced by the more electronegative carbon atom. It should be kept in mind that in Lewis

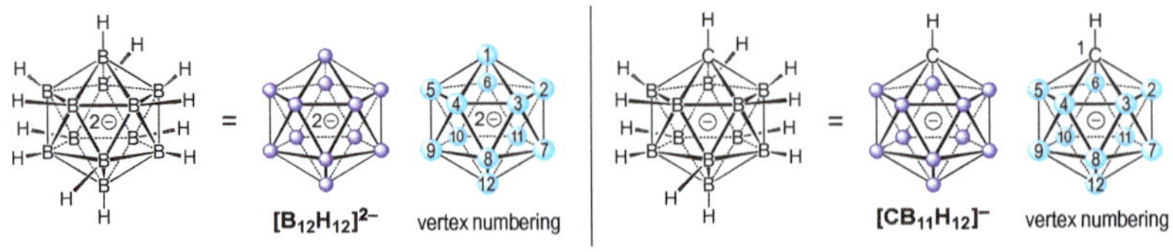

Fig. 1 Structure and vertex numbering of the icosahedral boron clusters $[B_{12}H_{12}]^{2-}$ and $[CB_{11}H_{12}]^{-}$.

structure-like representations of boron clusters, a bond line between two vertices corresponds to less than two electrons and is merely used to visualize the three-dimensional polyhedral geometry.

The σ aromaticity of the anions $[B_{12}H_{12}]^{2-}$ and $[CB_{11}H_{12}]^{-}$ results in exceptional chemical and thermal stability, which exceeds that of most organic molecules. The cesium salt $Cs_2[B_{12}H_{12}]$, for example, shows no decomposition at temperatures of up to 800 °C. Also, it is inert in strongly Brønsted-acidic and Brønsted-basic solutions. Even when treated with elemental fluorine in liquid hydrogen fluoride, the $\{B_{12}\}$ core remains unchanged, although vertex fluorination occurs. Similar properties are observed for the monocarborane cage. In terms of size, the volume of the unfunctionalized clusters, including hydrogen atoms, has been calculated to be *ca.* 150 $Å^3$ (Fig. 2) *(17)*. For comparison, the corresponding volumes of benzene and adamantane are approximately 100 and 135 $Å^3$.

Boron clusters become more inert upon halogenation, and poly-halogenated derivatives have primarily been applied as weakly coordinating anions *(18–23)*. They have made it possible to isolate and crystallize highly reactive cationic species such as silylium ions, borinium ions, chloronium ions and protonated arenes *(24–27)*. For this reason, chlorinated and brominated derivatives of $[B_{12}H_{12}]^{2-}$ and $[CB_{11}H_{12}]^{-}$ have also proved to be non-interfering counterions for highly active cationic catalysts.

The unique electronic and steric characteristics of icosahedral boron clusters have lead to various applications over the past decades. Research has not only demonstrated their utility as weakly coordinating anions, but they have also been applied in fields such as ligand design, supramolecular chemistry, fluorescence/phosphorescence studies and materials science (Fig. 3) *(28–34)*. In medicinal chemistry, the high boron content of such cages has made them ideal building blocks for boron neutron capture therapy and also as pharmacophores interacting with biological targets with high specificity, *e.g.*, causing enzyme inhibition. Generally, the toxicity of boron clusters is low *(35–41)*.

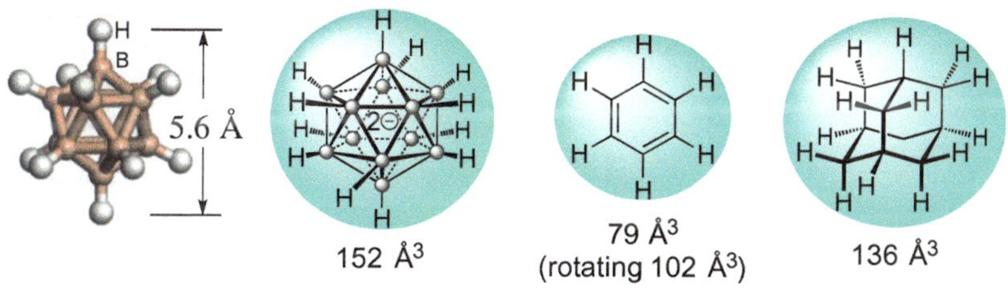

Fig. 2 Van der Waals volumes of and $[B_{12}H_{12}]^{2-}$, benzene and adamantane.

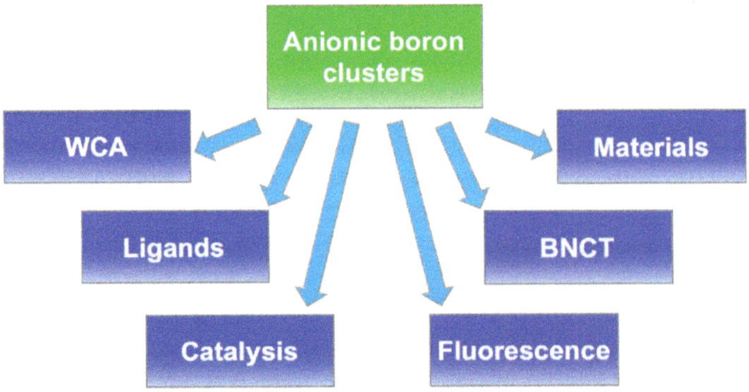

Fig. 3 Important fields of applications of anionic boron clusters; WCA = weakly coordinating anions, BNCT = boron neutron capture therapy.

Access to starting materials is an important prerequisite for a class of compounds to become and remain attractive in fundamental and applied research. The two parent anions $[B_{12}H_{12}]^{2-}$ and $[CB_{11}H_{12}]^{-}$ can both be prepared on a multi-gram scale in any standard chemical laboratory. Remarkably, $[Et_3NH]_2[B_{12}H_{12}]$ is obtained in a single step from $NaBH_4$ and elemental iodine, followed by cation exchange, on a 25 g scale *(21)*. High-yielding, reliable procedures for the synthesis of the monocarborane anion start from decaborane $B_{10}H_{14}$, but other routes have been explored as well *(6,42)*. Furthermore, alkali metal and ammonium salts of both parent anions are commercially available from several companies at reasonable costs on a per-mole basis.

In terms of characterization, anionic boron clusters are most conveniently analyzed by NMR spectroscopy and mass spectrometry. Both methods can be performed on small quantities of samples (micrograms to milligrams). With typical measurement times of minutes, both methods are not only useful in studies of isolated compounds but also in reaction monitoring. ^{1}H NMR spectroscopy typically affords indistinct B—H resonances that become significantly better-defined upon decoupling from ^{11}B. Such spectra still exhibit ^{1}H—^{1}H and ^{1}H—^{10}B coupling, which leads to relatively broad signals and inaccurate integration. Despite these complicating factors, valuable analytical information about the cluster is obtained, especially in combination with ^{11}B NMR spectroscopy.

The ^{11}B nucleus has a quadrupolar moment, and therefore the signals are not as well-resolved as those of, for example, ^{13}C, ^{29}Si or ^{31}P *(43–45)*. Depending on the structure and symmetry of the compound, observed line widths are between 20 and 200 Hz. On the other hand, the T1 relaxation times of ^{11}B are often in the range of 10–500 ms and allow for rapid pulsing. Thus, milligram quantities can be measured within minutes, furnishing spectra with high signal-to-noise ratio. While compounds of high symmetry and simple substitution patterns afford spectra that can be interpreted easily, the ^{11}B NMR spectra of multiply and unsymmetrically substituted clusters often exhibit overlapping signals and are less straightforward in terms of peak assignment. In such cases, ^{11}B—^{11}B COSY NMR spectroscopy and ^{11}B—^{1}H 2D measurements can be used to decipher the spectra, but complete assignment is not always possible *(46,47)*. It should also be mentioned that ^{11}B NMR spectra can be recorded in non-deuterated solvents, thus making the method highly economical especially in reaction monitoring.

Mass spectrometry is another particularly helpful method to characterize compounds based on the dodecaborate and monocarborane scaffolds. Because these clusters are already negatively charged (unless the substitution pattern leads to zwitterionic character), ionization is not an issue, and they can conveniently be detected by (−)-ESI mass spectrometry. Microgram quantities are sufficient for clear signals, and the high thermal stability of the cage results in little fragmentation so that relatively simple spectra are obtained. The isotopic distribution distinguishes boron clusters from other molecules with similar mass because such ions possess a markedly different isotope pattern.

In reaction monitoring, mass spectrometry not only offers the advantage of showing consumption of starting material and increasing concentration of product(s), but it also reveals the degree of substitution when multiple B—H vertices are functionalized. This is advantageous for B—H activation reactions in which one or several B—C bonds can be formed. The degree of substitution would be relatively difficult to determine by NMR spectroscopy unless reference spectra are available.

The negative charge of dodecaborates and monocarboranes can pose certain challenges in practical terms. Contrary to working with neutral dicarbaboranes, one has to deal with salts, whereby ammonium ions $[R_4N]^+$ and alkaline earth metals are the most frequently used countercations. The ionic nature sometimes limits the choice of solvent for reactions and work-up/purification, although a wide range of organic solvents of medium to high polarity, such as tetrahydrofuran, ethyl acetate, acetone, dimethyl sulfoxide and alcohols offer sufficient solubility. Remarkably, column chromatography can be applied in many instances for the purification of boron cluster salts.

2. B—H activation for the functionalization of anionic boron clusters

Established and potential applications of boron clusters have motivated chemists to develop various derivatization methods. Commonly used strategies to install new substituents at B—H corners rely on direct electrophilic substitution or a combination of halogenation and subsequent palladium-catalyzed cross coupling *(48–50)*. Challenges of these approaches are regioselectivity and control over the degree of substitution because of the high symmetry of the clusters. In this regard, the monocarborane anion offers a certain advantage. On the one hand, the C—H vertex can be functionalized by lithiation and subsequent reaction with an electrophile; on the other hand, the carbon atom causes a dipole moment in the cluster, leading to different reactivities of the upper belt B(2–6), lower belt B(7–11) and B(12) positions toward electrophiles *(6)*. This difference has been taken advantage of in halogenations and alkylations, but the vertex selectivity depends significantly on the specific transformation and reaction conditions.

In the past two decades, synthetic organic chemistry has benefitted tremendously from the direct activation and functionalization of C—H bonds *(51–57)*. In particular, the controlled introduction of new substituents to aromatic starting materials bearing a directing group has enabled the preparation of compounds that would otherwise require a higher number of synthetic steps and also be associated with the generation of more waste. A general outline of C—H activation/functionalization is shown in Scheme 1. The directing group DG contains one or severely Lewis-basic atoms which can coordinate to a transition metal TM. Frequently used directing groups are carboxylic acids and their derivatives or nitrogen heterocycles. Complexation of the metal by the organic starting material leads to a first intermediate in which the metal center is held in close proximity to the *ortho* hydrogen atom. This C—H position is activated with a low energy barrier, most often by a cyclometalation–deprotonation process.

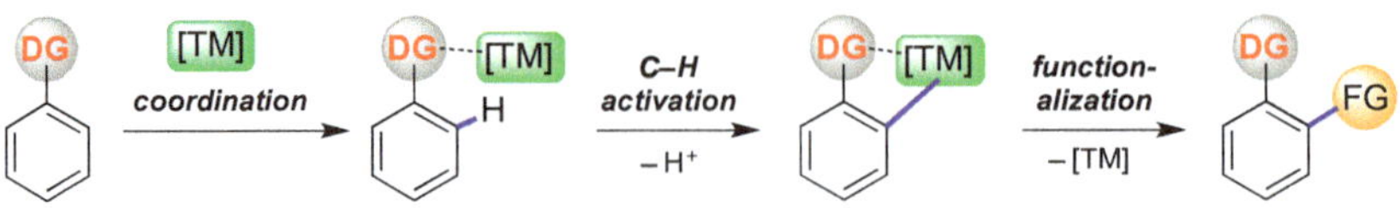

Scheme 1 Transition metal-catalyzed C—H activation of benzene derivatives assisted by directing groups (DG = directing group, TM = transition metal, FG = functional group).

The resulting species contains a direct carbon–metal bond, which is set to be functionalized with a coupling partner FG. Upon substitution of the *ortho* carbon atom, the metal fragment is released and can initiate modification of the next aromatic starting molecule and therefore acts as a catalyst.

Boron clusters exhibit spherical delocalization of electron density, as discussed in the introduction. Based on the analogy between 2D-aromatic benzene derivatives and 3D-aromatic boranes, the positional descriptors *ortho*, *meta* and *para* have been introduced for icosahedral clusters (Scheme 2). According to this convention, in a directing group-carrying dodecaborate or monocarborane, B(2–6) = *ortho*, B(7–11) = *meta* and B(12) = *para*.

As a consequence of the formal similarity between carbon- and boron-based aromaticity, transition metal-mediated B—H activation appears as an attractive alternative to address the general synthetic hurdles in the preparation of novel boron cluster derivatives. In a sequence closely related to that of C—H activation, *ortho* B—H activation affords the desired products step/atom-economically and with high regioselectivity

Early studies on the cyclometalation of neutral boron cages date back to the 1970s. In two publications from 1973 and 1975, Hawthorne and coworkers reported on the reactivity of phosphine derivatives of 1,2-dicarbadodecaboranes (parent structure 1,2-$C_2B_{10}H_{12}$, also referred to as *ortho*-carborane) *(58,59)*. They investigated the effect of a phosphine group attached to a carbon vertex on the cluster chemistry and demonstrated that treatment with iridium(I) complexes can induce generation of B–Ir(III) species by oxidative addition to B—H bonds. Following Hawthorne's seminal studies, neutral dicarbaboranes have remained the main focus of the organometallic chemistry of *closo* boron cages. Catalytic methodologies for their modification have mainly been developed by Xie's group over the past 10 years. Their work and also the result

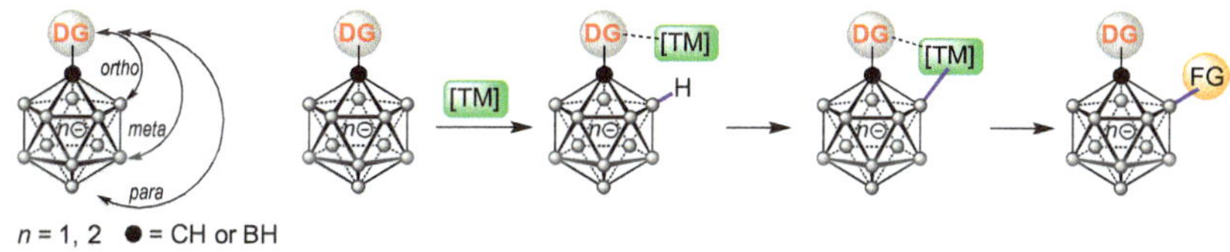

Scheme 2 *Ortho/meta/para* relationship in anionic boron clusters and directing group-controlled regioselective B—H functionalization (DG = directing group, TM = transition metal, FG = functional group).

from other laboratories have been summarized in several review articles *(60–64)*. On the other hand, the metal-mediated functionalization of anionic icosahedral clusters is a young research field *(65,66)*.

3. Formation of organometallic complexes of the monocarborane anion without directing groups

Prior to the stage of directing group control over B—H activation, findings on transition metal complexes of the monocarborane cluster were reported by Weller and coworkers. In 2005 and 2007, two publications appeared in which they investigated boron vertex functionalization of the parent anion $[CB_{11}H_{12}]^-$ *(67–69)*. It was demonstrated that the rhodium fragments $(PPh_3)_2Rh^I$ and $(PR_3)H_2Rh^{III}$ (R = isopropyl, cyclohexyl, cyclopentyl) combine with monocarborane anions to form complexes with agostic-like Rh/$H_{cluster}$ interactions (Fig. 4).

When placed in a vacuum, complexes $(PR_3)H_2Rh^{III}[CB_{11}H_{12}]$ underwent H_2 elimination, which resulted in the formation of dimeric Rh(I)/Rh(III) species that exhibited one direct B–Rh(III) and five Rh(I)/$H_{cluster}$ agostic-like interactions (Fig. 5) *(69)*. This represented the first example of monocarborane complexes with a $B_{cluster}$–transition metal bond, and full structural characterization was provided.

In solution at, the dimers showed fast internal isomerization on the NMR time scale at room temperature. In the solid-state structure of the cyclohexyl derivative, as determined by X-ray diffraction, the isomerization was frozen out, and the different environments of the two metal centers could be recognized clearly. Approximate coordination geometries were

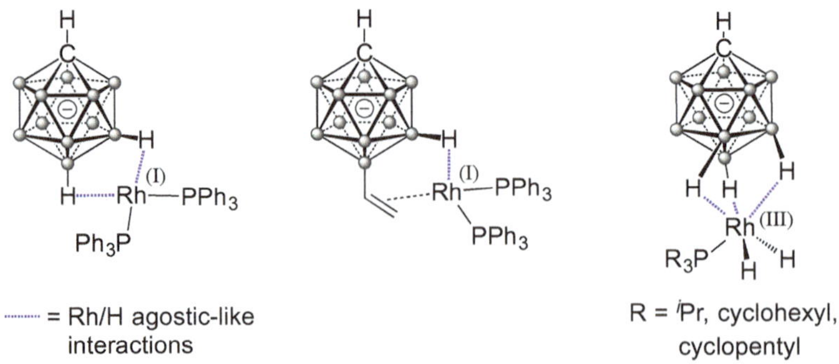

Fig. 4 Rhodium complexes of the monocarborane anion exhibiting agostic-like Rh/$H_{cluster}$ interactions.

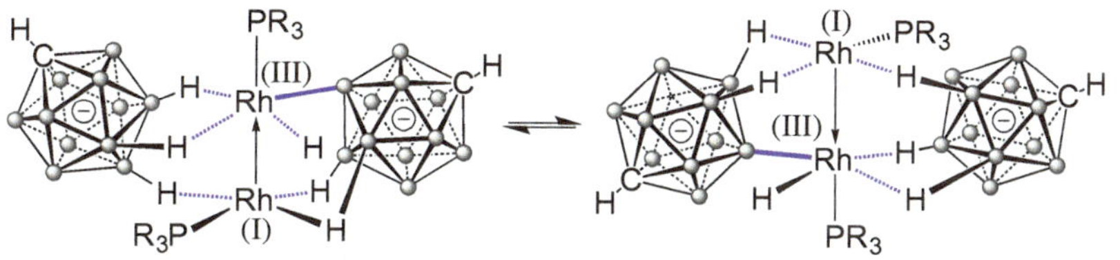

Fig. 5 Dimeric Rh(III)/Rh(I) monocarborane complex featuring direct B–Rh(III) and agostic-like coordination.

square pyramidal for Rh(I)/octahedral for Rh(III), and key distances were B–Rh(III) 2.061(3) Å and Rh(I)–Rh(III) 2.7962(2). Interestingly, the H_2 loss was reversible for all three derivatives (R = isopropyl, cyclohexyl, cyclopentyl), and exposing the dimers to hydrogen gas in deuterated dichloromethane regenerated the starting materials quantitatively.

After Weller's non-directed B—H activation of $\{CB_{11}\}$ cages, Lavallo, Maron and coworkers probed the organometallic chemistry of monocarborane anions in 2013 *(70)*. The authors started by incorporating a phosphine group at the carbon vertex of the cluster to form $[1\text{-}(^iPr_2P)\text{-}CB_{11}H_{11}]^-$ by coupling of the $C_{cluster}$-lithiate with $^iPr_2PCl$. The intention of this study was to evaluate the properties of the phosphine carborane as a ligand toward iridium. This compound could potentially interact with metal fragments by P coordination or at the B(7–12) positions through metal/$H_{cluster}$ agostic-like interactions.

Treatment of the monocarborane phosphine $[1\text{-}(^iPr_2P)\text{-}CB_{11}H_{11}]^-$ with the dimeric Ir(I) complex [Ir(*cod*)Cl]$_2$ (cod = cyclooctadiene) in benzene resulted in the formation of the zwitterionic complex $[1\text{-}(P^iPr_2)\text{-}CB_{11}H_{11}\text{-}Ir(cod)]$ (Scheme 3); it was fully characterized by multinuclear ($^1H/^{31}P/^{11}B$) NMR analysis in C_6D_6 and toluene-d_8 and by single crystal X-ray diffraction. At room temperature, NMR resonances from to B—H bonds interacting with Ir or to the potential product of oxidative addition, containing a B–Ir(III)–H moiety, could not be detected. On the other hand, upon cooling the sample to −80°C, the authors observed a signal at −1.42 ppm in the ^1H NMR spectrum, which was assigned to (B—H)$_2$–Ir, shifted upfield in comparison to the free phosphine carboranyl ligand (^1H–B(2–6) = 0.93 ppm). Infrared studies ruled out reversible B—H activation and Ir—B cyclometalation. Variable temperature $^{11}B\{^1H\}$ NMR experiments were conducted and suggested that a dynamic

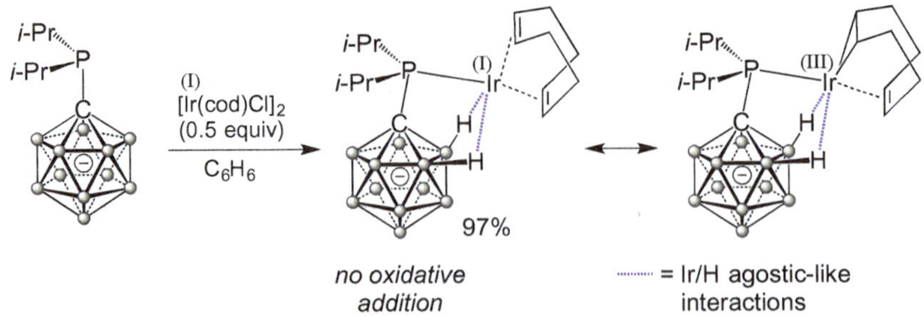

Scheme 3 Formation of a zwitterionic iridium complex by reaction of the mono-carborane phosphine $[1\text{-}(^i\text{Pr}_2\text{P})\text{-CB}_{11}\text{H}_{11}]^-$ with Ir(I).

process involving rotation of the metal fragment with respect to the B(2–6) positions took place. At room temperature, three ^{11}B{^{1}H} resonances in a 5:5:1 ratio were in agreement with the apparent five-fold symmetry of the complex at room temperature, indicating fast rotation on the NMR time scale. Upon cooling to −90 °C, the ^{11}B{^{1}H} NMR spectra became more complicated, consistent with a static structure of lower symmetry. The coalescence temperature was −75 °C, corresponding to $\Delta G^{\ddagger}$ = *ca.* 8 kcal/mol as the activation barrier to rotation. X-ray diffraction of a single crystal of $[1\text{-}(^i\text{Pr}_2\text{P})\text{-CB}_{11}\text{H}_{11}\text{-Ir(cod)}]$ showed that it was best described as a Ir(cod) fragment coordinated by the phosphorus center and two upper-belt B—H units. Interestingly, the authors observed an elongated distance of 1.451(4) Å for the coordinated cyclooctadiene C—C double bond *trans* to the $(\text{B—H})_2$–Ir motif. In comparison, the distance of the double bond *trans* to phosphorus was 1.362(4) Å, typical of a C—C double bond. Furthermore, the Ir—C distances *trans* to the B–H–Ir interactions were shorter than those found for the Ir—C bonds opposite to the phosphorus center. Based on these findings, it was concluded that the [Ir(η^2-CC)] moiety *trans* to B—H can alternatively be viewed as an iridacyclopropane, *i.e.*, as an Ir(III) complex with d^6 electronic configuration (Scheme 3, right resonance structure). This interpretation was additionally supported by DFT calculations. The study of $[1\text{-}(^i\text{Pr}_2\text{P})\text{-CB}_{11}\text{H}_{11}]^-$ thus revealed that tethering a phosphine substituent to the C(1) position of $[\text{CB}_{11}\text{H}_{12}]^-$ rendered metal coordination by upper-belt B—H vertices favorable, different from $[\text{CB}_{11}\text{H}_{12}]^-$ interacting with rhodium in the complex $[(\text{CB}_{11}\text{H}_{12})\text{Rh}(\text{PPh}_3)_2]$ described by Weller.

Toward a more complete understanding of the reactivity of [1-(^{i}Pr$_2$P)-CB$_{11}$H$_{11}$-Ir(cod)], the authors treated this complex with the chloride salt [Me$_4$N][Cl] in fluorobenzene (Scheme 4) *(71)*. Also in this experiment, no B—H activation was observed. Instead, chloride replaced the B—H "agostic ligands", and the product [(carborane-P^{i}Pr$_2$)Ir(*cod*)Cl] was obtained. Multinuclear NMR spectroscopy, IR spectroscopy and X-ray diffraction showed that the monocarborane was bonded to the iridium center only through its phosphorus atom.

In 2016, Lavallo's group reported on the reaction between the two-coordinate carborane Pd(0) complex [(1-(^{i}Pr$_2$P)-CB$_{11}$H$_{11}$)$_2$Pd]$^{2-}$ and chlorobenzene (Scheme 5) *(72)*. The main product, obtained in neat chlorobenzene at room temperature, was the formal Pd(II) species [(1-(^{i}Pr$_2$P)-CB$_{11}$H$_{11}$)$_2$Pd(Ph)]$^-$. Its formation was proposed to occur through oxidative addition of chlorobenzene and subsequent elimination

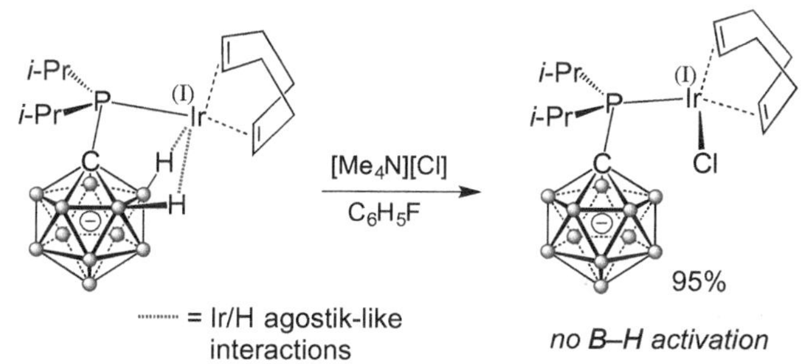

Scheme 4 Formation of the iridium(cyclooctadiene) complex coordinated by [1-(^{i}Pr$_2$P)-CB$_{11}$H$_{11}$]$^-$ and chloride.

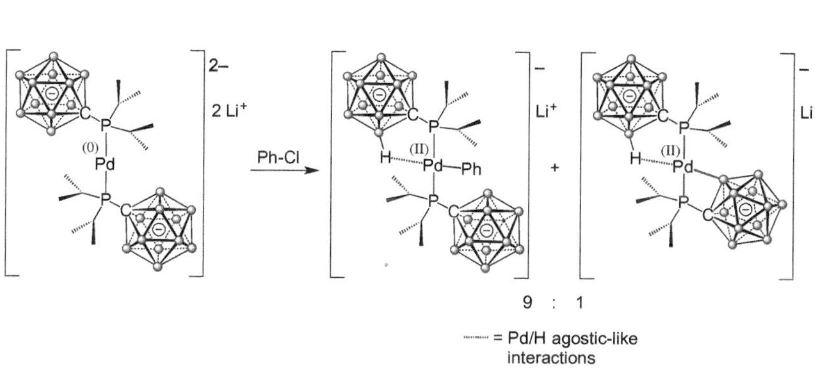

Scheme 5 Palladium(II) complexes formed from the bis(monocarborane phosphine) palladium(0) precursor.

of LiCl. In addition, a minor fraction of a second product was obtained, which was identified as a B–Pd(II) complex featuring a four-membered palladacycle.

This unusual kind of B—H activation was interpreted as initial oxidative addition of chlorobenzene, followed by σ bond metathesis involving loss of one B—H hydrogen atom and the Ph ligand, *i.e.*, overall elimination of LiCl and benzene. The structure of the minor product was confirmed by multinuclear NMR spectroscopy and single-crystal X-ray diffraction analysis. Several overlapping resonances were observed in the $^{11}B\{^{1}H\}$ NMR spectrum in THF at 25 °C, with a resonance at −21.6 ppm most likely stemming from the palladium-bound boron vertex. Furthermore, the authors presented calculations to investigate the reaction mechanism, and they also performed several Kumada cross coupling reactions between aryl Grignard reagents and substituted chlorobenzenes demonstrating the catalytic activity of the two-coordinate carborane Pd(0) complex.

4. Stoichiometric directing group-mediated synthesis of organometallic complexes

Weller's and Lavallo's results suggested that successful controlled B—H activation of anionic boron cages required the careful design of the cluster starting material. This was crucial for fundamental studies of organometallic species with a boron–metal interaction and also for catalytic processes leading to functionalization of the cage surface. In order to explore the intermediates of B—H bond activation, Duttwyler and coworkers started approaching these challenges by a directing group strategy related to C—H activation chemistry in 2017. Three criteria regarding the directing group were postulated: (1) Facile introduction to the cage, (2) ability to coordinate to the metal and favor B—H activation over agostic-like interactions, and (3) lack of facile competitive C—H activation. After an evaluation of several functional groups installed at the C(1) vertex of the monocarborane ion, the sulfonamide functionality turned out to fulfill these criteria. The tosyl amide of the monocarborane cage was synthesized from the C(1)-carboxylic acid by treatment with oxalyl chloride, followed by reaction with tosyl amine (Scheme 6). *(73)* The observed distances in the X-ray crystal structure were (Å) C—C 1.505(3), C═O 1.206(2), C—N 1.378(3) and N—S 1.657(2). The sum of angles around C═O was 360.0°, and the torsion angle C-C-N-S was 176.6°, showing coplanarity of the entire sulfonamido moiety. These values led the authors to anticipate

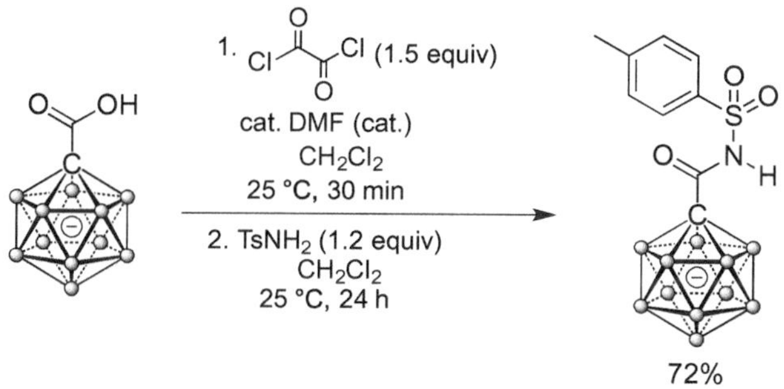

Scheme 6 Synthesis of the monocarborane C(1)-tosyl amide from the carboxylic acid precursor.

that N—H deprotonation, followed by N coordination to a transition metal, would enable B—H activation and formation of a five-membered metallacycle.

Using the tosyl amide directing group, three monocarbaborane-metal complexes were prepared selectively and fully characterized, involving iridium, rhodium and palladium. The reaction between the tosyl amide and a stoichiometric amount of IrCp*$(OAc)_2$ in acetonitrile lead to a yellow precipitate, which was identified as the corresponding anionic iridacycle (Scheme 7) *(73)*. Its structure was determined by X-ray crystallography and multinuclear NMR spectroscopy. The iridacycle consisted of the carborane amide coordinated as a bidentate B,N ligand to the IrCp* (MeCN) fragment. Key distances were B(2)—Ir 2.123(7) and N—Ir 2.147(4) Å, thus both values were consistent with direct B—Ir and N—Ir interactions. The five-membered metallacycle exhibited only slight puckering, with internal angles of B(2)-Ir-N 79.51°, B(2)-C-C 113.31°, C-C-N 112.01° and a torsion angle of B(2)-C-C-N 12.41°.

A DFT study on the reaction mechanism was carried out (Scheme 8). They started from a precomplex exhibiting N–Ir/acetate–Ir coordination, giving the acetic acid-coordinated product. According to computations at the B97-D/LanL2dz level of theory in acetonitrile continuum, product formation is exothermic by 7.86 kcal/mol, with a calculated barrier of 8.25 kcal/mol. The reaction mechanism proceeds through a six-membered ring transition state, where a hydrogen atom is transferred from B(2) to the carbonyl O atom of the acetate ligand. This hypothesis was consistent with proposed mechanisms for dicarbaborane activation and studies on C—H activation reactions occurring *via* deprotonation–cyclometalation. HOMO/LUMO calculations suggested that the HOMO of the iridacycle

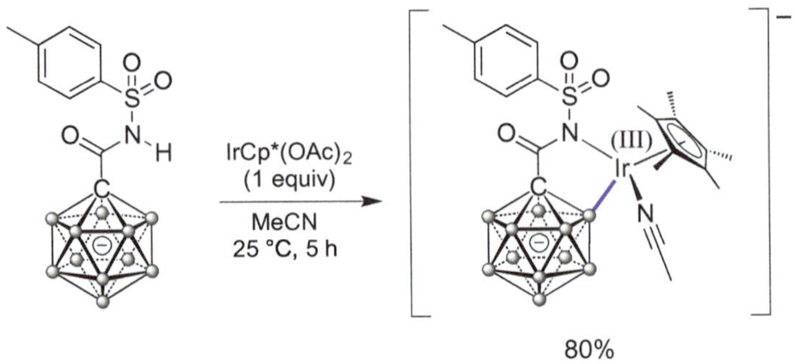

Scheme 7 Preparation of the monocarborane iridacycle from the tosyl amide precursor; overall negative charge highlighted by brackets only for the product.

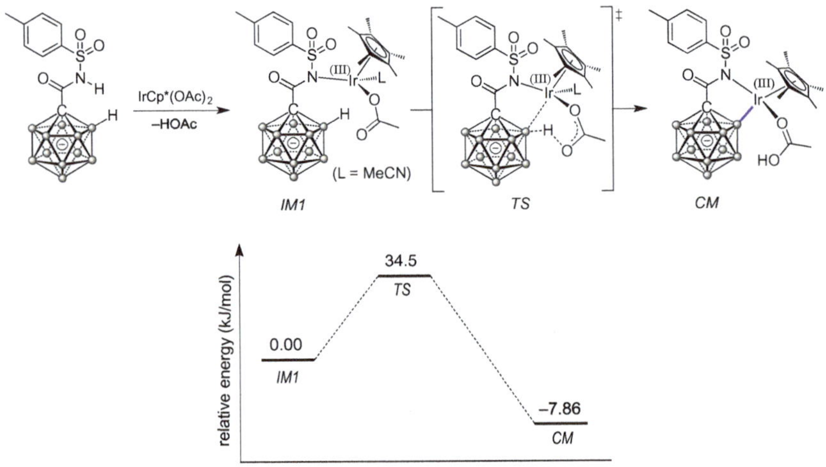

Scheme 8 Energy profile for the cyclometalation process calculated at the B97-D/LanL2dz level of theory in acetonitrile continuum (IM1 = intermediate 1; TS = transition state; CM = cyclometalated product).

is primarily localized around the iridium fragment as well as N and B(2), while the LUMO is localized around the tolyl-SO_2 moiety and the Cp* π-system. They indicated B(2) and Ir as the centers with the highest probability for reaction with an electrophile, while nucleophilic attack was predicted to occur at C(*ipso*) of the tolyl ring.

As a model reaction to test the reactivity of the cyclometalated complex, electrophilic chlorination was chosen. Substitution was anticipated to take place around the B—Ir moiety. Treatment with *N*-chlorosuccinimide

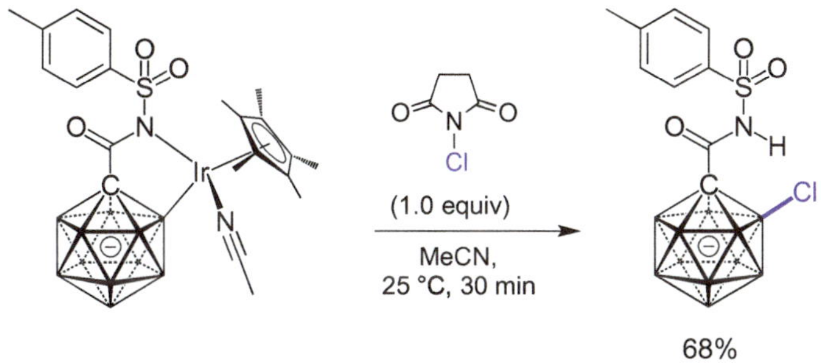

Scheme 9 Selective B(2)-monochlorination of the monocarborane tosyl amide iridium complex.

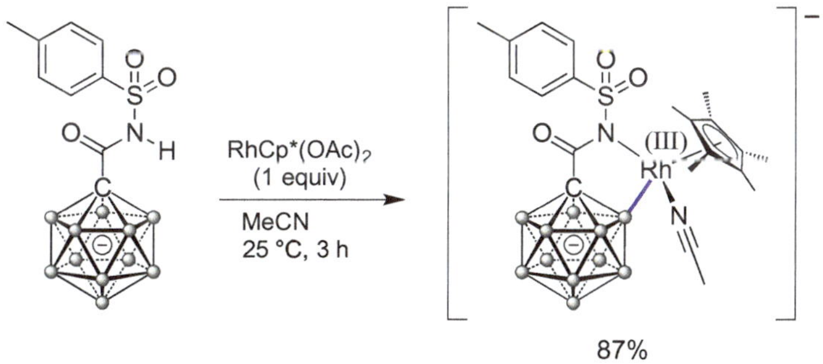

Scheme 10 Preparation of the monocarborane rhodacycle from the tosyl amide precursor; overall negative charge highlighted by brackets only for the product.

indeed afforded the B(2)-chlorinated product selectively in 68% isolated yield (Scheme 9); no other regioisomers were detected in this transformation.

The tosyl amide precursor was also treated with one equivalent of $Cp^*Rh(OAc)_2$ in a procedure similar to that involving iridium. The reaction afforded a new deeply red-brown product that was identified as the corresponding anionic rhodacycle (Scheme 10) *(74)*. X-ray crystallography showed that the complex consisted of the carborane moiety acting as a chelating B,N ligand toward the RhCp*(MeCN) fragment, comparable to the iridacycle. The key distances Rh–B(2) 2.111(6) and Rh—N 2.141(4) Å were similar to the ones found for the iridacycle and consistent with a direct B(2)–Rh interaction. The five-membered rhodacycle was relatively flat with an internal angle of B(2)-Rh-N 79.38° and a torsion angle of B(2)-C-C-N-9.54°.

Finally, reaction between the tosyl amide precursor and $Pd(OAc)_2$ resulted in the formation of the palladacycle, which was isolated in 80% yield as a colorless solid (Scheme 11) *(74)*. X-ray diffraction revealed a dimeric dianionic structure; the complex showed B,N chelation to one Pd(II) center and S=O–Pd coordination to the second Pd center. Observed distances were B(2)–Pd 1.999(6) Å, Pd—N 2.059(4) Å and Pd—O 2.435(3) Å. The ligand geometry around palladium was square-planar with a sum of angles around Pd of 360.07° and a distance between the best-fit plane through B-N-O-N and Pd of 0.126 Å.

Detailed NMR studies were carried out for the tosyl amide precursor and all three cyclometalated complexes. The tosyl amide exhibited ^{11}B NMR resonances at −14 ppm for B(2–6), −13 ppm for B(7–11) and–6 ppm for B(12), respectively. For the complexes, the B(2) NMR shifts were of key interest. Interestingly, no general trend was observed in terms of shift changes upon cyclometalation. The B(2) signal of the iridacycle appeared at −16 ppm, while for the rhodacyle and the palladacycle, values of −6 ppm and − 20 ppm were found.

All complexes showed remarkable inertness in solution and as solids. Solutions of the iridacycle in acetonitrile under a nitrogen atmosphere remained unchanged over weeks; also, heating to 70 °C in a sealed NMR tube for 2 days did not lead to any decomposition. In the solid state, either as a powder or in crystalline form, it remained unchanged in air for 2 weeks. Similarly, the rhodacycle and palladacycle were inert over weeks at room temperature and over days at 50 °C.

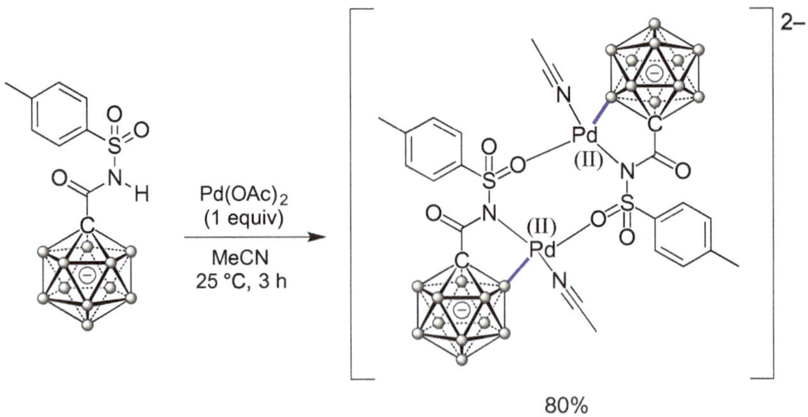

Scheme 11 Preparation of the dimeric monocarborane palladacycle from the tosyl amide precursor; overall double negative charge highlighted by brackets only for the product.

5. Functionalization of monocarboranes by transition metal catalysis

Shortly after the isolation of the tosyl amide-derived metallacycles, the first example of selective catalytic derivatization of the monocarborane anion was reported *(75)*. For these transformations, the pyrrolidine amide functionality served as the directing group. The amides with H, Br, and I at the B(12) position were prepared in yields of 88%, 76% and 87%. Initial screening to identify conditions for transition metal-catalyzed cage derivatization involved coupling with ethyl acrylate as the model substrate. Using $[RhCp^*Cl_2]_2$ in combination with $AgSbF_6$ and Cu(II) successfully led to B—C bond formation. Carrying out this transformation in dimethylacetamide at 100°C selectively afforded tetrasubstitution across the B(2–5) positions (Scheme 12). The product contained three reductively

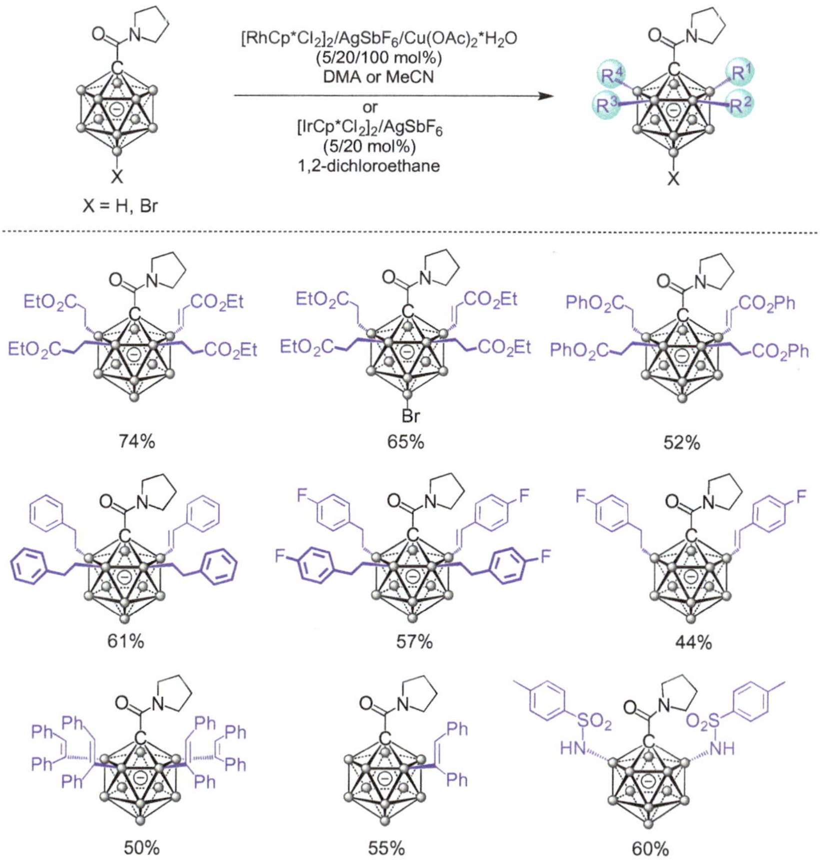

Scheme 12 Synthesis of monocarboranes functionalized at the B(2–5) positions.

coupled ester moieties and one ethyl acrylyl substituent with a remaining double bond. The reaction also worked with the B(12)–Br starting material without competitive side reactions.

Substrate scope was investigated by coupling to activated alkenes such as phenyl acrylate, styrene and 4-fluorostyrene to give tetrasubstituted clusters 57–74% isolated yields. Acetonitrile solvent provided access to products exhibiting a lower degree of functionalization. The B(2/4) di-substituted fluorophenyl compound could thus be obtained in 44% yield by monitoring the reaction by ESI-MS and stopping it at the point of di-substitution. Reaction with diphenylacetylene in dimethylacetamide resulted in tetra-alkenylation, showing that alkynes could be used as coupling partners. Ir(III) also effected B—H activation, however, generally resulting in a lower degree of substitution. Using $[IrCp^*Cl_2]_2/AgSbF_6$ in 1,2-dichloroethane, coupling with diphenylacetylene afforded the mono-alkenylated product in 55% yield. Under the same conditions, formation of B—N bonds was accomplished. Reaction with tosyl azide caused sulfonamidation at the B(2/4) positions. Characterization of the products was performed by multinuclear NMR spectroscopy, X-ray crystallography and mass spectrometry; additionally, several compounds were prepared on a gram scale

Selective monochlorination was achieved using Rh(III) and *N*-chlorosuccinimide as the chlorinating agent. Starting from pyrrolidine amides carrying H, Br or I at the B(12) position, treatment with three equivalents of *N*-chlorosuccinimide in the presence of $[RhCp^*Cl_2]_2$/Ag(I)/Cu(II) in 1,2-dichloroethane cleanly lead to B(2)-substituted products in 68–82% isolated yields (Scheme 13). Notably, no positional isomers were obtained. This finding was remarkable because electrophilic halogenation of monocarboranes generally favors substitution at the B(12) vertex,

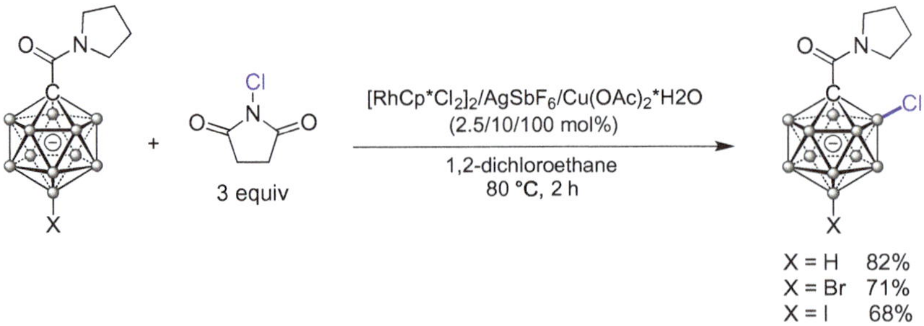

Scheme 13 Regioselective B(2) monochlorination of monocarboranes.

followed by further substitution at the B(7–11) positions. In the absence of Rh(III), the B(12)-chlorinated pyrrolidine product was formed. Prior to this study, no protocols had been described for the selective halogenation of upper belt boron vertices when B—H bonds at the B(7–12) positions were present. Therefore the methodology offered a useful strategy to prepare otherwise inaccessible derivatives of the monocarborane cluster.

Stoichiometric experiments with Ir(III) were conducted in order to probe the mechanistic pathway. Reaction of the pyrrolidine amide with one equivalent of $IrCp^*(OAc)_2$ in acetonitrile at room temperature afforded the neutral cyclometalated complex in 82% isolated yield (Scheme 14). X-ray crystallographic analysis revealed a direct cage–metal interaction with a B—Ir distance of 2.128 Å. Subsequent treatment of this species with diphenylacetylene afforded the mono-alkenylated product, suggesting that it was a key intermediate along the catalytic cycle.

In 2018, the Duttwyler group reported an approach to synthesize a series of isocoumarins in which the benzene ring is replaced by a monocarborane (Fig. 6) *(76)*. Isocoumarins are fused lactonic heterocycles that have attracted much attention in the field of medicinal chemistry. As natural products, they can be isolated from bacteria, fungi, plants and insects *(77,78)*. In terms of bioactivity, they have been studied with regard to their antiviral, antibacterial, anti-inflammatory and tumor growth-inhibiting properties. Replacement of a carbon atoms by boron allows for the investigation of boracoumarins, a novel class of compounds. The specific cluster/heterocycle fusion depicted in Fig. 6 leads to three-dimensional monocarborane isocoumarine analogues.

The strategy to access these analogs was based on B—H activation, followed by coupling with an alkyne and cyclization (Scheme 15). In this transformation, the C_{cage} carboxylic acid serves as a directing group and then turns into part of the oxygen-containing heterocycle by C—O bond formation. Screening of conditions showed that the desired one-pot

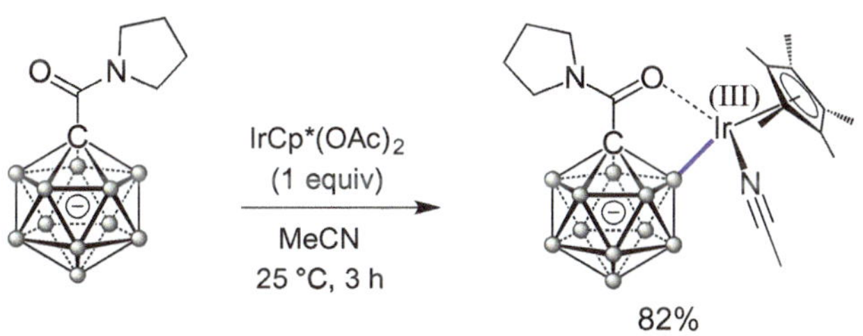

Scheme 14 Formation of the neutral monocarborane pyrrolidine amide iridacycle.

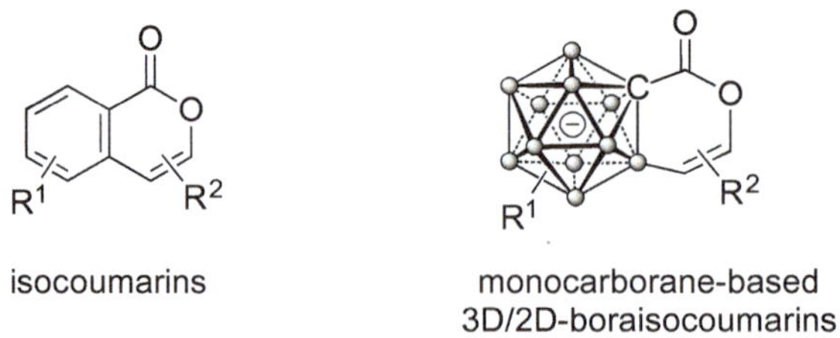

Fig. 6 General structures of isocoumarins and monocarborane-based boraisocoumarins.

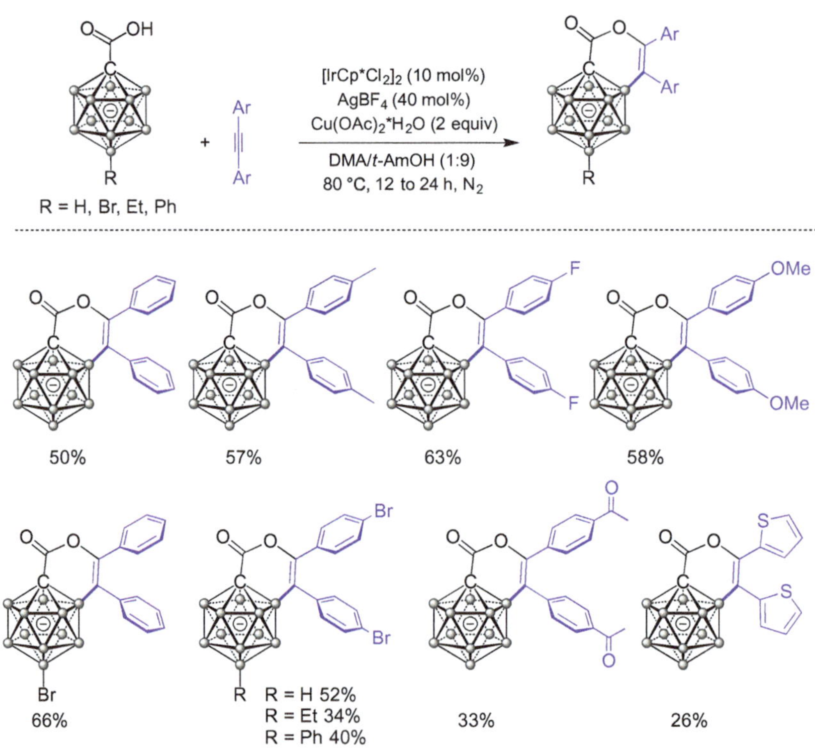

Scheme 15 Synthesis of boraisocoumarins by annulation between monocarboranes and diaryl-alkynes.

oxidative annulation starting from carborane carboxylic acids and diaryl alkynes occurred in the presence of an Ir(*III*)Ag(I)/Cu(II) catalytic system.

Subsequently, the established reaction conditions were applied to the generation an array of boraisocoumarins, and selected products are displayed in Scheme 15. Variation of the diaryl alkyne included electron-donating and electron-withdrawing groups, furnishing the products in 50–66% isolated yields, *e.g.*, with phenyl, methoxyphenyl and fluorophenyl substituents. Protecting group-free ketone and thiophene moieties could be used as well. Although the yields were reduced in these cases, the methodology highlighted the possibility of step-efficient preparation of boraheterocycles which had not been accessible by traditional cluster chemistry. Furthermore, monocarborane starting materials with substituents such as bromo, ethyl and phenyl at the B(12) position furnished the isocoumarin analogs in 34–66% isolated yields.

Starting from the fused bora-heterocycles, the feasibility of additional B—H functionalization was investigated. In this reaction, the lactone moiety was anticipated to act as the directing group and give rise to a second B—C bond formation. Under Ir(III) catalysis, completely regioselective alkenylation at the B(4) position was observed (Scheme 16). The transformation was then applied to the synthesis of symmetrically and unsymmetrically substituted butterfly-shaped products in moderate to high yields. This study demonstrated the versatility of the carboxylic acid moiety as a directing group. On the one hand it enables initial B—H activation, but also offers additional utility by participating in C—O bond formation and eventually inducing a second B—C coupling process.

Shortly after the publication of the isocoumarin study, Uchiyama and coworkers reported on a one-pot process to access fused cluster–biaryl products from the parent monocarborane $[CB_{11}H_{12}]^-$ *(79)*. Their strategy involved copper-mediated coupling of the C(1) vertex with cyclic diaryliodonium salts to afford 2,2′-carborane-iodobiaryl intermediates

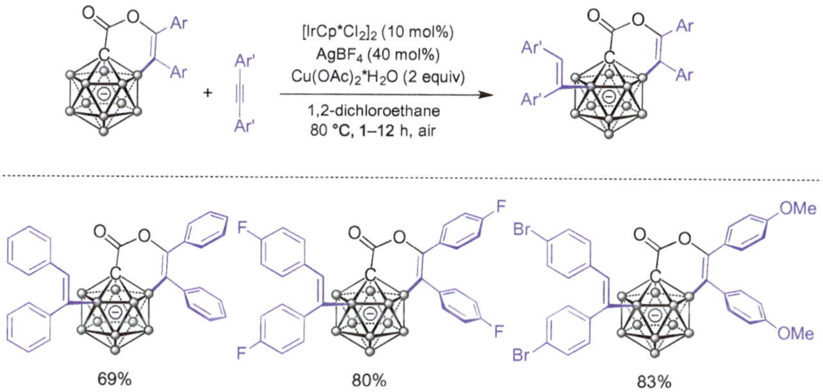

Scheme 16 Alkenylation of the boraisocoumarins by B—H activation and B(4)–C bond formation.

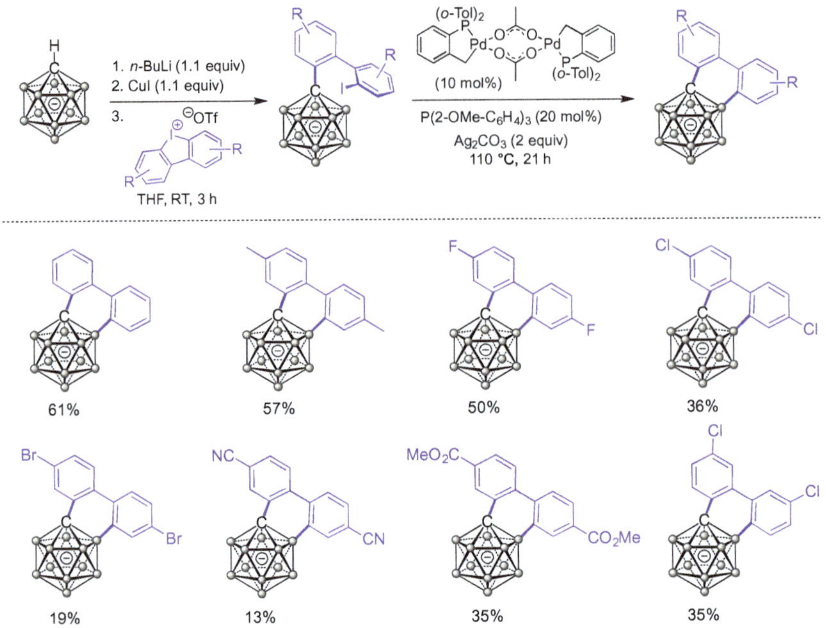

Scheme 17 Synthesis of fused monocarborane–biaryls (yields based on ^{1}H NMR spectra).

(Scheme 17). This step was immediately followed by palladium-catalyzed B—H bond activation and B—C bond formation. The authors also examined one-pot conditions under which the palladium catalyst was present from the beginning, but this resulted in significantly reduced yields.

A series of disubstituted diaryliodonium triflates was evaluated in terms of substrate scope. Yields for electron-deficient to moderately electron-rich starting materials were in the range of 13–61%, including fluoro, chloro, cyano, ester and methyl substituents. Importantly, however, most products were not isolated in pure form; instead NMR yields were reported. Diaryliodonium salts whose structure could give rise to regioisomers furnished inseparable mixtures of products except for the 5,5′-dichloro starting material. The most remarkable feature of this methodology was that no pre-functionalization of the parent monocarborane cage was necessary. It was directing group-free and relied on the C(1) lithiation/cupration to afford regioselective C(1)/B(2) annulation.

The five-fold rotational symmetry of the $[CB_{11}H_{12}]^{-}$ anion offers the opportunity to directly functionalize of the entire upper or lower boron belt, *i.e.*, the B(2–6) or B(7–11) positions. The B—H vertices within each belt feature equal reactivity, and in principle this allows for one-pot, five-fold substitution, leading to products in which the C_5 rotational axis is retained. The resulting compounds represent appealing building blocks for applications in areas such as ligand design and supramolecular chemistry. Early studies on the derivatization of monocarboranes showed that under harsh conditions, $[CB_{11}H_{12}]^-$ can be hexahalogenated at the B(7–12) positions. These reactions afford multiply substituted C_5-symmetrical products in a single laboratory step, and this has been an important method to access extremely weakly coordinating and redox-inert anions. On the other hand, direct symmetrical introduction of five substituents by B—H activation was developed until recently.

The first demonstration of regioselective catalytic derivatization of the monocarborane cluster at the B(2–6) positions was reported in 2018 *(80)*. The corresponding synthetic strategy relied on the C(1) carboxylic acid directing group and iodoarenes for coupling with the upper belt B—H vertices (Scheme 18). The anion $[1\text{-COOH-CB}_{11}H_{11}]^-$ has the advantage that it is an easily accessible starting material and that the COOH group can be used as a functional group handle for ester or amide formation, among others.

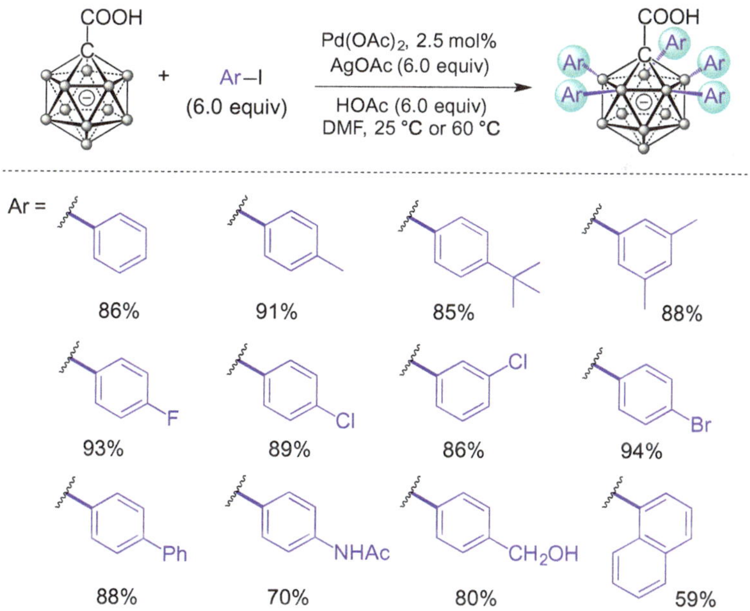

Scheme 18 Synthesis of penta-arylated carboranes by Pd-catalyzed coupling with iodoarenes.

Successful one-pot, five-fold B—H activation was achieved using a catalytic system that involved $Pd(OAc)_2$, AgOAc, and HOAc plus six equivalents of the iodoarene. The reactions were air-tolerant and could therefore conveniently be set up under ambient conditions and required either no or only slight heating. The products were purified by silica gel column chromatography and isolated in generally high yields. Simple phenyl, alkylphenyl and halophenyl coupling partners were tolerated well. Oxygen- and nitrogen-containing substituents were also compatible with this procedure, among them even unprotected OH groups. All transformations proceeded with complete regioselectivity at the B(2–6) positions, thus furnishing novel C_5-symmetrical derivatives of the $[CB_{11}H_{12}]^-$ anion. Furthermore, a five-fold coupling/decarboxylation cascade took place with 1,3-bromoiodobenzene and other electron-deficient iodoarenes for coupling at a reaction temperature of 60°C (Scheme 19). This was the first demonstration of the use of a traceless directing group in monocarborane chemistry.

The reaction mechanism of this five-fold B—H activation was studied in detail by DFT calculations. They favored a merry-go-round substitution sequence, *i.e.*, the first coupling is followed by arylation at an adjacent B position, and this process continues until the entire upper belt is functionalized. Also, the efficient one-pot penta-arylation/directing group removal could be rationalized. Electron-poor aryl substituents lower the energy of the dianions $[:CB_{11}Ar_5H_6]^{2-}$ more favorable, species that are generated along the decarboxylation pathway.

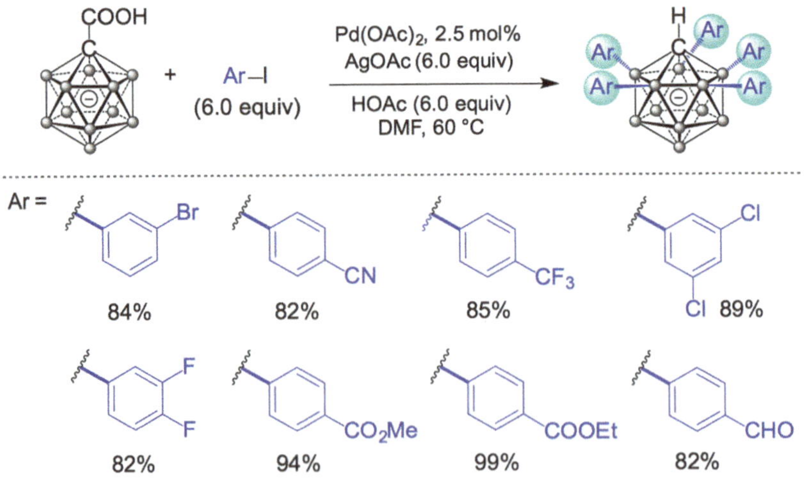

Scheme 19 Synthesis of penta-arylated carboranes by one-pot Pd-catalyzed coupling with iodoarenes and directing group removal.

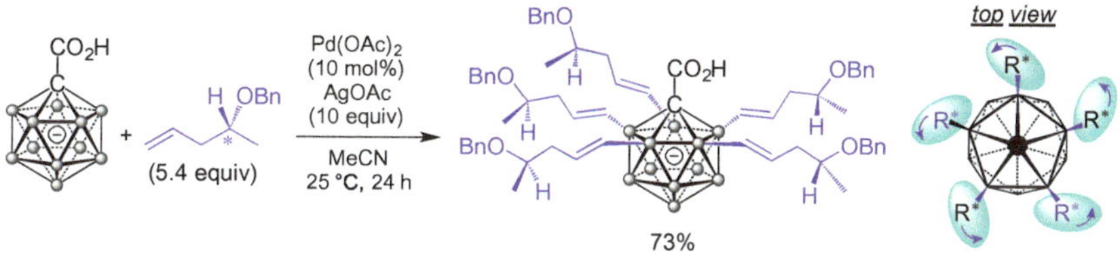

Scheme 22 Formation of homochiral, penta-alkenylated monocarboranes.

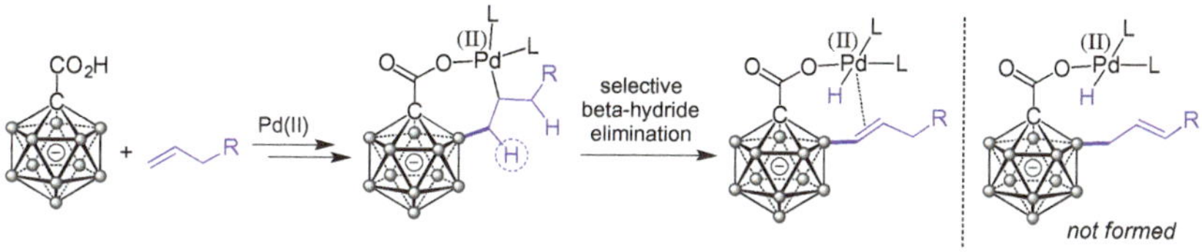

Scheme 23 Putative selective hydride elimination to give the vinylic products.

the ether and chiroptical measurements, showed that the transformation occurred without racemization at the stereogenic carbon centers.

Selective β-hydride elimination was proposed as an explanation for the formation of only vinyl-type (*E*)-penta-alkenylated products. According to this mechanistic proposal, a B–C–C–Pd intermediate is formed along the catalytic cycle after B—H activation/B–Pd bond formation and insertion of the alkene (Scheme 23). Hydride elimination then takes place selectively at one of the saturated carbon centers, leading to the B-vinyl-type alkene–Pd complex, which is assumed to be more stable than the allyl-type intermediate. Hydride ligand removal assisted by Ag(I) and continuation of the substitution process eventually affords the observed products.

6. Functionalization of the dodecaborate cluster by transition metal catalysis

Until 2016, the selective catalytic B—H activation of dodecaborates had remained unexplored due to the stability of the $[B_{12}H_{12}]^{2-}$ dianion and the challenge to form B–metal bonds under mild conditions. Traditional methods to access functionalized dodecaborates primarily rely on reactions with electrophiles or halogenation/cross coupling. Reactions using electrophiles under harsh conditions often afford several products from multiple degrees of substitution or from formation of regioisomers. Palladium-mediated cross coupling is possible but requires the presence of a B–halogen bond in the starting material, introduced by an electrophilic halogen source *(50,82,83)*. Dodecahydroxylation of $[B_{12}H_{12}]^{2-}$ for the one-pot synthesis of $[B_{12}(OH)_{12}]^{2-}$ was reported by Hawthorne and coworkers using H_2SO_4/SO_3 in the presence of platinum and palladium catalysts such as $[Pt(bpym)]Cl_2$, $[Pt((NH_3)_2)Cl_2]$ or $[PdCl_2]$ *(84)*. A reaction pathway involving an B–M–H hydride intermediate (M = Pd, Pt) in the B—H activation process was suggested, but no mechanistic evidence was presented. In 2016, Schleid and coworkers reported on the first B–metal intermediate for B—H bond activation of the $[B_{12}H_{12}]^{2-}$ anion *(85)*. An aqueous mixture of Bi^{3+} cations and $[B_{12}H_{12}]^{2-}$ furnished charge-neutral solvated $Bi–B_{12}H_{11}$ complexes by formal metalation-deprotonation. The complexes were characterized by X-ray diffraction and NMR spectroscopy. Schleid's results proved that B—H bonds in dodecaborate clusters can be activated in a controlled manner similar to organic compounds and carboranes. However, further functionalization of the B—Bi bond or transformations with substoichiometric amounts of Bi(III) were not explored.

In 2016, Duttwyler and coworkers for the first time presented transition metal-catalyzed regioselective B—H activation of the dodecaborate cage to synthesize 1,2,3-trisubstituted derivatives *(86)*. The dimethylureido functionality served as a directing group for metal coordination and subsequent B—H activation. Various reaction conditions were screened to optimize coupling of the ureido precursor with internal alkynes in the presence of a rhodium catalyst. Using 2.5 mol% $[RhCp^*Cl_2]_2$ and $Cu(OAc)_2$ as an oxidant afforded the product with regioselective *ortho*-monoalkenylation and concomitant B—O cyclization to provide 1,2,3-trisubstituted clusters (Scheme 24).

Under the established conditions, the methodology was compatible with various electron-rich and electron-deficient aryl substituents moderate to high yields. Unprotected ketones and thiophene rings were tolerated without competing side reactions with Rh(III). Unsymmetrical aryl alkyl alkynes such as MeC≡CPh and dialkyl alkynes such as EtC≡CEt also afforded the corresponding products with high regioselectivity but in reduced yields.

The strategy to use *N*-acyl directing groups for selective B—H activation of the dodecaborate cluster was extended by the same group in 2018 *(87)*. Specifically, the starting materials were amides of the type $[B_{12}H_{11}–(NH(CO)Ar)]^{2-}$. In synthetic organic chemistry, aryl amides not only serve as directing groups for C—H activation, but they are also easily accessible and modifiable pharmacophores in drug discovery. An important question at the outset of this study was whether the aromatic C—H bonds or the cluster B—H vertices would be activated preferentially by transition metals.

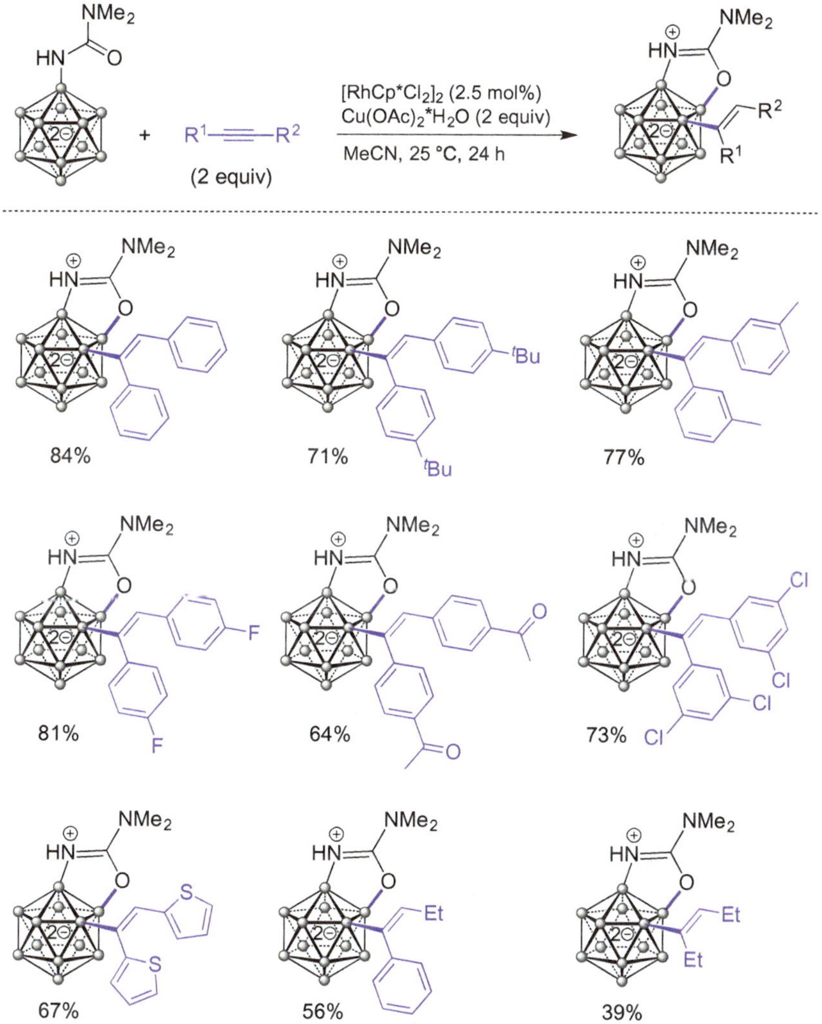

Scheme 24 Rh(III)-catalyzed alkenylation/annulation of dimethylureido dodecaborate.

Rh(III)-catalyzed coupling of electron-rich and electron-poor as well as naphthyl and thiophenyl amides worked well in combination with diphenylacetylene as the alkyne (Scheme 25). The transformations were carried out in acetonitrile under ambient conditions. Notably, all dodecaborate starting materials underwent exclusive B—H activation without competitive aryl C—H activation. Furthermore, the diarylalkyne coupling partner could be varied with very good functional group tolerance.

Until 2018, the selective B-alkylation of boron clusters under mild conditions had remained an unsolved challenge; methylation and ethylation

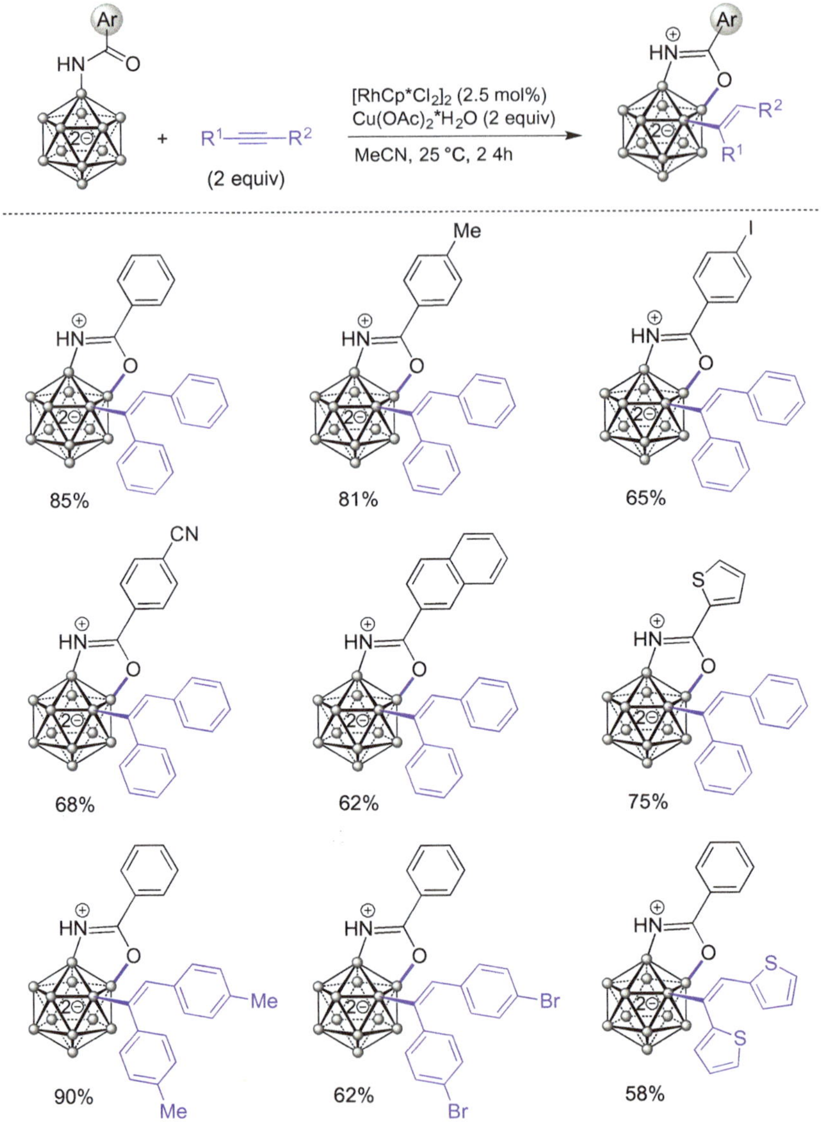

Scheme 25 Rh(III)-catalyzed alkenylation/cyclization of dodecaborate amides.

using strongly electrophilic alkylation agents is possible but is associated with poor control over the degree of substitution and regioselectivity. To overcome this synthetic limitation, the alkylation by reductive coupling with different alkenes was probed. The reaction with different olefins under Rh(III) catalysis afforded the desired 1,2,3-trisubstituted alkylated/cyclized

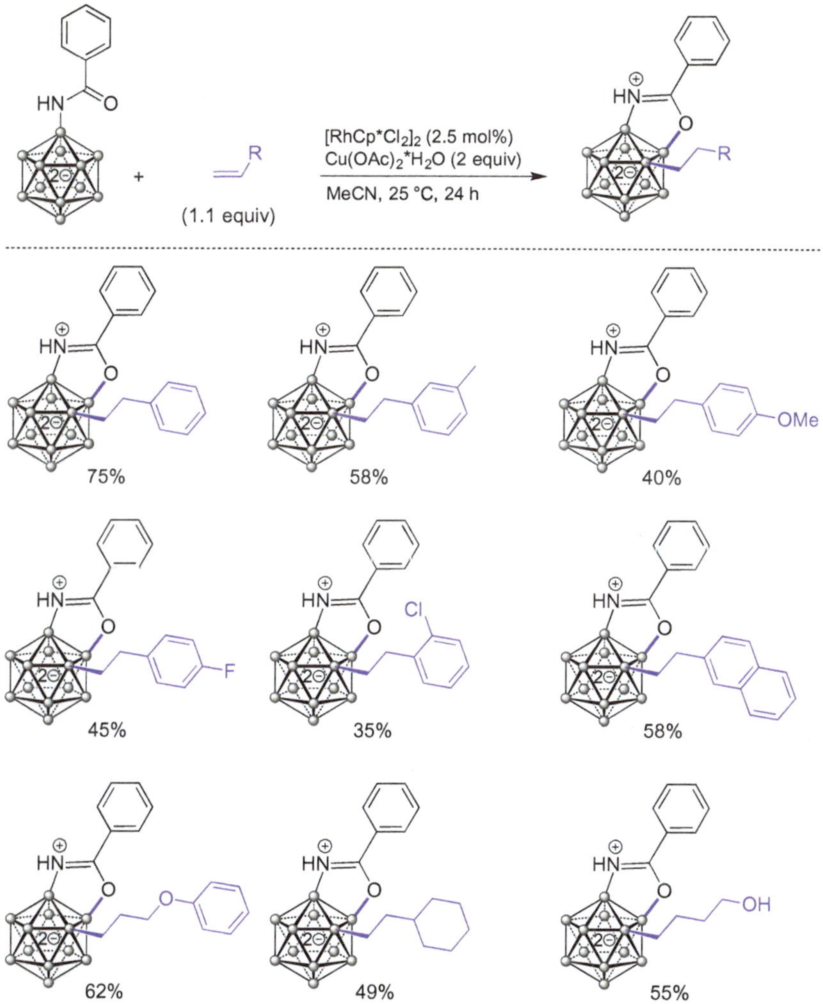

Scheme 26 Rh(III)-catalyzed alkylation/cyclization of the dodecaborate benzamide.

clusters (Scheme 26). Electron-rich and electron-deficient styrenes as well as 2-vinylnaphthalene furnished the corresponding products in moderate to good yields. The reaction also tolerated non-aromatic alkenes, notably unprotected alcohols. Additionally, different aryl amide directing groups (aryl = *e.g.*, *p*-tolyl, 4-fluorophenyl, 1-naphthyl, 2-naphthyl, 2-thiophenyl) were tested in combination with styrene as the coupling partner, providing phenethylated products in yields of 54–70%.

The reaction mechanism of the two-fold B—H activation of the ureido- and amide-substituted dodecaborates was studied by an investigation of

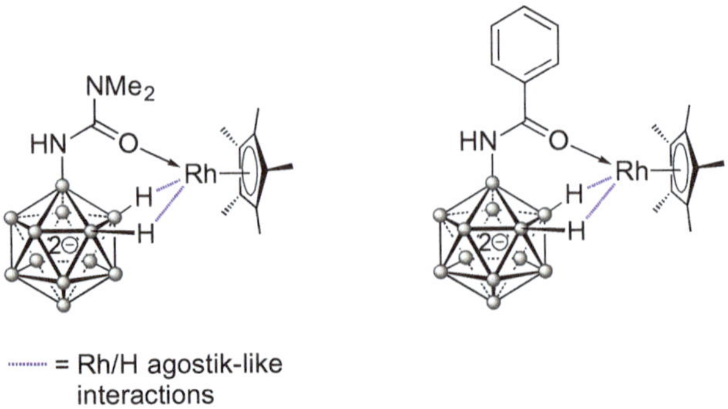

Fig. 7 Isolated Rh(III) complexes from the studies on B—H activation of the dodecaborate cluster.

early key intermediates of the catalytic cycle. Rhodium complexes with Rh/$H_{cluster}$ agostic-like interactions were isolated and fully characterized (Fig. 7). In both cases, the intermediates were best described as a Rh(*III*) Cp* fragment coordinated by the carbonyl group and two B—H units of the cluster. NMR and X-ray diffraction data clearly suggested that all B—H bonds were still intact, *i.e.*, the complexes represented a snapshot of the catalytic cycle just prior to the deprotonation/cyclometalation step. Isolation of rhodacycles with a direct B—Rh bond was attempted, but clean formation and capture of such species was not successful. In terms of the sequence of coupling events, further control experiments suggested that B—C bond formation occurs first, followed by B—O cyclization, decomplexation and release of the final product.

7. Dodecaborate clusters with antimicrobial activity

In 2018, Duttwyler and coworkers reported a metal-free methodology for the synthesis of fused dodecaborate-diboraoxazoles by B—O bond formation *(88)*. Starting materials were dodecaborate amides, which were obtained by mono-acylation of $[B_{12}H_{11}(NH_3)]^-$ with acyl chlorides under mild conditions (Scheme 27). This protocol involved deprotonation of the NH_3 functionality by NaH and required only 1.1 equivalent of acylation agent in dry THF solvent. The reaction tolerated various acylating reagents with different steric and electronic properties to afford corresponding *N*-acylation products in high yields. These products were subsequently utilized for the formation of the fused dodecaborate-diboraoxazoles.

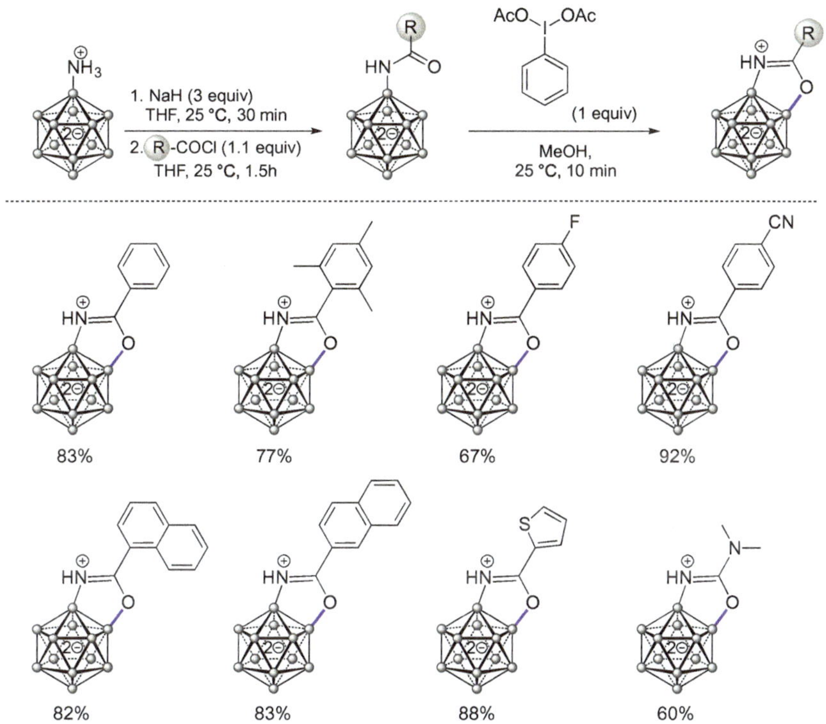

Scheme 27 Synthesis of dodecaborate amides and cyclization to give fused diboraoxazoles.

Cyclization of the acylation products was explored using different reaction conditions involving $[RhCp^{*}Cl_2]_2$ along with various oxidants such as Ag^{I} or Cu^{II} and also metal-free oxidants such as iodine(III) reagents. A key observation during the screening was that the use of less transition metal resulted in better yields. Remarkably, adding only the reagent $PhI(OAc)_2$ in MeOH and using no Rh at all, cyclization occurred smoothly to afford the desired fused diboraoxazoles within 10 min.

The hybrid 3D/2D-fused diboraoxazoles resemble benzoxazoles, which are flat aromatic organic N,O-heterocycles. Benzoxazoles are constituents of many synthetic pharmaceutical and agrochemical compounds as well as natural products. They also feature a wide range of bioactivities, such as anti-inflammatory, antiproliferative, and antimicrobial properties. Thus, the dodecaborate-diboraoxazoles can be viewed as 3D analogs of classical benzoxazoles. In order to probe their bioactive features, the antimicrobial activity of the diboraoxazole compounds was studied against *Neisseria gonorrhoeae*, *Staphylococcus aureus* and *Enterococcus faecalis*. Some of the synthesized compounds displayed minimal inhibitory concentrations (MIC) of 4–16 μg/mL against *N. gonorrhoeae*, which was found to be more efficient than some of the marketed antibiotics such as azithromycin and ciprofloxacin. Interestingly, control compounds involving only benzoxazoles without the boron cluster moiety displayed lower activity, which highlighted the importance of the cage for the antibacterial activity. The borane–heterocycle hybrids are therefore promising candidates for further assays in which diverse biological targets are probed. In addition, they are novel examples of the concept "escape from the flatland," which is based on the organic synthetic strategy to start from a flat aromatic molecule and modify a part of it in such a way that it becomes more three-dimensional, *e.g.*, by introducing saturated carbon centers.

The strategy to synthesize dodecaborate-based diboraoxazoles with potential antibiotic activity was extended toward trisubstituted derivatives as shown in Fig. 8A. These results were disclosed in 2022 *(89)*. The incentive was to explore directing groups featuring a non-aromatic amide scaffold and known inherent bioactivity. To that end, amides of camphanic acid and menthyl carbonate were selected. Their suitability as directing groups to afford products with antimicrobial properties was anticipated on the basis of the coordination ability of the carbonyl moiety and the reported bioactivities of camphanic acid and menthol. Furthermore, these natural products possess an alicyclic structure with non-racemizing stereogenic centers. Generally speaking, 1,2,3-trisubstitution on an icosahedral cage surface leads to R_{cage} and S_{cage} stereoisomers (Fig. 8B). In the case of the targeted trisubstituted natural product-bearing dodecaborates, the given chirality within the directing group, combined with the R_{cage} and S_{cage} stereochemistry,

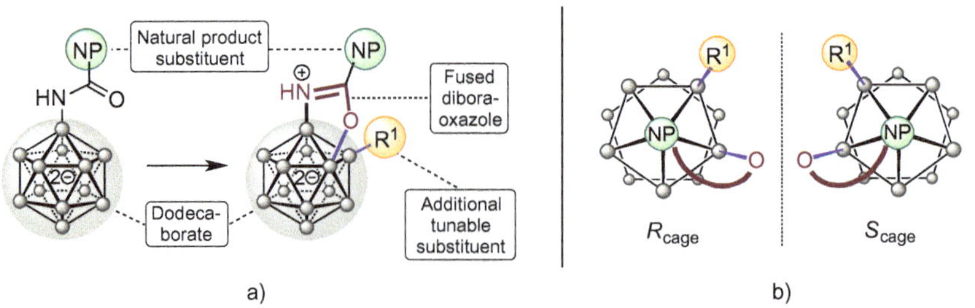

Fig. 8 (A) Strategy toward trisubstituted dodecaborates mediated by natural product-based directing groups; (B) R_{cage} and S_{cage} stereoisomers arising from cage surface 1,2,3-trisubstitution.

was expected to lead to diastereoisomers that would be formed with a diastereomeric ratio $dr \neq 1{:}1$.

The required amides were prepared by treatment of $[B_{12}H_{12}(NH_3)]^-$ with three equivalents of NaH in THF for complete deprotonation of the NH_3 moiety, followed by combination with 1.1 equivalents of (−)-camphanic acid chloride or (−)-menthyl chlorocarbonate. This procedure was similar to the first step depicted in Scheme 27.

Successful B—H activation/B–C/B–O bond formation was observed using styrene coupling partners under Rh(III) catalysis in acetonitrile (Scheme 28). The coupling–cyclization consistently occurred in moderate to high yields of 63–92% under ambient conditions. Substituted styrenes with electron-withdrawing and electron-donating functionalities furnished the products with excellent regioselectivity and control over the degree of substitution. Typical diastereomeric ratios were in the range of 60:40–80:20 while, while increased values of up to 91:9 were observed for styrenes with 4-*t*Bu and 4-OC(O)Me substitution. Coupling with non-aromatic acrylic esters CH_2–CH–R (R = CO_2Me, CO_2tBu, $(CH_2)_9CH_3$) proceeded

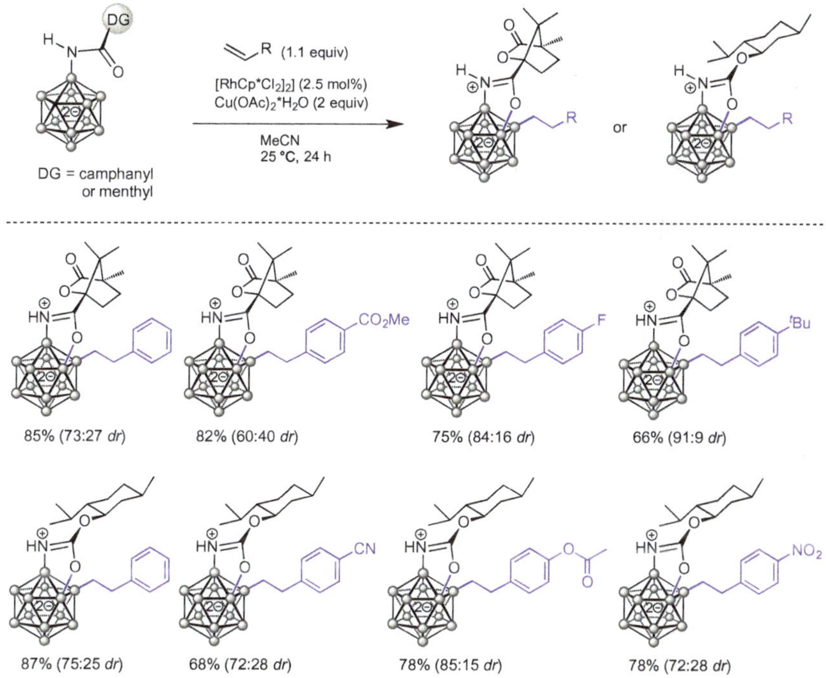

Scheme 28 Selected products of the camphanyl and menthyl amide-controlled alkylation/cyclization of the dodecaborate cluster (*dr* = diastereomeric ratio).

in high yields of 85–92%. However, in these cases the diastereomeric ratio was 1:1, suggesting that the nature of the alkene coupling partner plays a decisive role in the diastereo-discriminating step. For all new compounds, two sets of signals were observed in the NMR spectra, consistent with diastereomer formation due to unchanged absolute configuration of the directing group and R_{cage}/S_{cage} configuration on the cluster surface. Even though an investigation using 2D NMR techniques and a single crystal X-ray diffraction study were conducted, unambiguous determination of the preferred surface stereochemistry could not be achieved.

The antimicrobial activity of the trisubstituted compounds was evaluated against several Gram-positive and Gram-negative antibiotic-resistant bacterial strains such as *Neisseria gonorrhoeae* ATCC 49226, *Staphylococcus aureus* ATCC 25923, *Enterococcus faecalis* ATCC 29212, *Pseudomonas aeruginosa* ATCC 27853, *Escherichia coli* ATCC 25922 and *Shigella sonnei* SD10053. All camphanyl derivatives showed strong anti- microbial activity against *N. gonorrhoeae* with best MIC values of 4 μM. Most products of the camphanyl series also exhibited activity against *S. aureus* and *E. faecalis* with MICs of 8–16 μM. Similarly, all menthyl compounds showed strong activity against *N. gonorrhoeae*, with the best activity observed for the nitro derivative at an MIC of 2 μM. As a general trend, phenyl substituents containing the *tert*-butyl group, halides or polar functionalities such as –OR and $–NO_2$ showed higher effectivity.

The menthyl/nitrophenyl compound, which showed the lowest MIC values, *i.e.*, highest activity, was subsequently assessed against *N. gonorrhoeae* and *S. aureus* in time-kill analyses. It displayed rapid bactericidal activity against both pathogens. In control experiments, the currently recommended first-line bactericidal antibiotics ceftriaxone and vancomycin were evaluated under identical conditions. Exposure of *N. gonorrhoeae* to the menthyl/nitrophenyl compound at $4 \times$ MIC resulted in >100,000-fold reduction in bacterial counts within the first hour, whereas ceftriaxone required 8 h to achieve similar inactivation. In the case of *S. aureus*, exposure at $4 \times$ MIC caused >100,000-fold reduction within the first 2 h, whereas vancomycin could not effectively achieve inactivation, even after 8 h of exposure. The time-kill assays, corroborated by additional microscopic bacterial staining, underscored the strong potential of the trisubstituted dodecaborates as leads for novel antimicrobial compounds to treat bacterial infections.

8. Outlook

For inorganic and organic synthetic chemists, B—H activation remains a fascinating field to explore. The preceding sections demonstrate the rapid progress with regard to selective functionalization of anionic boron cages over the last few years. The methodologies enable the synthesis of novel derivatives in a small number of steps, and they also provide access to compounds that can not be prepared by electrophilic substitution or halogenation/cross coupling. Unique characteristics of the B—H activation procedures include excellent control over regioselectivity and degree of substitution as well as a broad variety of substituents that can be installed at the boron vertices. Compared to C—H activation, in which coupling to a vast number of other elements with high stereoselectivity and also theoretical understanding has been accomplished, transition metal-mediated B—H functionalization is a research area where efforts in several directions are still necessary. In terms of fundamental studies, details of the mechanistic manifolds must be better investigated, and to this end, state-of-the-art calculations are essential. An improved insight into catalytic cycles will enable researchers to establish optimized conditions more efficiently. This in turn will enable the identification of novel coupling partners to form B–X bonds (X = main group elements). Furthermore, a shift to using earth-abundant metals such as manganese, iron, cobalt and nickel is an important goal. A diversification of the synthetic toolbox will have a considerable impact on the preparation of tailor-made boron clusters and their applications.

Acknowledgments

Financial support by the National Natural Science Foundation of China (No. 21871231 and 21850410451) and the Special Funds for Basic Scientific Research of Zhejiang University (No. K20210335, 2019QNA3010 and 2018QNA3011) is gratefully acknowledged.

References

1. Hosmane, N. S., Eagling, R., Eds. *Handbook of Boron Science: With Applications in Organometallics Catalysis, Materials and Medicine*; World Scientific Publishing: London, 2018.
2. Grimes, R. N. *Carboranes*, 3rd ed.; Elsevier: Amsterdam, 2016.
3. Hosmane, N. S. *Boron Science: New Technologies and Applications*; CRC Press: Boca Raton, 2011.

4. Davidson, M., Hughes, A. K., Marder, T. B., Wade, K., Eds. *Contemporary Boron Chemistry*; Royal Society of Chemistry: Cambridge, 2000.
5. Grimes, R. N. *Dalton Trans.* **2015**, *44*, 5939–5956.
6. Douvris, C.; Michl, J. *Chem. Rev.* **2013**, *113*, PR179–PR233.
7. Grimes, R. N. *J. Chem. Ed.* **2004**, *81*, 657–672.
8. Sivaev, I. B.; Bregadze, V. I.; Sjöberg, S. *Collect. Czechoslov. Chem. Commun.* **2002**, *67*, 679–727.
9. Štíbr, B. *Chem. Rev.* **1992**, *92*, 225–250.
10. Commission on the Nomenclature of Inorganic Chemistry. *Nomenclature of Inorganic Chemistry*, 2nd ed.; International Union of Pure and Applied Chemistry/ Butterworth & Co.: London, 1971. see also: Beckett, M. A.; Brellochs, B.; Chizhevsky, I. T.; Damhus, T.; Hellwich, K.-H.; Kennedy, J. D.; Laitinen, R.; Powell, W. H.; Rabinovich, D.; Viñas, C.; Yerin, A. *Pure Appl Chem.* **2020**, *92*, 355–381.
11. Poater, J.; Solà, M.; Viñas, C.; Teixidor, F. *Chem. Eur. J.* **2016**, *22*, 7437–7443.
12. Otsuka, M.; Takita, R.; Kanazawa, J.; Miyamoto, K.; Muranaka, A.; Uchiyama, M. *J. Am. Chem. Soc.* **2015**, *137*, 15082–15085.
13. McKee, M. L. *Inorg. Chem.* **2002**, *41*, 1299–1305.
14. Aihara, J. *Inorg. Chem.* **2001**, *40*, 5042–5044.
15. King, R. B. *Chem. Rev.* **2001**, *101*, 1119–1152.
16. Zhoa, M.; Gimarc, B. M. *Inorg. Chem.* **1993**, *32*, 4700–4707.
17. Assaf, K. I.; Ural, M. S.; Pan, F.; Georgiev, T.; Simova, S.; Rissanen, K.; Gabel, D.; Nau, W. M. *Angew. Chem. Int. Ed.* **2015**, *54*, 6852–6856.
18. Riddlestone, I. M.; Kraft, A.; Schaefer, J.; Krossing, I. *Angew. Chem. Int. Ed.* **2018**, *57*, 13982–14024.
19. Knapp, C. In *Comprehensive Inorganic Chemistry II*; Vol. 1; Elsevier: Amsterdam, 2013; pp. 651–679.
20. Reed, C. A. *Acc. Chem. Res.* **2010**, *43*, 121–128.
21. Geis, V.; Guttsche, K.; Knapp, C.; Scherer, H.; Uzun, R. *Dalton Trans.* **2009**, 2687–2694.
22. Krossing, I.; Raabe, I. *Angew. Chem. Int. Ed.* **2004**, *43*, 2066–2090.
23. Reed, C.; Carboranes, A. *Acc. Chem. Res.* **1998**, *31*, 133–139.
24. Kim, K.-C.; Reed, C. A.; Elliott, D. W.; Mueller, L. J.; Tham, F.; Lin, L.; Lambert, J. B. *Science* **2002**, *297*, 825–827.
25. Shoji, Y.; Tanaka, N.; Mikami, K.; Uchiyama, M.; Fukushima, T. *Nat. Chem.* **2014**, *6*, 498–503.
26. Stoyanov, E. S.; Stoyanova, I. V.; Tham, F. S.; Reed, C. A. *J. Am. Chem. Soc.* **2010**, *132*, 4062–4063.
27. Reed, C. A.; Kim, K.-C.; Stoyanov, E. S.; Stasko, D.; Tham, S. T.; Mueller, L. J.; Boyd, P. D. W. *J. Am. Chem. Soc.* **2003**, *125*, 1796–1804.
28. Axtell, J. C.; Saleh, L. M. A.; Qian, E. A.; Wixtrom, A. I.; Spokoyny, A. M. *Inorg. Chem.* **2018**, *57*, 2333–2350.
29. Núñez, R.; Tarrés, M.; Ferrer-Ugalde, A.; de Biani, F. F.; Teixidor, F. *Chem. Rev.* **2016**, *116*, 14307–14378.
30. Núñez, R.; Romero, I.; Teixidor, F.; Viñas, C. *Chem. Soc. Rev.* **2016**, *45*, 5147–5173.
31. Fisher, S. P.; Tomich, A. W.; Lovera, S. O.; Kleinsasser, J. F.; Guo, J.; Asay, M. J.; Nelson, H. M.; Lavallo, V. *Chem. Rev.* **2019**, *119*, 8262–8290.
32. Fisher, S. P.; Tomich, A. W.; Guo, J.; Lavallo, F. *Chem. Commun.* **2019**, *55*, 1684–1701.
33. Sivaev, I. B.; Stogniy, M. Y.; Bregadze, V. I. *Coord. Chem. Rev.* **2021**, *436*, 213795–213889.
34. Zhao, X.; Yang, Z.; Chen, H.; Wang, Z.; Zhou, X.; Zhang, H. *Coord. Chem. Rev.* **2021**, *444*, 214042–214068.

35. Hey-Hawkins, E., Viñas Teixidor, C., Eds. *Boron-Based Compounds: Potential and Emerging Applications in Medicine*; John Wiley & Sons: Oxford, 2018.
36. Stockmann, P.; Gozzi, M.; Kuhnert, R.; Sárosi, M. B.; Hey-Hawkins, E. *Chem. Soc. Rev.* **2019**, *48*, 3497–3512.
37. Leśnikowski, Z. J. *J. Med. Chem.* **2016**, *59*, 7738–7758.
38. Gabel, D. *Pure Appl. Chem.* **2015**, *87*, 173–179.
39. Issa, F.; Kassiou, M.; Rendina, L. M. *Chem. Rev.* **2011**, *111*, 5701–5722.
40. Scholz, M.; Hey-Hawkins, E. *Chem. Rev.* **2011**, *111*, 7035–7062.
41. Valliant, J. F; Guenther, K. J.; King, A. S.; Morel, P; Schaffer, P.; Sogbein, O. O.; Stephenson, K. A. Coord. Chem. Rev. 2002, 232, 173–230.
42. Berger, A.; Buckley, C. E.; Paskevicius, M. *Inorg. Chem.* **2021**, *60*, 14744–14751.
43. Štíbr, B.; Tok, O. L.; Holub, J. *Inorg. Chem.* **2017**, *56*, 8334–8340.
44. Hermánek, S. *Chem. Rev.* **1992**, *92*, 325–362.
45. Mason, J., Ed. *Multinuclear NMR*; Plenum Press: New York, 1987.
46. Venable, T. L.; Hutton, W. C.; Grimes, R. N. *J. Am. Chem. Soc.* **1984**, *106*, 29–37.
47. Venable, T. L.; Hutton, W. C.; Grimes, R. N. *J. Am. Chem. Soc.* **1982**, *104*, 4716–4717.
48. Olid, D.; Núñez, R.; Viñas, C.; Teixidor, F. *Chem. Soc. Rev.* **2013**, *42*, 3318–3336.
49. Dziedzic, R. M.; Spokoyny, A. M. *Chem. Commun.* **2019**, *55*, 430–442.
50. Al-Joumhawy, M. K.; Marei, T.; Shmalko, A.; Cendoya, P.; La Bordea, J.; Gabel, D. *Chem. Commun.* **2021**, *57*, 10007–10010.
51. Colobert, F., Wencel-Delord, J., Eds. *C–H Activation for Asymmetric Synthesis*; Wiley-VCH: Weinheim, 2019.
52. Sambiagio, C.; Schönbauer, D.; Blieck, R.; Dao-Huy, T.; Pototschnig, G.; Schaaf, P.; Wiesinger, T.; Zia, M. F.; Wencel-Delord, J.; Besset, T.; Maes, B. U. W.; Schnürch, M. *Chem. Soc. Rev.* **2018**, *47*, 6603–6743.
53. Wu, X.-F., Ed. *Transition Metal-Catalyzed Heterocycle Synthesis via C–H Activation*; Wiley-VCH: Weinheim, 2016.
54. Dixneuf, P. H., Doucet, H., Eds. *C–H Bond Activation and Catalytic Functionalization I (Topics in Organometallic Chemistry 55)*; Springer: Heidelberg, 2016.
55. Dixneuf, P. H., Doucet, H., Eds. *C–H Bond Activation and Catalytic Functionalization II (Topics in Organometallic Chemistry 56)*; Springer: Heidelberg, 2016.
56. Chen, Z.; Wang, B.; Zhang, J.; Yu, W.; Liu, Z.; Zhang, Y. *Org. Chem. Front.* **2015**, *2*, 1107–1295.
57. Li, J. J., Ed. *C–H Bond Activation in Organic Synthesis*; CRC Press: Boca Raton, FL, 2015.
58. Hoel, E. L.; Hawthorne, M. F. *J. Am. Chem. Soc.* **1973**, *95*, 2712–2713.
59. Hoel, E. L.; Hawthorne, M. F. *J. Am. Chem. Soc.* **1975**, *97*, 6388–6395.
60. Qiu, Z.; Xie, Z. *Acc. Chem. Res.* **2021**, *54*, 4065–4079.
61. Zhang, X.; Yan, H. *Coord. Chem. Rev.* **2019**, *378*, 466–482.
62. Quan, Y.; Xie, Z. *Chem. Soc. Rev.* **2019**, *48*, 3660–3673.
63. Quan, Y.; Qiu, Z.; Xie, Z. *Chem. Eur. J.* **2018**, *24*, 2795–2805.
64. Yu, W.-B.; Cui, P.-F.; Gao, W.-X.; Jin, G.-X. *Coord. Chem. Rev.* **2017**, *350*, 300–319.
65. Kanazawa, J.; Kitazawa, Y.; Uchiyama, M. *Chem. Eur. J.* **2019**, *25*, 9123–9132.
66. Duttwyler, S. *Pure Appl. Chem.* **2018**, *90*, 733–744.
67. Molinos, E.; Kociok-Köhn, G.; Weller, A. S. *Chem. Commun.* **2005**, 3609–3611.
68. Molinos, E.; Brayshaw, S. K.; Kociok-Köhn, G.; Weller, A. S. *Organometallics* **2007**, *26*, 2370–2382.
69. Molinos, E.; Brayshaw, S. K.; Kociok-Köhn, G.; Weller, A. S. *Dalton Trans.* **2007**, 4829–4844.
70. El-Hellani, A.; Kefalidis, C. E.; Tham, F. S.; Maron, L.; Lavallo, V. *Organometallics* **2013**, *32*, 6887–6890.
71. Estrada, J.; Lee, S. E.; McArthur, S. G.; El-Hellani, A.; Tham, F. S.; Lavallo, V. *J. Organomet. Chem.* **2015**, *798*, 214–217.

72. Chan, A. L.; Estrada, J.; Kefalidis, C. E.; Lavallo, V. *Organometallics* **2016**, *35*, 3257–3260.
73. Shen, Y.; Pan, Y.; Liu, J.; Sattasathuchana, T.; Baldridge, K. K.; Duttwyler, S. *Chem. Commun.* **2017**, *53*, 176–179.
74. Shen, Y.; Pan, Y.; Liu, J.; Sattasathuchana, T.; Baldridge, K. K.; Duttwyler, S. *Eur. J. Inorg. Chem.* **2017**, 4420–4424.
75. Shen, Y.; Pan, Y.; Zhang, K.; Liang, X.; Liu, J.; Spingler, B.; Duttwyler, S. *Dalton Trans.* **2017**, *46*, 3135–3140.
76. Lin, F.; Shen, Y.; Zhang, Y.; Sun, Y.; Liu, J.; Duttwyler, S. *Chem. Eur. J.* **2018**, *24*, 551–555.
77. Saeed, A. *Eur. J. Med. Chem.* **2016**, *116*, 290–317.
78. Saddiqa, A.; Usman, M.; Cakmak, O. *Turk. J. Chem.* **2017**, *41*, 153–178.
79. Akimoto, G.; Otsuka, M.; Miyamoto, K.; Muranaka, A.; Hashizume, D.; Takita, R.; Uchiyama, M. *Chem. Asian J.* **2018**, *13*, 913–917.
80. Lin, F.; Yu, J.-L.; Shen, Y.; Zhang, S.-Q.; Spingler, B.; Liu, J.; Hong, X.; Duttwyler, S. *J. Am. Chem. Soc.* **2018**, *140*, 13798–13807.
81. Shen, Y.; Zhang, K.; Liang, X.; Dontha, R.; Duttwyler, S. *Chem. Sci.* **2019**, *10*, 4177–4184.
82. Himmelspach, A.; Finze, M.; Vöge, A.; Gabel, D. *Z. Anorg. Allg. Chem.* **2012**, *638*, 512–519.
83. Peymann, T.; Knobler, C. B.; Hawthorne, M. F. *Inorg. Chem.* **1998**, *37*, 1544–1548.
84. Bondarev, O.; Hawthorne, M. F. *Chem. Commun.* **2011**, *47*, 6978–6980.
85. Zimmermann, L. W.; Van, N.-D.; Gudat, D.; Schleid, T. *Angew. Chem. Int. Ed.* **2016**, *55*, 1909–1911.
86. Zhang, Y.; Sun, Y.; Lin, F.; Liu, J.; Duttwyler, S. *Angew. Chem. Int. Ed.* **2016**, *55*, 15609–15614.
87. Zhang, Y.; Wang, T.; Wang, L.; Sun, Y.; Lin, F.; Liu, J.; Duttwyler, S. *Chem. Eur. J.* **2018**, *24*, 15812–15817.
88. Sun, Y.; Zhang, J.; Zhang, Y.; Liu, J.; van der Veen, S.; Duttwyler, S. *Chem. Eur. J.* **2018**, *24*, 10364–10371.
89. Varkhedkar, R.; Yang, F.; Dontha, R.; Zhang, J.; Liu, J.; Spingler, B.; van der Veen, S.; Duttwyler, S. *ACS Cent. Sci.* **2022**, *8*, 322–331.

About the author

Simon Duttwyler studied chemistry at the University of Zurich, Switzerland. In 2010 he obtained his PhD degree from the University of Zurich under the supervision of Prof. Jay S. Siegel. In 2007/2008 he worked for Prof. Christopher A. Reed at the University of California, Riverside. From 2010 to 2013 he carried out postdoctoral research with Prof. Jonathan A. Ellmann at Yale University. In 2013 he started his independent career at Zhejiang University, China, where he was promoted to full professor in January 2019. His research focuses on the synthesis and applications of boron cluster compounds.

FSC
www.fsc.org
MIX
Papier | Fördert
gute Waldnutzung
FSC® C083411

Kolibri 360 GmbH
Ferdinand-Jühlke-Straße 7
99095 Erfurt, Deutschland
produktsicherheit@kolibri360.de